STEEL IN THE 80s

PARIS SYMPOSIUM
FEBRUARY 1980

**ORGANISATION FOR ECONOMIC
CO-OPERATION AND DEVELOPMENT**

The Organisation for Economic Co-operation and Development (OECD) was set up under a Convention signed in Paris on 14th December 1960, which provides that the OECD shall promote policies designed:
— to achieve the highest sustainable economic growth and employment and a rising standard of living in Member countries, while maintaining financial stability, and thus to contribute to the development of the world economy;
— to contribute to sound economic expansion in Member as well as non-member countries in the process of economic development;
— to contribute to the expansion of world trade on a multilateral, non-discriminatory basis in accordance with international obligations.
The Members of OECD are Australia, Austria, Belgium, Canada, Denmark, Finland, France, the Federal Republic of Germany, Greece, Iceland, Ireland, Italy, Japan, Luxembourg, the Netherlands, New Zealand, Norway, Portugal, Spain, Sweden, Switzerland, Turkey, the United Kingdom and the United States.

Publié en français sous le titre :
L'ACIER DANS LES ANNÉES 80

THE OPINION EXPRESSED AND ARGUMENTS EMPLOYED
IN THIS PUBLICATION ARE THE RESPONSIBILITY OF THE AUTHORS AND
DO NOT NECESSARILY REPRESENT THOSE OF THE OECD

*
* *

OECD Symposium on
"THE STEEL INDUSTRY IN THE 1980s"
Paris, 27th and 28th February, 1980

PROGRAMME

27th February   Chairman/Moderator   Mr. Luther Hodges, Jr.
United States Deputy
Secretary of Commerce

10.15 a.m.   Opening address   Mr. E. van Lennep
Secretary-General, OECD

10.30 a.m.
Theme   **Current Trends in the Steel Industry and Developments in the 1980s**

a)   *The Economic Background; Short and Medium Term Outlook in the OECD Area and in the World Generally*

"The World Economy in the 1970s and 1980s"   (Document No. 1)
by
Dr. Sylvia Ostry, Head of Economics and Statistics Department, OECD

b)   *The Probable Growth of Steel Demand and Trade Through the Decade 1980-1990*

"Outlook for World Steel Industry up to 1985: Demand, Trade
and Supply Capacity"   (Document No. 2)
by
Mr. Tsutomu Kono, General Manager, Corporate and Economic Research Department, Nippon Steel Corporation

"World Trends in Steel Consumption and Production to 1990" (Document No. 3)
by
Mr. Helmut Wienert, Research Economist, Rheinisch-Westfälisches Institut für Wirtschaftsforschung

Panellists' and floor discussion

1.15 p.m.   Luncheon - offered by the Secretary-General
Château de la Muette

3

3.00 p.m.   *c)*   *Changes in Capacity and the Supply Potential*

"The Growth of Steel-Making Capacity in the 1980s"   (Document No. 4)
by
Mr. Kimiro Suzuki and Mr. Tudor Miles, Secretariat Group on Steel, OECD

"Global Scenario of World Steel Industry Growth particularly
up to 1985"   (Document No. 5)
by
Dr. B.R. Nijhawan, Senior Interregional Adviser, UNIDO

Panellists' and floor discussion

d)   *The Future Supply/Demand Balance and its Implications, including Developments in International Trade in Steel*

"The Economics of the Current Steel Crisis in OECD
Member Countries"   (Document No. 7)
by
Dr. Robert W. Crandall, Senior Fellow, The Brookings Institution

"How Steel-Making Enterprises can become Internationally
Competitive"   (Document No. 6)
by
Dr. Franco Peco, Assistant to the President, ASSIDER

Panellists' and floor discussion

Panellists for 27th February:

Mr. D.F. Anderson	Director, Department of Economic Affairs, IISI
Mr. J. Brunner	Chief Economist, Broken Hill Proprietary Co. Ltd., Melbourne, Australia
Dr. P.W. Marshall	Director, Putnam, Hayes and Bartlett Inc., Newton, Massachusetts, USA
Prof. E. Ruist	Stockholm School of Economics, Sweden

5.00 p.m.   Sir Charles Villiers, Chairman, British Steel Corporation, will present a paper - "Job Creation by the British Steel Corporation in Major Steel Closure Areas" - (Document No. 8) relating to the theme of the second day's meeting.

Discussion.

5.45 p.m.   Chairman's summary.

6.30 p.m.   Reception - offered by the Secretary-General
Château de la Muette

4

28th Februar   Chairman/Moderator   Viscount E. Davignon
                                                            Member, Commission of the European
                                                            Communities responsible for Internal
                                                            Market and Industrial Affairs

10.00 a.m.
   Theme   **The Policy Responses to the Problems of the World Steel Industry**

            a)   *Policies for Adjustment, Modernisation and Adaptation of the Steel
                 Industry in the Light of Expected World Developments*

            b)   *The Problems of the Labour Force and Re-Adaptation and
                 Re-Employment Policies*

      Papers to be presented:

            "Policy Responses to the Problems of the World Steel
            Industry"   (Document No. 10)
                                   by
            Mr. Edward S. Florkoski, OECD Consultant

            "The Problems of the Labour Force and Re-adaptation and
            Re-employment Policies"   (Document No. 9)
                                   by
            Mr. William Sirs, President of the International Metalworkers Federation's
            Iron, Steel and Non-Ferrous Metal Department

            "The Steel Industry in the International Context"   (Document No. 11)
                                   by
            Mr. L. McBride, President, United Steelworkers of America

            "Restructuring of the Steel Industry and the Reemployment and
            Readjustment of Labour"   (Document No. 12)
                                   by
            Mr. K.A. Sanden, Editor, Norsk Jern-og Metallarbeiderforbund

      Others who will present papers or make statements are:

                     Mr. L.W. Foy   Chairman, Bethlehem Steel Corporation,

                     Ms. K. O'Reilly   Executive Director,
                                         Consumer Federation of America

                     Mr. E. Tesch   President, ARBED
                                         President, EUROFER

            Representative Charles Vanik House of Representatives,
                                         Washington

                     Mr. A. Wolff   Verner, Liipfert, Bernhard and McPherson,
                                         Washington

      Discussion

1.00 p.m.   Luncheon Interval

3.00 p.m.   Discussion

4.30 p.m.   Chairman's summary

5

SPEAKERS, PANELLISTS AND PARTICIPANTS
WHO TOOK THE FLOOR ON WEDNESDAY, 27th FEBRUARY

SUBJECT : CURRENT TRENDS IN THE STEEL INDUSTRY AND DEVELOPMENTS IN THE 1980s

The Secretary-General 11
The Chairman: Luther Hodges, Jr. 15

Dr. Sylvia Ostry, Head of Economics and Statistics Department (Speaker) 17-61
Mr. Tsutomu Kono, General Manager, Corporate and
    Economic Research Department, Nippon Steel Corporation, Japan (Speaker) . 18-68
Mr. Helmut Wienert, Research Economist, Rheinisch-Westfälisches Institut, Germany
    (Speaker) 20-29-83
Dr. Paul Marshall, Director, Putnam, Hayes and Bartlett Inc., US 21-48
Mr. D.F. Anderson, Director, Department of Economic Affairs, IISI 22-31-39
Mr. John Brunner, Chief Economist, Broken Hill Proprietary Co. Ltd.,
    Australia 23-49
Mr. Alan Wolff, Verner, Liipfert, Bernhard & McPherson, US 24
Dr. B. R. Nijhawan, Senior Adviser (Interregional), UNIDO (Speaker) 25-37-113
Mr. Herman Rebhan, General Secretary, International Metalworkers' Federation,
    Geneva 26
Mr. Tudor Miles, Secretariat Group on Steel (Speaker) 27-35-97
Dr. Ing. E.H. Willy Korf, President, Korf-Stahl AG, Germany 28-52
Representative Charles Vanik, US House of Representatives, Washington 28
Dr. Robert W. Crandall, Senior Fellow, The Brookings Institution, US (Speaker) 29-43-140
Professor Erik Höök, Managing Director, Jernkontoret, Sweden 30
Mr. Thomas Atkinson, Director, General Motors Corporation, US 31
Professor Erik Ruist, Stockholm School of Economics, Sweden 31-47
Dr. Franco Peco, Assistant to the President, ASSIDER, Italy (Speaker) ... 32-45-127
Mr. Odd Göthe, Deputy Secretary-General, Ministry of Industry, Norway 33
Mr. Steffen Møller, Chief Economist, Danish Matalworkers Union 33
Monsieur J. Doyen, Secrétaire Général, Centrale Chrétienne des Métallurgistes,
    Belgium 40
Mr. Aleksander Cavié, Technical and RD Manager, Yugoslav Iron and
    Steel Federation 42
Mr. Michael Marshall, MP, Parliamentary Under-Secretary of State for Industry,
    Department of Industry, United Kingdom 50
Dr. Ruprecht Vondran, Hauptgeschäftsführer, Wirtschaftsvereinigung Eisen und
    Stahlindustrie, Germany 51
Mr. Richard F. Schubert, President, Bethlehem Steel Corporation, US 53
Mr. Alfredo Acle, Director General, Comision Coordinadora de la
    Industria Siderurgica, Mexico 54
Monsieur Bernard Mourgues, Secrétaire Général, Fédération des
    Industries Métallurgiques, Force Ouvrière, France 55

Conclusions of Chairman 57

SPEAKERS, PANELLISTS AND PARTICIPANTS
WHO TOOK THE FLOOR ON THURSDAY, 28th FEBRUARY

SUBJECT : THE POLICY RESPONSES TO THE PROBLEMS OF THE WORLD STEEL INDUSTRY

The Chairman: Viscount Etienne Davignon 149

Mr. Edward S. Florkoski, OECD Consultant, (Speaker) 154

Mr. William Sirs, President of the International Metalworkers Federation's Iron, Steel and Non Ferrous Metal Department, (Speaker) 168

Mr. L. McBride, President, United Steelworkers of America, (Speaker) 187

Mr. Knut Arne Sanden, The Norwegian Iron and Metalworkers' Union (Speaker) . 193

Mr. G.H. Sambrook, Managing Director Commercial, British Steel Corporation, (Speaker) 197

Mr. G. Elliot, Office of General Relations, Department of Industry, Trade and Commerce, Canada 205

Mr. Rudolf Judith, Chairman, Consultative Committee, European Coal and Steel Community, Vorstand IG Metall, Germany 207

Mr. Herman Rebhan, General Secretary, International Metalworkers Federation, Geneva 208

Mr. Olof Rydh, Secretary, Sv. Metallindustriarb. förbundet, Sweden 210

Mr. Steffen Møller, Chief Economist, Danish Metalworkers Union 211

Mr. Michael Marshall, MP, Parliamentary Under-Secretary of State for Industry, Department of Industry, United Kingdom 212

Mr. J. Doyen, Secrétaire général, Centrale Chrétienne des Métallurgistes, Belgique 212

Dr. Paul W. Marshall, Director, Putnam Hayes and Bartlett Inc., US 213

Mr. Lewis W. Foy, Chairman, Bethlehem Steel Corporation, US (Speaker) 215

Ms. Kathleen O'Reilly, Executive Director, Consumer Federation of America, (Speaker) 219

Mr. Emmanuel Tesch, President, ARBED, President, EUROFER, (Speaker) 225

Mr. Naohiro Amaya, Vice-Minister for International Affairs, Ministry of International Trade and Industry, Japan 229

Representative Charles Vanik, U.S. House of Representatives, Washington, (Speaker) 232

Mr. Alan Wolff, Verner, Liipfert, Berhnard and McPherson, US, (Speaker) 243

M. Pierre Gadonneix, Directeur des Industries Métallurgiques, Mécaniques et Électriques, Ministère de l'Industrie, France 251

Mr. Kurth Orban, President, American Institute for Imported Steel 253

Dr. Ruprecht Vondran, Hauptgeschäftsführer, Wirtschaftsvereinigung Eisen-und Stahlindustrie, Germany 254

Mr. Thomas Atkinson, Director, International Economics Group, General Motors Corporation, US 255

Mr. Staffan Sohlman, Deputy Director General, Board of Commerce, Sweden ... 255

Mr. Richard Simmons, President, Allegheny Ludlum Steel Co., Former President,
Specialty Steel Industry Committee, US 256

Dr. Manfred Caspari, Directeur Général adjoint des Relations Extérieures,
Commission of the European Communities 257

Mr. Alfredo Acle, Director General, Comision Coordinadora de la
Industria Siderurgica, Mexico 258

H.E. Mr. Robert D. Hormats, Deputy US Trade Representative,
Office of the US Trade Representative 259

M. Robert Altman, Directeur général, Société Lorraine des
Produits Métallurgiques, France 260

M. Mario Castegnaro, Membre du Comité Exécutif, OGBL, Eschlalzette,
Luxembourg 260

Conclusions of Chairman 262
Closing Remarks of Secretary-General 264

List of Participants in the Symposium 265

9

OPENING ADDRESS
BY THE OECD SECRETARY-GENERAL

Mr. Emile van Lennep

Mr. Chairman, Excellencies, Ladies and Gentlemen,

I would like to welcome here today all the participants in this Symposium on the Steel Industry in the 1980s. The fact that so many of you have been able, in spite of your many commitments, to accept my invitation underlines the acute nature of the problems confronting this basic sector in modern industrial and industrialising economies and the range of the interests involved. It is a cause of particular satisfaction that participants have come not only from OECD countries but from a number of others in whose economic development steel is playing an important role. We have participants from 30 countries and from six international organisations. There are participants from government, from national administrations and from parliamentary bodies, from the industry itself, from the ranks of leaders of steel enterprises and the trade unions; there are also those from steel-using industries and steel stock-holders and traders; and we have participation from university circles and research institutes.

I would like to express my thanks to all those who by their efforts and their advice have made this Symposium possible. I should first, Sir, express my appreciation to you for having agreed to chair the meeting today on the economic problems of the steel sector and to Mr. Davignon who will be in the chair tomorrow for the discussion of the policy responses to these problems. Particular thanks also go to the authors of the various papers and to those who have been able to indicate in advance the statements they will be making. A small group of panellists will be making comments and observations on the papers and we are grateful for the role they will play in leading off the discussions. Above all, I hope that this Symposium will give all participants the opportunity to exchange views in a frank and direct manner.

In seeking to contribute in the OECD to the urgent consideration of how the problems of the steel sector might best be resolved we have adopted a rather innovative approach and tried to be as flexible as possible. As you know, after the preliminary work carried out by an ad hoc Working Group, the Steel Committee of the OECD was set up on 26th October, 1978, and has been meeting at frequent intervals. We are continuing to develop the means by which, through the advisory bodies of the OECD, inputs may be made to the work in steel from the industry and trade union sides which will contribute to the formulation of appropriate policies.

As a further step, we thought it essential to give an opportunity for a rather free-ranging discussion between the interest involved. It is for this reason that I have invited you, on a personal basis, to come together to exchange views, during two days, on the fundamental nature of the problems facing the industry and the policy responses required.

I should like to take the opportunity of this opening statement to make some remarks on the problems in the steel industry and on the challenges which the industry presents to policy-makers in the present international context.

In virtually all modern industrial economies steel occupies an important place. So do quite a number of other industries, such as petrochemicals and electronics. These are as basic or vital to our economies as the steel industry. In market economies, governments normally confine themselves to helping maintain the conditions for an optimal allocation of resources without assuming direct responsibility for formulating policies for individual industries. Why then have governments and the OECD paid special attention to steel?

From the mid-1970s, the combination of excess capacities in many countries, resulting from the unanticipated weakening in the trend growth of steel demand, together with the shift in competitive positions, created a very serious situation. A number of governments began to introduce measures to stabilize their steel markets and preserve the position of their steel industries — measures with significant international effects. It was therefore agreed that the problems posed by increasing unilateral government intervention should be considered within the framework of international co-operation in the OECD.

Let me now turn briefly to the interrelationship between steel and the general economy. The demand for steel is determined by the overall level of economic activity. In trying to foresee what general economic conditions are likely to prevail we have to take into account, after the experiences of 1973-74, a second series of sharp increases in the price of oil over the past twelve months. Inflationary pressures continue, as do high rates of unemployment. There are considerable uncertainties as to the future course of our economies — uncertainties on the rate of growth over the next twelwe to eighteen months and what growth rate it may be possible to sustain over the coming decade.

There is also much uncertainty concerning the longer-term relationship between economic growth and demand for steel or, to put it differently, about the trend changes in the steel intensity of overall demand. Evidence over the last decade or so suggests a significant decline in this relationship in the OECD countries. On the other hand, there are growing requirements for higher quality steels and for more sophisticated products. Further, a number of the developing countries are at a particularly steel-intensive stage of industrialisation, and their steel consumption is growing strongly year by year.

In its turn, the steel industry plays a significant macro-economic role. I would mention first the relation to inflation. Adequate supplies of steel at the right economic price are essential to our economies. While price controls and restraints applied to steel have proved to be counter-productive, even in respect of the long-term development of steel prices themselves, measures to restrict competition and raise prices well above equilibrium levels would run counter to the priority task of containing and reducing inflation. The steel industry is in general a large user of energy. In the present energy situation the industry will have to concentrate considerable effort on containing cost pressures from the recent rises in the price of energy.

Realistic investment programmes have an essential part to play in revitalising the steel industry. The cost and conditions of that investment will be determined partly by the general economic situation and partly by specific policies. Like other highly capital-intensive activities the steel industry needs large funds to maintain its productive potential and to carry out the modernisation of its products and processes.

12

In this area, two extremes have to be avoided. On the one hand, in view of the long lead times involved, financing difficulties reflecting short-term market conditions can lead to loss of competitiveness and even to inadequate investment in the steel industry worldwide. On the other hand, there is the danger that to generous investment incentives, which appear justified from the point of view of individual enterprises, can lead to excess capacities and serious misallocation of resources, nationally and/or internationally.

Although the steel sector does not employ the large numbers of people of many processing industries, it plays an important role as an employer. In a number of countries it tends for historical reasons, to be located in areas where it is either the only source of industrial employment or it shares the labour market with other declining industries. The unavoidable reduction in the labour force of many older steel industries therefore raises very serious social problems in a regional context. These cannot be solved by the steel industry alone.

What then might the response be in the area of manpower and social policies? It is clear that measures will be needed in the countries where steel production is expanding rapidly to ensure the availability of an adequate trained labour force. In some of the older steel-making centres where excess capacity has to be removed the socially tolerable rate at which the labour force can be reduced will depend both on the overall employment situation and on the success of specific manpower policy measures to improve the geographical and occupational mobility of the labour force and to create alternative work opportunities.

I have already stressed the need for an international and co-operative approach in tackling the problems of the world steel industry. There is a broad consensus that the answer to the industry's difficulties will have to lie in accepting a gradual shift of the productive potential in line with comparative advantage. This implies a combination of modernisation and scrapping in some of the older centres of production and moderation in expansion elsewhere. Progress is being made with regard to the acceptance and application of the necessary policy measures; but this progress has been slow. The industry is passing through a difficult transition period and further steps must be taken to avoid disruptive trade movements and destructive price competition.

There is a very real danger that the essential process of adjustment could be interrupted; any new slowdown in economic activity would increase this danger. It is essential therefore that there is a greater mutual understanding between trading partners and that a proper balance of policies is pursued. Self-restraint is needed to avoid recourse to new or more restrictive trade measures, even if this appears legitimate in terms of international trade rules. Increasing restrictive measures and confrontations in the trade field would have grave consequences not only for the steel industry itself but for international economic co-operation in general.

Despite the seriousness of the short-term situation, we must have the courage to lift our eyes beyond the immediate horizon and develop choherent longer-term policies. In doing so, we must recognise that to attempt to resist the unavoidable gradual shift in location patterns reflecting competitive strenghts and the emergence of new dynamic markets for steel can only have a detrimental effect on the free market system to which we all attach so much importance. OECD Ministers adopted, in June 1978, some General Orientations for Policies for Adjustment, and, in June 1979, they again underlined the importance of the work of this Organisation on Positive Adjustment Policies. Ministers recognised that a constructive approach is needed to promote adjustment to new conditions, relying as much as possible on market forces to encourage mobility of labour and capital to their most productive uses.

Though it is difficult in the economic circumstances I have described for countries to move away from defensive action nevertheless, if we are to achieve sustained growth, a progressive shift to more positive adjustment policies together with the appropriate macro-economic policies is essential. Should special intervention be required, Ministers at the June meeting last year indicated important criteria which should be followed, ensuring, inter alia, that the action be temporary, reduced progressively according to a pre-arranged timetable, and fully linked to the implementation of plans to phase out obsolete capacity and re-establish financially viable entities. Further, the direct and indirect costs of assistance to an industry should be made as evident as possible to decision-makers and the public at large. Ministers emphasized that purely defensive measures would, over the longer run, be detrimental to the interests of the world economy as a whole and to the interests of the steel industry itself.

During these two days of discussions, varied viewpoints will be expressed on the difficult and complex problems of the steel sector. It is not, therefore, envisaged that this Symposium would arrive at any specific recommendations as to the course of action to be followed. What I do hope is that each of you, in tackling your own problems, will do so with a better understanding of the viewpoints of others and that as a result, a greater degree of consensus might be reached in the search for solutions to the problems of the industry. I have no doubt that for our part our work in the OECD will benefit considerably from the views of those responsible for guiding the fortunes of the steel industry.

I wish you every success in your deliberations.

I

CURRENT TRENDS IN THE STEEL INDUSTRY AND DEVELOPMENTS IN THE 1980s

Chairman/Moderator: Mr. Luther Hodges, Jr.,
United States Deputy Secretary of Commerce

Thank you, Mr. van Lennep, for your very appropriate remarks. Ladies and gentleman, may I again welcome you to this first OECD Symposium on the steel industry. On behalf of all the participants and indeed a very distinguished audience, may I express appreciation to the OECD for sponsoring the Symposium.

I am especially honoured to chair this important panel today on the outlook for the coming decade for this most basic industry. Steel is of unparelleled importance to the individual economies of all nations and to world trade generally. Clearly, the direction and structure of the international steel sector is central to the future of the international economy.

We are all very aware that since 1975 the steel industries of the industrial nations have experienced very slow growth and depressed financial performance. The enthusiasm which accompanied the world steel boom of 1973/74 has been supplanted by a sombre, more pessimistic mood.

True enough, the problems of the industry are quite serious and, I would add, long-range in nature. When the OECD Steel Committee was established in October, 1978, a review group identified several key factors: excess steelmaking capacity internationally; lower growth in demand with pronounced cyclical behaviour; low prices on world markets; depressed financial performance, inhibiting needed investment for modernisation; shifts in traditional trade patterns, forcing socially and politically difficult re-adjustment; and increasing governmental intervention, especially in foreign trade. The unpleasant reality is that today, twenty-five months later, these problems remain largely unresolved. Globally, the steel industry is only slightly better off than it was two or three years ago and the problems may well get worse in the future. Even though world and OECD crude steel production both grew some four to five per cent in 1979, the capacity utilisation rate for the OECD was only some 76 per cent of nominal capacity.

It is in this context that we gather here today to assess the course of the steel industry over the next decade. Our charge is to examine likely changes in supply, demand, capacity and capacity utilisation; and to assess the implications of these trends for international trade.

Our work today will provide the basis for tomorrow's important session which will address the policy implications of the outlook that we develop in this session.

The importance of the judgments that we make on the future course of demand and supply is underlined by the significance of international trade to the world steel economy and to all the OECD nations.

I would like to emphasize a few of the key points of the relationship between the steel industry and the development of the international economy as a whole. These will affect our conclusions and will receive attention in detail as we proceed today.

1. With the demand for steel so closely linked to overall growth and given the uncertainty in general about the future of the international economy, we must approach our task with caution and with respect for the inherent difficulties. Our search is for a possible or probable range of the demand for steel products;

2. The trends of steel use per unit of gross domestic product in the industrialised nations and the LDCs are moving in opposite directions. Steel intensity in the industrialised world is declining and consumption by industrialising LDCs is growing rapidly. We need to weigh the trade-offs between these trends as they affect demand;

3. The importance, institutionally and socially, of the industry as an employer presents serious problems for adjustment where there is overcapacity. We need to keep these adjustment problems in mind, not only for the policy implications, but also for inclusion in our judgments on the supply side;

4. The difficulties inherent in the adjustment process have clear implications for trade. We must also consider in our assessment of supply trends a realistic understanding of the current transition period.

As an official of the United States Government, let me take the liberty of speaking for just a moment about the importance of the future of the steel industry to the United States. The realities facing this critical industry prompt us to be especially mindful of the domestic and global factors affecting our industry at this point in time. I should mention, however, that as a matter of historical and philosophical preference we have followed to the greatest possible extent the discipline of the market place to determine resource allocation.

It is our desire to leave trade flows and other economic decisions to market forces in the future. But we are keenly aware that the United States cannot carry out its role in the world without a strong and financially sound steel industry. As the world's largest net importer of steel, judgments made on the future global supply, demand, capacity and trade in steel are of particular concern to the United States. That is why I am especially pleased to be here and very proud of this fine panel and very distinguished audience.

As I have mentioned, our session today focusses on the outlook, with as indicated in your programme a general economic forecast and specific discussions on demand and capacity and subsequent discussion on the implication of these forecasts. Tomorrow you will focus more on policy implications arising out of the environment which we identify today.

I have chosen to present our three speakers this morning in rather rapid succession with no more than ten minutes for each speaker to summarise his or her paper, which has previously been distributed to you and which is available to you in its total content. Following the three presentations we will hear a brief statement from the members of the panel, but I know that you did not come to hear just the head table, so to speak, and I urge your own participation and will make every effort to be certain that we minimise a dialogue from the head table and that you participate in these deliberations.

a) The Economic Background; Short and Medium Term Outlook in the OECD area and in the World Generally

"The World Economy in the 1970s and 1980s" (Document No. 1)

Dr. Sylvia OSTRY:

I first want to say how pleased I am to be participating in this very important occasion and I am looking forward to lively debate and discussions from the floor. So I will try to carry out the Chairman's instructions and be brief.

As we all know too well, the economic developments of the 70s presented problems which exceeded in magnitude and complexity any the world has faced since the great depression and the last war. After a prolonged period of stagflation in the 70s for most of the OECD, the upswing finally began to show signs of life when the second oil shock hit in 1979. As previous speakers have already stressed (and I am sure you will hear this over and over again this morning) this is a particularly difficult time to be peering into the future; the customary caveats which are always attached to any projection are augmented today by the number of risks and uncertainties prevailing both in the economic and the political environment. Nevertheless, that is my task and the task of my fellow speakers today: to peer into the future. When I was thinking of this last night I was uncomfortably reminded of an adage — a forecaster's adage — that he who lives by the crystal ball must learn to eat ground glass.

The paper which has been distributed illustrates the hazards of prediction by comparing the outlook for the 70s, which was prepared by the OECD Secretariat at the onset of that decade, with the actual outcome of the period just closed. Some figures are presented in Table 2, but I think perhaps more significant than the difference in statistical performance measures is the difference in language and tone of the assessment of prospects in 1970: the confident optimism of the Sectretariat (and they were not alone) which contrasts so sharply with the pervasive uncertainty and what someone has called "the visceral malaise" which marks the mood on the eve of the 1980s.

The series of shocks which struck the world economy during the last decade not only dampened the rate of economic activity and created a deeply entrenched inflationary behaviour, but heightened scepticism — indeed in some quarters even provoked hostility — to traditional tools of stabilization policy in a world which had become so integrated as an economic system and so polycentric as a political system. Further, while a great part of the actual slowdown in growth over the past decade did stem from depressed demand, there seems little doubt that there have also been long-run negative effects on the course of potential output. For example, the rate of andogamous technical change appears to have slowed, although the reasons for this are by no means clear. Quantity and quality investment has been impaired by the unhealthy inflationary climate. The rise in energy prices has caused an acceleration in technical obsolescence and factor substitution in the production process, leading to a decline in actual and potential labour productivity.

I will not refer here to the OECD 1980 outlook, which is presented in the paper and summarised in Table 3. It can best be summed up by saying that, barring another apocalypse, the overall outlook is pretty flat. For the medium term the crucial factor is of course the track of oil prices. The landscape of the 80s, I need hardly say, is dominated by the shadow of the oil producers. On one scenario which is described in the paper, after an inauspicious beginning, the course of growth in the OECD area could achieve quite respectable rates of around 4 per cent by mid-decade — that is quite respectable by recent experience. But the assumptions lying behind the scenario must be spelt out:

i) absence of major world-scale disruptive factors such as bad harvests, raw material shortages and sharp cutbacks in oil production;

ii) annual real increases in crude oil prices of no more than 4 per cent;

iii) a successful winding-down of average OECD inflation to about half its present level by mid-decade;

iv) relatively strong growth performance of non-OECD countries.

This scenario cannot be ruled out as not feasible or unrealistic; but I think it must be described as optimistic. It would require as a *sine qua non* effective energy policies adopted without delay which, however, in themselves would involve substantial growth-inducing investment, thus promoting a virtuous circle of improved supply and conservation and diminishing inflationary pressures. It would also crucially depend on income claims brought into line with the trend increase in productivity and adverse movements in the terms of trade. No less importantly, it would require the successfull resistance of protectionist pressures and the adoption of policies to ease and sustain structural adaptation-policies which you will be talking about. Given the backlog of unrealised technical progress, the unused potential for improved international division of labour and the scope for a strong expansion of investment capacity the outlook for the 1980s need not alarm us. On the contrary, one might say with confidence that it would merit at least two cheers. However, whether this desirable outcome will actually occur will as always be dependent on the will and wisdom of decision-makers in both the private and the public sectors to face up to and to solve the structural problems which are the unhappy legacy of the 1980s.

b) The Probable Growth of Steel Demand and Trade Through the Decade 1980-90

"Outlook for World Steel Industry up to 1985:
Demand, Trade and Supply Capacity" (Document No. 2)

Mr. KONO:

I deeply appreciate the opportunity which this OECD Symposium affords me to express my opinion on the outlook for steel demand and supply for the next decade. I am fully aware of the hazards involved in making such a forecast for the 1980s, particularly in view of the multitude of uncertainties surrounding the political and economic situation of the world today.

I would like to beg your indulgence and limit my present discussion to the world economy and steel supply and demand to the first half of the 1980s. I predict that world economic growth up to 1985 will continue at the annual rate of 4 per cent. I am assuming also that during the period up to 1985 there will be no sharp downturns in economic growth similar to that experienced in 1974/75.

Next, I would like to discuss my projection for steel demand through 1985. What I am going to present today is my personal view on this subject. I expect that under the projected economic conditions which I have just indicated that world apparent crude steel consumption in 1985 will probably reach the 900 million tonne level. As shown in Table 4 world apparent steel consumption will increase at an average annual growth rate of approximately 3 per cent during the period 1978-1985. A breakdown of the 900 million tonnes figure shows that the industrialised countries as a group will consume 445 million tonnes and developing countries 135 million tonnes. This gives us a total of 580 million tonnes and an annual growth rate of 3.2 per cent for the Western world. The corresponding consumption figure for centrally planned economies is 320 million tonnes. Putting the two figures together brings world consumption to 900 million tonnes. However, if world steel demand remains stagnant the total figure may be as low as 850 million tonnes.

In my forecast of world steel demand, the steel demand in developing countries and in centrally planned economies warrants special attention. Figure 1 shows steel consumption by country group up to 1978 using 1973 levels as 100. Steel consumption in developing countries in 1978 was remarkably high — 153 — and consumption in centrally planned economies, reached a high 124, in contrast with industrialised countries consumption, which was only 86 per cent of 1973 levels. This high rate of growth in steel consumption in developing countries was primarily caused by the high growth in steel use among both the oil-producing countries and the newly industrialised countries. As shown in Figure 2, I predict that the lower economic growth rate and the tendency of oil-producing countries and newly industrialised countries to review and trim their development plans will reduce the growth in steel consumption of developing countries to around 5.8 per cent. I believe that the growth rate of steel consumption in centrally planned economies up to 1985 will also fall back to somewhere about 3 per cent a year.

On the other hand I anticipate that the steel consumption of industrialised countries will continue to grow at 2.4 per cent if their annual economic growth rate is 3.8 per cent. This is because the growth of steel demand heavily depends on the level of fixed capital formation (see Figure 3). For the foreseeable future I do not expect high growth in fixed capital formation in the industrialised countries because of the lower return on investment and the high cost of capital.

Now let us look at the steel supply capacity through 1985. I would like to point that there is no uniform formula for calculating steel capacity in all steel-producing countries in the Western world. At present three different ways are used to calculate steel capacity: nominal capacity, effective capacity and production capability. (See Figure 4 for more details.) The results of these supply capacity assessments differ significantly one from another. We need to establish an internationally acceptable method of assessing capacity. In my paper the steel supply capacity of the Western world has been calculated on the basis of effective capacity.

Now, let us look at steel capacity by country groups. As can be seen in Table 5 the steelmaking capacity of industrialised countries in 1985 is predicted to be 562 million tonnes, a level which shows little increase over present capacity. In contrast, the capacity of developing countries will be 114 million tonnes, a large increase commensurate with the magnitude of the growth in demand and honest desire of these countries to increase their degree of self-sufficiency. The total capacity of the Western world will be 676 million tonnes. On the basis of an effective capacity of 676 million tonnes and past production experience, the figure for possible production in 1985 can be calculated as around 600 million tonnes.

Now, I would like to compare the projected steel demand of the Western world in 1985 with the production capacity in the same year. As I mentioned earlier, steel demand will be about 580 million tonnes. Adding to this approximately ten to 15 million tonnes for the net export to countries with centrally planned economies we find that the steel demand supply position in 1985 will be tight as shown in Figure 5. This means that the present surplus capacity in the Western world will diminish steadily and will almost disappear by the year 1985. Furthermore, due to the structural change which has taken place in steel demand since the 1973 oil crisis, the production capacity of some specific steel products may fail to meet demand because of a lack of rolling capacity.

Today I have no time to discuss the trend of world steel trade. However, let me just summarise my views: the growth of world steel trade between 1978 and 1985 will be relatively slower than the level attained from 1973 to 1978 because the growth in steel demand in centrally planned economies and in developing countries will fall below previous levels.

19

Now let me summarise my forecast for steel demand, trade and supply through 1985: first, I foresee steel consumption in the Western world growing to 580 million tonnes at a growth rate of about 3 per cent; secondly, I predict that possible steel production in the Western world will remain about 600 million tonnes. This will mean that supply and demand will be balanced with a possibility of some supply shortages Thirdly, I expect world steel trade to grow at a rate somewhat lower than that attained up to 1978.

Before closing my presentation I would like to touch briefly upon two particular problems facing the steel industries of industrialised countries: one is the steel industry's profitability and the other is co-operation with the steel plant construction plans of developing countries. The foremost task facing the steel industry, particularly in industrialised countries, is not just to resolve demand/supply problems but more importantly to carry out basic structural counter-measures. It is imperative for the steel industries of industralised countries to re-establish profitability by improving labour productivity and international competitiveness. It is also important that the industrialised countries accommodate developing countries' desire to be more self-sufficient in meeting their steel needs. We believe that a reasonable approach to this problem is for industrialised countries to provide suitable economic and technical co-operation for steel plant construction projects in developing countries. I am convinced that the best way to ensure the profitability and employment potential of the steel industry is through expanding steel demand by sustaining the growth of world economy and by making efforts to uphold the workings of international co-operation.

World Trends in Steel Consumption and Production to 1990
(Document No. 3)

Mr. WIENERT:

The subject of our Symposium is development of steel markets in the 80s. In one way or another we therefore have to deal with the forecasting of future developments. In a retrospective view the accuracy of former predictions is not always encouraging, because forecasts have often proved incorrect. In addition to this empirical fact there are also theoretical arguments for a certain distrust of economic forecasts. Nevertheless people go on producing them. The objective of economic forecasts is not to change uncertainty of future developments into certainty. This they cannot achieve. They are much more meant to draw a consistent pattern form available facts and theoretical relations given certain data and assumptions. They should also point out the consequences of past develop-ments for the future and estimate their quantitative significance. It is unavoidable, however, that the subjective expectations of the forecaster also influence the figures.

Let us first look at the development of steel consumption. This depends on the growth of GDP and on the development of the steel intensity of GDP. Steel intensity depends mainly on the level of economic development reached by a country or region. In countries which are still within the process of industrialisation steel intensity increases, in countries already highly industrialised it decreases. Therefore if there is no overall economic growth, steel consumption diminishes in the highly developed countries but expands in the less developed ones. The considerable range in the stages of development of the various countries of the world and its slow rate of change has an important influence on the average growth rate of world steel consumption. I would now like to give a rough picture of its development in the past two decades and of its future up to 1990.

In the 60s a relatively strong growth of world steel consumption was determined by the highly industrialised countries of today, hence average steel intensity was not yet declining. The consumption of the less developed countries had only reached such a low level that it had hardly any influence upon the level of world steel consumption.

In the 70s steel intensity in today's highly developed countries started to decline noticeably and with it the growth rate of steel consumption. Moreover this decrease was intensified by the slackening overall growth rate of their economies. In contrast to this steel intensity in the LDCs increased strongly and in addition the overall economic growth in these countries was less weak than in the developed countries. In consequence the high growth rates of steel consumption in LDCs led to a considerably strengthened share of these countries in world steel consumption. However, it was still not large enough to prevent the marked slowdown of the average growth rate of world steel consumption in the 70s.

In the 80s the LDCs will probably reach a share in world steel consumption large enough to stabilize the average growth rate of world steel consumption at almost 3 per cent per year, although the growth rate in the highly developed countries may continue to decline. Given an average annual growth rate of world GDP of 3.5 to 4 per cent, world steel consumption in 1990 might reach 1,000 to 1,100 million tonnes crude steel equivalent. Taking into account the steel-saving continuous casting process this would mean an effective crude steel consumption of 950 to 1,030 million tonnes. The share of the LDCs including China in world steel consumption would increase from 14 per cent in 1975 to nearly 25 per cent in 1990, while the total share of the EEC, North America and Japan would at the same time decline from nearly 50 per cent to only 40 per cent.

Let us now have a short look at production. By and large the regional distribution of steel production should of course follow the development of consumption both in time and space. I assume that the obsolete excess capacity in the highly industrialised countries will be shut down hesitantly and that the capacity of the LDCs will continue to expand markedly and operate at high rates due to an excess of internal demand over supply. Therefore the probability of a low operating rate of facilities in today's steel exporting regions, that is to say, the EEC and Japan, is rather strong. On the other hand the likelihood of threatening worldwide steel shortage is, I believe, relatively small. The LDCs share in world crude steel production could amount to 20 per cent in 1990 compared with only 9 per cent in 1975, whilst the total share of the EEC, North America and Japan would go down at the same time from 56 per cent to only 43 per cent.

To sum up, we can say that the steel market in the 80s will be characterised by first, a slower growth in consumption and production and second, a considerable shift of emphasis from the highly industrialised to the less developed regions. The highly industrialised regions will be mainly confronted with problems of adjusting their capacity whereas the LDCs will have their problems when setting up new production plants. Subsidies for obsolete and inefficient steel plants and restrictions on external steel trade may be able to delay this process; they cannot however prevent it in the long run.

CHAIRMAN:

You have now heard one economic forecast — there could be many, but we have set a tone for the environment — and two very professional analyses of consumption, production and capacity. Before turning to your questions I would like to ask for brief comments from several of our panellists, particularly focussing on the demand side.

Dr. MARSHALL:

This is indeed a fortunate position, to be able to comment on forecasters without having to forecast. I listened carefully: all of them presented a large amount of caveat to the effect that there is increasing difficulty in forecasting over the next five years.

The problem, I fear, is that the forecasters solve their problem by giving wider ranges and the policy makers still listen to the optimistic end. I sense in a meeting like this the desire to co-operate. That is important, for one of the easiest ways for us to leave here feeling we have co-operated is for all to agree that demand will pick up and that our problems will be solved by some one other than ourselves. So my reaction to the forecasts is: they are optimistic. It would be nice if operating rates would come up because of increased demand without having to face the painful side of the equation which is the capacity side.

The other point which I would make is that demand will be determined by people other than policy makers in steel and private steel companies; the world economy has a lot of actors and a lot of players, so we can treat demand as an outside factor and try to forecast it.

Capacity on the other hand will be controlled by decision-makers, many of whom are in this room. To assume that some trend in capacity will continue because it has in the past assumes that people will act in the way they have in the past. I think that the reason for this meeting is to try to encourage people not to act in the way they have in the past in terms of trying to let someone else solve the problem.

One final comment. All the forecasts are aggregate. If I had been asked to forecast I would have done an aggregate forecast too, but the real problem is a sectoral problem and I don't mean industry-wise but region-wise. Mr. Kono said that on the low side there would be possibly 850 million tonnes of demand as opposed to 900 million tonnes. If that 50 million tonnes is not spread evenly around the world all of you realise the tremendous consequences for any individual region. That is the level of uncertainty with which we are dealing here and it is very large.

Mr. ANDERSON:

Dr. Marshall thinks that these forecasts are optimistic, but I am afraid that I cannot even say whether they are optimistic or not. Most probably like most forecasts they will turn out to have been wrong. I say this because I belong to the fraternity of forecasters and we all know that you run great risks of exposing yourself with a forecast. The reason why this is so is that the long-term relationship between steel consumption and gross domestic product or whatever you take has been distorted over the last ten or fifteen years in a way that makes the use of global models for making steel demand forecasts, let alone steel production forecasts, really very questionable. That is why my institute, for instance, has refrained since 1972 from coming up with any forecast of steel demand, because we believe that these figures do not carry a reasonable expectency of being accurate and that they might therefore be more misleading than beneficial for the decision-makers in industry.

If you wish I will just briefly say what are the main reasons for this distortion in the relationship between steel consumption and any macro-economic indicator. We have undertaken a study of what we call the causes of the recession in steel demand of the mid-1970s. We have not just made this study in order to satisfy the curiosity of research economists, but rather to identify those factors which are of a long-term character and will therefore also influence the development of steel demand in the 1980s. We found of course that the factors which have caused the crisis in steel are also those which have caused the crisis in steel forecasting. Naturally there are many of them and economists will always tell you that it is very difficult to identify the effect and the impact of a single factor.

But if I am to identify particular factors, they are mostly concerned with developments in the field of investment, notably of gross fixed capital formation. Not only has the rate of investment slowed down already since the mid-1960s, which is very important for steel because two-thirds of steel demand in an industrialised country is investment-related; but also there is an important change in the expenditure pattern of investment away from investment for capacity expansion to investment for productivity increase; and the investment for capacity expenditure is much less steel-intensive than that for productivity increase. That is one of the long-term factors. To overcome the problem you would need to have not only good past data on capital formation, broken down into construction and all sorts of other things such as plant and machinery, but you also would want to have a good forecaster in this area; but then the same troubles that arise for steel economists in making steel consumption forecasts have also hit our colleagues the generalists because they also seem not to know really what is going to happen in the future.

So I must say that although as the French playwright said it is easy to criticise but to do it yourself is very difficult, I think that we should really try to improve our tools of forecasting before we come up with such figures as have been presented here.

Mr. BRUNNER:

I would like to pursue Mr. Anderson's comments perhaps a little further and even more sceptically. I believe that the problems of long-term economic forecasting are even more fundamental than he is suggesting. When all is said and done and however sophisticated the mathematics and the model-building, long term economic forecasting is really not much more than a reflection of the mood of the moment.

Ten years ago it was perhaps permissible to think that in some way the future was an extrapolation of the past. This is no longer the case. Even if one believed that it was, there is the problem of deciding which period in the past is relevant to the future. Dr. Ostry said that it is a particularly difficult time to be peering into the future. I would suggest it is always a particularly difficult time to be peering into the future. I believe that the steel industry can therefore only proceed on the basis, to quote an English expression which will probably defy translation, that anything can happen and probably will. I must say that I also feel, to rub salt in the wound, that as much restructuring is required of my profession as of the steel industry. It is time we got out of obsolete and non-productive activities.

But if all this sounds rather negative I am in fact trying to draw out a lesson which is possibly more relevant to tomorrow's discussion. Let me say that forecasts are getting no better. The Interfutures Study which has been referred to thought it was a reasonable assumption to make that the real price of oil would double by the end of the century, but it had in fact doubled by the time copies reached Australia.

I genuinely don't want to be negative so much as to draw out a point that I think is worth making by reference to a particular forecast which I remember very vividly. Ten years ago, when I first interested myself in this area, Japan had just announced that she was going to produce 160 million tonnes of crude steel in 1975. This threw the North American and Western European steel industries, in my experience, into near panic as they realised what this meant in terms of Japan taking up the entire growth of world trade. As we know, in 1980 the Japanese are battling to produce 50 million tonnes less than that. I am not saying this in order to have a dig at my friend Mr. Kono; the reason I am saying it is to point out that the Japanese have learnt how to respond to that appalling forecasting error and to operate profitably at a breakeven point of 70 per cent. In the last half-year they have turned in profits 36 per cent higher than in their record half-year in 1973, and they have done this despite the facts of their high debt/equity ratios and their lifetime employment which were once seen as unfair advantages and may have been when working at 100 per cent capacity utilisation but manifestly are not an advantage when working at 70 per cent.

I would like to suggest, therefore, that we probably have a great deal to learn from the Japanese experience, because I do believe that the steel industry worldwide has got to be able to learn to live with a much wider range of possible outcomes than we have looked at in the past. In other words we have all got to be able to operate at breakeven points round about 70 per cent.

CHAIRMAN:

I think at this early point in our deliberations we have concluded that economists are fallible and that forecasting leaves something to be desired. I don't know that I would go so far as to recommend that the economists eat that broken glass as they go about restructuring their industry, but do look forward to continuing to hear from you. I now turn to the floor for discussion. I have not known the steel industry in my country to be particularly inhibited or shy, so I trust that you will speak out.

Mr. ALAN WOLFF:

I have a question for Mr. Kono. If you take Dr. Ostry's assumptions for growth in GNP in the short-term outlook and say that that might be the pattern for the coming four or five years, what then would be the relationship of capacity to demand look like in 1985?

Mr. KONO:

As I stated earlier and also mentioned in my paper there is a great deal of marginal capacity which has already lost international competitiveness, therefore it is desirable that these obsolete facilities be closed as soon as practicable. The figures which I mentioned earlier might be deemed optimistic and if the situation in the future demands it the closure of obsolete facilities must be accelerated to an even faster rate than I predicted.

Mr. WOLFF:

If one assumes relatively little growth or no growth in the OECD countries and relatively flat growth in the developing countries obviously there would be a balance in capacity and demand if you phase out enough capacity. But knowing what capacity exists I wondered what kind of shortfall might be anticipated in steel demand.

Dr. OSTRY:

I think, if you look at Mr. Kono's paper and mine, that his projection for the OECD up to 1985 — the average growth up to mid-decade, not the immediate outlook in the next 18 months — is just about the same, 3.8 against 4 per cent, so that the demand side insofar as it is determined by the assumptions about average growth would not be very different; but if you look at his Table 3, which breaks down the GDP forecasts, he assumes a much lower rate of growth, 4.8 per cent for the non-OECD countries, than is contrained in the 6 per cent projection on my part. So I just don't know how that would affect his total outlook.

Mr. SOHLMAN:

In his paper Mr. Kono states that if world steel demand remains stagnant the total figure of steel consumption may decline to 850 million tonnes. I would like to ask Mr. Kono how he arrived at this figure because I think that a decrease in the forecast from 900 million tonnes to 850 million tonnes is relatively small assuming that steel demand remains stagnant.

Mr. KONO:

When I said 850 million tonnes it should be considered in relation to what Dr. Ostry said. The World Bank GDP forecasts suggest 3.5 per cent growth in industrialised countries but I believe that demand would not go as low as this. I also assume that the demand from the communist bloc would fall from 30 to 20 million tonnes.

Dr. MARSHALL:

One of the issues which was very difficult for me to sort out in the forecast and it might be worth some discussion was what is the net import forecast for non-OECD members. One of the tendencies, I think, is to assume that part of the problem, the OECD Member country problem, will be solved by exporting a large amount of steel to developing countries. One of the reasons I had a difficulty in understanding what the forecasters felt here is there is not a consistent definition of LDCs. A couple of people on the panel of speakers this afternoon have done work in this area. Could we get those people who have looked at the non-OECD consumption in capacity forecasts to give us an idea of how much net imports in non-OECD countries are going to go up by 1985?

Dr. NIJHAWAN:

I would like to make a few comments about the fallacy of forecasting. The forecasts normally do not follow any dynamic models but have to be based on a pragmatic basis. If you look at the year 1900 world steel production was about 40 million tonnes. In 1950 there was production of the order of 200 million tonnes going up by 1967 to about 500 million tonnes. If somebody had forecast in the year 1900 would he have come up with a figure of 710 million tonnes in the year 1974? What model would he have followed? What would be the basis of his forecasting? Forecasting, even though we follow sensitivity models and intensity studies, very often overlooks some of the largest and most important international events such as the two world wars between the years 1900 and 1974. The figures which have emerged as a result of these international events have raised steel production from about 40 million tonnes to 710 in 1974 and 745 in the year 1979. Japan increased steel production after the war from possibly a million tonnes to 120 in 1974 while hoping to achieve something of the order of 160 million tonnes by 1985.

Nobody could have forecast the oil price increases and nobody could have forecast also the possibilities of global conflagrations of the order that we have witnessed during this century. The forecasts which you see in the documents have been made more on a pragmatic basis based on underlying current depression of economic models due to the oil price increase and to many other factors which are not directly related to any models or forecasting. For example, when the Japanese were not permitted after the war to expand their defence industries and concentrated their efforts on the steel industry, what model did they follow to raise their production from one million to 120 million tonnes by the year 1974? I believe there were no real models and that is the fallacy of forecasting. If you look at the figures which have been given I believe some might say they are optimistic, others might say they are pessimistic, but they are governed by factors very different from technical and economic factors and follow international trends which are not easy to analyse and depict in a forum such as the one we have assembled here.

Mr. REBHAN:

The Secretary-General this morning stressed the importance of the steel industry. It was a British politician from a steel constituency, the well known Mr. Bevan, who said that steel represented the commanding heights of the economy. I think that everyone has stressed here how important steel is in modern life and that it is a cornerstone of development in the Third World. Although it represents economic strength everywhere in the world, it was allowed to slip from boom to bust and the fat years and the lean years determined the investment with little concern for market conditions of far-sighted modernisation of the works.

Let me speak for a moment from the point of view of the workers in the industry. Under such circumstances, workers are the first to be taken for granted as they pay the real price of boom and bust. They pay the price in insecurity and, as you know, they are involved in dangerous working conditions. The present prolonged steel crisis which we are going through is a challenge to all of us to develop the tools for well determined and socially responsible — let me underline — steel policy. We cannot simply talk about restructuring for the next crisis. We must make provisions and adjustments for a healthy steel industry with viable jobs. Restructuring, as you well know, not only means the physical remodelling of plants and workplaces, but must start and begin in the decision-making process where it is absolutely essential that the voice of the employees or their unions is heard.

We welcome the recognition by the OECD of these facts and I would like to throw some bouquets for the excellent economic reports to this conference. Despite the gloomy statements made by the other panellists here I think it is important that they predict the course of developments and that alone is the proof of the need for such a symposium. Dr. Ostry has said that the economists — those people who make the predictions — have to eat ground glass. Looking at the future was always difficult: Sir Winston Churchill used to tell journalists that he much preferred to consider the future after the event because he could contemplate the future much better with the benefit of hindsight. Unfortunately in this deep crisis we cannot be that indulgent. We cannot just wait and see thousands upon thousands of steel workers out of work. Contrary to what Mr. Anderson said, we regard forecasts, whatever their limits — and we know what their limits are — as essential and we believe that they can be made in such a way as to increase their usefulness in working out effective policies and in making intelligent decisions. We need orders of magniture which represent the most probable approach on the basis of a thorough ongoing study of developments.

Forecasts are not an exercise merely to assess the supply and demand of the various markets, but should also take into consideration policies such as the need to maintain employment in a competitive steel industry. Therefore consultations among governments, steel employers and the trade unions like those practised in establishing the General Objectives of steel in the European Community are just one illustration of the social and economic analysis that we have in mind. In this respect we share the views of Mr. Kono's paper that "the forecasting of world steel trade is one of the most important tasks of the OECD Steel Committee"; Mr. Kono further added that "we need in any case to establish an internationally acceptable method of assessing capacity".

We believe that we cannot take a static view of the world economy. Forecasts with all their limitations are necessary, but will never tell us how we are going to create the 60 million jobs that have to be found between 1980 and 1987 in the industrialised North alone, and what is more crucial — and I think as OECD countries we should be very much concerned about it — is the desperate need for the 600 million jobs in the South — just to quote a most recent statistic of the ILO.

Dr. Ostry's paper on the world economy speaks of manifestations of an intensified income distribution struggle at a national and international level and rightly recommends more investment, which is essential in our opinion and will increase the overall demand for steel. But investments, whether they are made for expansion or for the improvement of the economic infrastructure, cannot be at the price of depressing wages. Faster economic growth than is at present being experienced must be achieved and this can only be done through imaginative social policies, all the more since as Mr. Anderson pointed out a good deal of this investment will be for modern technology.

There is an enormous backlog of demand in developing countries, as we all know, if the people in those regions are to enjoy a better and fuller life. This must be done through their own internal development in which steel has to play such an important part. The forecasts which have been submitted to this conference recognise that there will be an enormous need for steel in the Third World for the next five or ten years, and this will maintain to some degree an export potential for industrialised countries; and as the developing countries develop more steel industries there will be an exchange of different steel products.

The metal unions in the industrialised countries joined together with metal unions in the developing countries, are aware — and I want to stress this — of the need for adaptation in view of the growing steel industry in other parts of the world. At the same time, it must be pointed out, there is no advantage in building massive basic steel complexes in developing countries when these resources would be better employed in establishing steel processing and manufacturing plants spreading employment throughout the economy. I am sometimes very irritated when I hear people say that the steel industry ought to go to developing countries alone because steel uses a lot of energy, uses a lot of raw materials etc. The next thing will be we will send our modern industry to those developing countries. What are the people in the industrialised countries going to do then? Compose music? Become Nobel prize winners? Write poems? How are 50 million people in Great Britain going to live? I hear this in Great Britain quite often — I think we should contemplate that problem for a moment.

I am sure that it is our common view that developing countries do not want to produce steel for its own sake, but rather for employment of the people; and this requires a development of an internal market. We are pleased that developing countries are represented here, and I think the Secretary-General should be congratulated for having arranged this.

We agreed with Dr. Marshall when he said that OECD should therefore investigate the possibilities for an international sectoral approach in steel in order to make these predictions and forecasts a little more orderly. Let me stress just in conclusion, that the social problems which we will be discussing tomorrow cannot be solved in a vacuum. We need the right measures at the general economic and industry level guided by social objectives.

Mr. MILES:

At this point I just wish to respond to a question put by Dr. Marshall and also raised by the last speaker as to what is the size of the LDCs requirements of steel from the developed countries? In 1978 their net import of steel from the developed countries — that is their imports less their exports, was 26 million tonnes. This has been a very fast growing trend. At the beginning of the decade in 1970 it was only 12 million tonnes. There is also the requirement of the centrally planned economies, which took net imports of 10 million tonnes in 1978 from the developed countries.

I think I should just respond to Mr. Kono's doubts as to whether the strong growth trend in these countries will continue into the 80s. This very strong growth trend is a little more robust than perhaps he is hinting. When I come on to my presentation shortly, I will say that in fact we had assumed that these countries will go on requiring an increasing tonnage of steel from the developed countries. In Mr. Kono's view, I think, their import requirements will stay the same at about the 25/26 million tonnes level.

Dr. KORF:

This discussion on capacities needs some clarification. We are talking about crude steel capacity. But you cannot sell crude steel. Only rolled products are relevant for the market. Between the crude steel capacity and the rolled product capacity there is a very big difference depending on whether you use ingot casting or continuous casting, to which the whole steel industry is in process of changing. The yield of a tonne of crude steel through continuous casting is 20 per cent better than through ingot casting. I think that in future people should look more to finished steel capacity than to crude steel capacity.

Future capacity also depends on availability of raw materials and energy. Do we know that all the iron ore is there to produce 900 million tonnes of steel? If you talk to the large iron ore companies of the world about their future plans, they have the same problems as the steel industry; they have had a major recession in recent years and they are also fighting for the money for new investment. If these investments are not made very soon maybe there will be a lack of steelmaking capacity because of a lack of iron ore.

Or look at energy — at coking capacity. Last year the United States imported 5 or 6 million tonnes of coke, because due to environmental problems they could not produce the coke in their own country. If this continues, there will soon be a limitation on the United States blast furnace output due to lack of coke.

Also, since the recession in 1975/76 a lot of people have been laid off by steel plants and even if the technical capacity is still there in some industrialised countries there will be problems in getting the people back to use all the technical capacity. In the Federal Republic of Germany, we produced in 1974 — the peak year — 54 million tonnes of crude steel. Last year we were back to 46 million tonnes, yet in some publications you read that there is a capacity of 69 million tonnes of crude steel. I have great doubts whether in Germany or in other countries all the capacity which is shown in the publications is really there and can be used.

Let me say a final word on obsolete plants. I think there are a lot of obsolete steel plants in the industrialised countries. They have to be closed, because if not the companies which own them will either go bankrupt or will be dependent on governments. This means subsidies; but to have a healthy steel industry in the future it must be the target to have all steel companies run by the standards of the free enterprise system. I think it is a wrong philosophy to see a steel company as an employment agency, as is done in many cases.

REPRESENTATIVE CHARLES VANIK:

As a political observer, I am least qualified to participate in projections of steel demand; but in all of this morning's discussion excepting a footnote in Dr. Ostry's statement, I heard nothing about the new steel demand which will develop through the dynamics of energy substitution. Substituting for oil will be driving the world, or should be, into energy alternatives almost all of which will demand huge supplies of steel: oil from coal, and coal gasification consumes steel in some computable proportion to the coal converted. These massive installations require enormous utilisation of steel. The inducements and subsidies which America is providing for alcohol fuels should be almost as dynamic as those which provide for conversion from coal. In America we are contemplating now spending $20 billion for synthetic fuel development. In our windfall tax bill we are contemplating incentives which ought to generate about $100 billion worth of activity both in conservation and in alternative energy sources. If the rest of the world is as deeply involved in energy alternatives to oil, I am concerned as to whether we will have enough steel to produce and take care of conservation and the synthetic fuel processes.

Dr. OSTRY:

I think that is implicit in and perhaps it should have been more explicit in the paper. The so-called optimistic scenario of growth reaching 4 per cent by mid-decade after a recovery from the present recession — whenever that is going to take place — does indeed involve this kind of investment-led activity — a good deal of which will stem from either capital replacement which is more energy efficient or the requirement of alternative energy sources in the OECD countries. That is why I say that that scenario is a possible scenario, but depends crucially on undertaking a coherent energy policy.

Dr. CRANDALL:

I think that Mr. Vanik only looks at one aspect of the substitution which is now going on, which is the forthcoming substitution of new forms of energy for older forms of energy in the United States. In fact, there is considerable evidence in the economics literature, which you may choose to trust or to distrust, which suggests that energy and capital are complements, the package of which are substitutes against labour. When you increase the price of energy — as is happening right now — you substitute labour for capital cum energy, a result which could well offset the increase in the requirement for steel. I would see also in the consumption of goods and services that there is likely to be a substantial substitution against heavier materials in favour of lighter materials, and I do not think that is likely to augur well for steel.

Mr. WIENERT:

It is true that energy investment requires steel, but whether such quantities of steel sould be required as we imagine today is a point which we should study in greater detail. Our major nuclear power plants need 50,000 to 60,000 tonnes of steel. Compared with the overall domestic steel consumption in Federal Germany, for example, with a rate of two or three nuclear power plants per year, these are quantities which are not really terribly important. The same applies to major facilities. Of course, for major industrial facilities we also need steel — 100,000 or 150,000 tonnes perhaps. This is one aspect. There will be a demand for steel but will the demand be as great as we imagine and as we could hope?

A second point, this additional demand for steel replaces a downward trend in other areas. We have already heard, for example, that rising energy prices create a situation where we have to save weight. This applies to the motor manufacturing industry. If we consider that the motor industry in the coming five years will reduce the weight of the average car 15 per cent, this involves a considerable loss in steel volume which will be compensated only partially through the investment necessary because of the energy situation. I do not think that we should be over hopeful about developments of that sort. From the mere changes in the use of energy we cannot expect a major recovery for steel. There will be a major recovery for steel demand if energy investments have a kind of cumulative effect, if there is an expansion of economic growth which would mean that additional investments in highly industrialised countries would become more important as part of the overall investment budget. Then we would have to reconsider the whole question of steel intensity.

I would also like to refer to the subject of the net steel imports of developing countries and whether these could save or at least support the steel sectors of developed countries. In the past, it is true that developing countries were an ever-growing market for steel exports from the European Community and from Japan. But if domestic markets in certain developing countries grow so much that there is no risk involved in building up to one million tonnes maximum capacity mini-mills, then this will happen; it is in fact a process which is already under way. We cannot forecast the volume of net imports. We can only say that they will not increase as fast as they did in the past since steel capacity in developing countries is rising and will be expanding as their consumption levels rise. Net imports depend on consumption levels and capacity levels. I gave two ranges for both in my paper. The problem is that, depending on whether I use the low consumption figures and the high capacity figures or vice versa, there is obviously a considerable impact on net imports. Thus we have the forecast that the net import needs of developing countries can either fall or rise. This is a forecast which is not of much use to steel capacities in industrialised countries. Net imports from developing countries will not save industrial countries. I think they will increase the risk since any extension of capacity in developing countries would have been covered by imports. If expansion in developing countries is insufficient then we are faced with a new situation.

CHAIRMAN:

Dr. Marshall, do you still feel optimistic?

Dr. MARSHALL:

My basic feeling is that net imports by the developing countries has been helpful in the past in solving the problem of excess capacity in the developed world. What I worry about is that we will continue to rely on them as a means of avoiding other policy choices confronting developed countries. I think Mr. Kono believes net imports will stay constant and I think that Mr. Miles believes they will go up. The CIA Report last summer believes that net imports will decline because capacity will grow very rapidly in the LDCs. What I worry about is that people will first decide what they want to do with their own capacity and then justify that decision on the basis of what LDC forecasts are going to pick out of the hopper. That is why I brought the issue up. I think that such forecasts will be used to rationalise whatever decisions various developed countries want to make, and I would warn you to worry about that. I would worry about it as I look at the various decisions that will be made in the next year or two.

PROFESSOR HÖÖK:

Mr. Vanik said that he was possibly not the right man to take part in a discussion about the forecasts. In a way I think he is exactly the right man to answer questions about the uncertainties in our forecast. Dr. Ostry refers in page 63 of her paper to the huge forecasting errors made at the beginning of the 70s — I am not quite sure that there were such big errors. The forecasts made in OECD at that time were about potential growth made from the capacity side; but analysing what has happened in practice you can see that it has been mostly a failure in demand management policy, resulting in unused capacity, high unemployment and so on. The forecasts in a way were quite good technically; and the same applies to most of the steel forecasts. Due to failures in governmental policies there has been a really big fall in investment. The basic problem here is in forecasting the efficiency and capability in governmental policy and whether governments will make deliberate efforts to pursue and achieve stable growth policies. I think that is the real question we have for Mr. Vanik and other politicians. If we shall not be able to have a steady growth policy and a good economic policy then we can give up most of forecasting about steel and about capacity demand and so on for other products.

I think that in this Organisation especially, where you have all governments discussing economic policy, there should be the right people to give some idea about the future outlook from an economic point of view and whether governements are going to be able to pursue a rational economic policy for the future five or ten years.

CHAIRMAN:

That is a very good point. My limited experience would indicate that politicians are as imperfect as economists and therefore it is going to be very hard to resolve that one. Do any of the panel want to respond at this moment?

Mr. ATKINSON:

I would like to respond in part to a line of reasoning which Representative Vanik brought up and was taken up by a later speaker, namely, energy and more specifically the use of steel in automotive production. It is true, of course, that the manufacturers of automobiles are greatly decreasing weight and that this takes a lot of steel out of the vehicle. On the other hand it is not quite so simple as that. As energy costs change we must realise that the lighter materials have a high energy content; indeed someone has characterised aluminium as practically pure energy. Moreover the petroleum feed stocks for plastics have gone up greatly in price, so that I think we can be unduly crude in our forecasts on a poundage basis as to how much of steel and steel products will be coming out of the car. I see quite a great elasticity of substitution at work in price, in weight and in strenght so that the quality and the value of the steel components in an automobile may well be much higher than one would anticipate simply on the basis of the reduction in weight which we are anticipating.

PROFESSOR RUIST:

This last intervention brought up a point which is very important. We are accustomed to express our forecasts in tonnages, but when we find that the specific steel consumption counted as a number of tonnes per dollar of GNP or something like that is declining, we do not take into account that one part of that decrease is due to a higher quality of part of the steel. The average quality of steel is certainly increasing in time and that means that the output of the steel industry in terms of dollars in constant prices is not decreasing at the same rate as the tonnages. We should take this into account in our forecasts.

A second point which was also mentioned is the price of steel. Dr. Marshall said that we always take demand as exogenous in our discussions. I think we try to forget that there is a price elasticity of steel even if it does not operate within a few months or one year. It is a long-term tendency, but it is quite certain that in the long run steel demand does react to changes in relative prices. That is also something that we should take into account.

Mr. ANDERSON:

The steel used in the field of energy is relatively little in terms of quantity but it has a very high value because all these new alternative processes, for instance, liquefaction, gasification of coal need very high quality steel because they have to resist very high pressures and temperatures.

The overall effect of the impact of future energy price increases is very difficult indeed to assess. We are at present trying to do something of the sort and we have asked around among our members what they feel will be the impact. At present I think that the steel industry in several countries is divided. Some think it will lead to higher use of steel and of heavier products, others think that it will lead to lower use of steel. Both groups explain their point of view very well; for instance, higher investment to open up new energy sources or to economise on energy will divert resources from investment for ordinary growth purposes; that is to say consumers' disposable income and other things will grow at a slower pace and this will not fail to have an impact on the demand for steel. There are others on the opposite side. Coal people, for instance, say that the major energy reserve of the world is coal and therefore the only hope we have is to develop its use further. This of course would lead to massive investment in ships and pipelines by which to transport the coal and the building of new power stations or replacement of domestic heating by district heating which needs beautiful quantities of steel pipes. So it is very difficult indeed to answer this question as to what the overall impact will be.

Now to return just for a while to the development of the imports of the new countries — the newly industrialising countries of the developing world in general. There is no doubt that what we called years ago the deficit-covering part of net imports into the developing countries has been shrinking historically in relative terms. However, the value and quality of the steel which still continues to be imported into these economies as they grow along the path of industrialisation has of course increased, but the problem here (and this is one of the shortcomings of these forecasts) is what will happen not only to direct, but also to indirect trade in steel, that is to say the imports and exports of steel-containing manufactured goods.

There are already changes underway which will affect the future steel consumption in the traditional exporting countries. Just take the case of motor vehicles produced in Brazil; they use the steel produced in Brazil rather than in any of the former exporting countries, and they are not only kept in Brazil but they are exported to other countries that used to be traditional markets of the long-standing automobile producers.

This whole problem brings out one point: these forecasts in tonnage terms and in crude steel equivalent should be changed because they don't really mean what they used to mean; they should also be corrected for the impact of indirect trade in steel; this will I think be an absolute must for anybody who wants to look at steel demand developments in the 1980s.

Dr. PECO:

I think that we ought to look at all of these reports and consider the whole of the discussion that is taking place at today's proceedings. At this stage of the discussion as far as I am concerned as regards the behaviour of companies in the steel market in the difficult years foreseen, I should like to stress certain aspects contained in the three documents that were presented this morning and which have been discussed up till now.

I would like to begin by what you yourself, Mr. Chairman, and the Secretary-General of the OECD said at the beginning of this meeting. You gave us a sort of leitmotiv concerning studies to be made and results to be achieved. You spoke of the search for or establishment of an economically fair price on the market as a result of free market forces combined with maximum self-discipline in international trade. Within this framework we have been discussing the balance between supply and demand this morning in terms of general economics and more specifically in connection with steel markets in the coming years. There has been argument about optimistic or pessimistic forecasts. Dr. Ostry told us, that we had a tendency to be optimistic in the 50s and 60s and it has already been pointed out that in forecasting we tend to be influenced by circumstances prevailing at the time the forecasts were made. I would not like to be amongst those who still consider that our forecasts for the 80s were optimistic. It is difficult to know what is really going to happen, but Mr. Kono in his paper spoke of a gradual re-absorption of excess capacity. It is true that this re-absorption will probably be simpler and easier at the level of products; it will be a more difficult business in terms of crude steel. Mr. Anderson a moment ago, and Dr. Korf earlier, spoke about the effects of continuous casting and we will hear more about this in connection with the reports this afternoon.

In fact, some re-absorption is possible and in the years to come we may manage to strike a new balance on the steel market but what is important is the different developments taking place at the geographical level of consumption and production. Another argument that we have just heard concerns net imports into developing countries and in international trade generally. Mr. Wienert in his paper has discussed the distinction to be made between trade which is necessary to balance out production and supply in various countries and more competitive forms of trade. What I am interested in is the presence on international markets of offers of steel. It is not so much the actual trade done as the pressure on the markets, the desire to export, the desire to sell which, given the existence of over-capacity will impose a very heavy burden on international markets and will make it very difficult to get an economically fair price established.

Mr. Wienert has also spoken of the necessity to assess the development in steel consumption according to income per capita in the various countries. There are old studies in this field — I myself dealt with it ten years ago. I would like simply to stress that each country, each market has its own development curve concerning developments in consumption. It is fast to begin with and can even at the end of a period of time be negative. The structure of each country is different so the possibilities of consumption vary too. Those are some considerations that it seems to me will link up the other reports which we shall have a chance to discuss this afternoon.

Mr. ODD GÖTHE:

Coming from Norway, I have a specific interest in the future of shipping and therefore the shipbuilding industry's activity. As I understand, the Norwegian shipowners' views on the future of shipping vary but there is something of an optimistic view that the shipping will have a slight upturn from 1983 perhaps up to 1985. If this is true, this will certainly have an effect on shipbuilding and therefore on the steel consumption in shipbuilding. So my question goes to the panel: is there any specific view on shipbuilding activity around 1985?

Mr. KONO:

Since last year the production of the Japanese shipbuilding industry has been increasing and in 1980 the production capacity is going to increase to 5.5 million tonnes. Mr. Vanik referred to the increasing utilisation of coal; there is going to be a shift into the use of smaller tonnage ships. There might also be some transition from oil to diesel so that as far as the shipbuilding industry is concerned the situation is going to be better.

Mr. STEFFEN MØLLER:

Two very brief comments, first about forecasting. I think we have heard a lot about forecasting problems, overcapacities, demands, imports and so on, both from the optimistic and from more pessimistic points of view, but we have only in Dr. Ostry's paper seen some indications of changes in general productivity; we have not seen much about steel productivity. Could you give any forecast of productivity within the steel producing companies?

Secondly, some of us are responsible for employment, and whether we like forecasting or not we need dimensions for the future; but if we only know about

33

tonnages and percentage increases in tonnages we do not know much about the employment situation. So it is very necessary for us who are going to monitor the employment side of it to know a bit about the dimensions of employment in the future.

Mr. KONO:

In the iron and steel industry the improvement in productivity must be dealt with in a different way. First of all the energy consumption must be considered. Secondly, how to improve the yield; this is also very important since the yields can be improved by the use of continuous casting. Thirdly, Mr. Anderson has pointed out that the simplification of processes can improve productivity. These three points will be very important factors in the future in assessing future prospects.

I can mention an example in the Japanese iron and steel industry. From 1973–1978 there was a 10 per cent saving in energy. During this period the yield improved by about 4 to 5 per cent; and this is a very good way to improve productivity. As regards employment, I am not a specialist so my answer might not be sufficient but I can tell you that in Japan it is very important to maintain the level of employment. This is what the industry is trying very hard to do.

Mr. WIENERT:

Productivity increases in the steel industry are still possible on the scale that we have known up till now. On the one hand this will result from the replacement of old plant by new plant and the replacement of outdated processes by more modern processes. But it seems likely that modern plant will also offer opportunities for productivity gains — a more rational process than up till now could give rise to such gains. In terms of aggregates we can optimise the way in which production processes take place, but it will be difficult to think in terms of aggregate processes within a single works. It seems to me that there are considerable possibilities through electronic control mechanisms. This will also have repercussions on the restructuring of employment: new jobs will be created, in particular in the field of supervision of electronic equipment; and there may be a reduction in less qualified work.

In the case of the Federal Republic of Germany I might briefly mention the following. We had a reduction in employment in steelworks at a time when production was still going up in terms of tonnage. It is clear that if we expect a reduction of production in the 1980s this must have repercussions on the number of people employed. There are already agreements on shift work and working hours and so on, and it seems likely that these may have some stabilizing influence, but the problem of the reduction of numbers of jobs in the highly industrialised countries is something that we will inevitably have to face up to. Certain processes will give rise to the creation of new jobs but only in those industries which are competitive from the international point of view.

Perhaps I could very briefly take up the argument about higher earnings and higher quality. Better earnings through more know-how does play some role. It is true to say that you cannot speak purely in terms of tonnages. But our problem is this — the higher consumption in the developing countries is for mass-produced steel and for steel profiles and we have considerable capacity of this sort in the highly industrialised countries. It is important to have high quality steel but these high quality steels would have to gain in importance in the developing countries. We find that the special steels are becoming more important in the highly industrialised countries, but this doesn't always correspond to consumption structures in other countries. We have capacity in our countries to produce types of steel for which there is more demand in developing countries.

Dr. CRANDALL:

I wonder if I could come back to the first question that was asked by Mr. Wolff and try to pose it a little more starkly. The premise of his question was, how would Mr. Kono's projection for the 1985 situation differ if his projections on GDP growth were the same as Dr. Ostry's? I would like to ask Mr. Kono how his Table 5 would differ if for instance the projected GDP growth in the world were 30 per cent greater or 30 per cent less than he assumed. That is, how would he see the capacity growing in the developing versus the developed countries if GDP is somewhat substantially higher or substantially lower than we had projected?

Mr. KONO:

In the advanced industrial countries, when we forecast we have to compare the growth of gross domestic capital formation (GDCF) with the growth of GDP. As regards developing countries we cannot really integrate them in one category, but as Mr. Miles pointed out in his paper on page 107 Table 5 must divide them into industrial countries, oil-producing countries and primary goods producing countries. The industrial countries and oil-producing countries account for 70 per cent of the demand of developing countries and have the key to the demand of the developing countries. We have been told 7 or 8 per cent growth of steel consumption in developing countries is possible, but I regard the growth of demand as probably 6 per cent considering the fragmentation of the situation in developing countries.

c) **Changes in Capacity and the Supply Potential**

"The Growth of Steel-making Capacity in the 1980s" (Document No. 4)

Mr. MILES:

First I want to make it clear that this paper is a co-operative effort by the OECD Secretariat, principally the research work of my colleague Mr. Suzuki. Capacity figures are some of the least trustworthy statistics we have in the area of steel. When I was trained as a statistician many years ago I was told first to say that the figures were so unreliable that one should disregard everything, but from then onwards to assume that everything was absolutely reliable. Well, I will proceed on those lines, but a healthy element of scepticism is appropriate here. We have got some faith in these figures but we certainly know their limitations. We have been collecting statistics of capacity of the steel industry from Member countries of OECD for something like 20 years and we have those records which we have scrutinised. We have also looked at the countries outside the OECD area and we have scrutinised the information which we have plant by plant and come up with the estimates set out in our paper. All I can say is that these are the best estimates we can make. But I repeat that some element of scepticism is appropriate.

The first point in our paper is how inflexible the decisions on capacity have been. We have the remarkable situation in the 1970s, when there was a very marked change in the market situation in the first half of the decade, yet capacity in the OECD countries went on growing year by year right up to 1978. I think that should be a lesson to us to make up realise how long the time lags are between decision and fulfilment in the steel industry. The reverse side of that is that, if there is some future change in the market situation, then again we should expect some considerable time lag before there is a response in capacity. If we do not foresee this, if in fact there is a better demand situation in a year or two, then it will take a long time to convince the decision-makers that they should respond to it with increased capacity.

The capacity estimates which we have put forward in our paper for the OECD area are based on the output of the OECD countries' steel industries in 1973/74, a peak time where everyone was trying to get maximum output. We have added to that the developments which we know about which have been notified to us, and we have subtracted such closures as we know about. On that basis in 1979 there was something like a surplus of capacity in the OECD countries of about 85 million tonnes, which is nearly four times the annual output of a country like France or Italy. In terms of current utilisation most of the spare capacity is in the EEC countries and in Japan.

The next point to which I wish to draw attention is the quite dramatic fall in investment expenditure in the OECD countries that has occurred since 1977, and also the change in the character of the investment. The investment that is taking place is fuel-saving investment, particularly pushing on with continuous casting, investment in product improvement areas and on the finishing end and also, in some countries, a heavy commitment of investment expenditure to meet current environmental requirements. Continuous casting increases capacity but most of the investment which has been taking place has been of a kind that does not do so. There has also been something of a switch in investment: where there is new capacity comming along, it is in investment-saving or capital-saving investment. There has been a marked preference for the electric furnace which does not require an iron-making base to it.

We are in the position in these estimates of having a book surplus of about 85 million tonnes but having said that I would wish to set about questioning it. I doubt very much whether the investment and research expenditure levels that we have had in the last few years have been adequate to keep the investment intact. I very much question whether some of the plant which is still in our records is capable of producing at the full rate achieved in 1973/74. Is it acceptable today in terms of today's manning rates, in terms of today's fuel efficiency rates and perhaps in terms of some of today's quality standards?

Another point we bring out is whether in fact you get as much out of your capacity as you used to. There have been some social trends — some of the labour is unwilling to work the overtime or the weekend schifts which used to be the practice — and also I do not believe there has been adequate maintenance and investment expenditure to get full output. When demand in fact did pick up a bit in 1978 and 1979 in many countries there were still some obstacles to increasing output. We soon ran into bottlenecks — I have only got to mention coke for all of you to know what I mean.

Because I have cast doubts on some of this surplus I don't wish to go to the other extreme and suggest that we are running into a steel shortage situation. In the economic environment which Dr. Ostry described, I think that there is a considerable risk of quite large cyclical swings in the market. The cycle is synchronised in many countries because it is tied to the oil supply and price situation. So I think we must expect a fairly strong cyclical situation in the 1980s. Also, as has been referred in some of the papers, we have a backlog of investment which some day somebody will do something to catch up on.

Moving on to the developing countries, I do not wish to spend long on these because I am to be followed by Dr. Nijhawan. The figures we portray show a very strong growth in demand. I quoted some figures in answer to Dr. Marshall's inquiry this morning. As I said, my personal view is that there is a lot of robustness in the strength of this demand. Mr. Kono has given a slightly different view, but our paper shows that there are substantial plans for increased capacity in the developing countries and our own conclusion is that if their economic development is not to be held up they will require substantial imports of steel. One's whole experience over several decades is that when a country reaches a fast industrialising stage its steel making capacity does not keep pace with the growth of consumption. I still hold to that view.

We also draw attention to an interesting point on technology: the very high proportion of the new capacity going up in developing countries which uses the direct reduction process and not the traditional blast furnace/oxygen furnace route. The explanation for this may partly have been offered by Mr. Wienert this morning; he talked of low risk investments in small size units and I find that a very plausible explanation.

There is however a dilemma: continuous casting is a reasonably capital-saving route of making steel, but some day someone is going to run up against a shortage of scrap for all these new electric furnaces, and I am not entirely convinced that the direct reduction process will be able to fill the gap.

Global Scenario of World Steel Industry Growth Particularly up to 1985 (Document No. 5).

Dr. NIJHAWAN:

I would like to make a few general comments before coming to the summary of the document. There has been some discussion about the role of developing countries in the growth of the steel industry, considered in the context of global steel industry expansion and growth. The document which I prepared covers chiefly the growth of the steel industry in the developing world. In this connection I submit that the steel industry is a highly capital-intensive industry; its expansion or establishment of growth in the developing world is *not* designed to promote employment potential. It is designed primarily to provide the basic platform in basic industry to promote a chain reaction growth of medium, light and heavy engineering industry and to assist in the overall economic and industrial development of the developing world. There are many other avenues such as the expansion of growth of the agricultural industry and the growth of the rural industries which have a much higher employment potential.

The developing countries in the years to come — 1985 to 1990 — will in our opinion play a leading part in a relative sense in the expansion of the steel industry considered in the global context. Much of this expansion in the developing countries will take place through the sponge iron route based on direct reduction, as many of the developing countries are endowed with resources of natural gas. The many direct reduction processes which have proven their commercial industrial acceptability will be the basis on which the developing countries will expand and are expanding their steel industry. The forecast of the capacity increase which is likely to take place in the decade ahead will not necessarily be based on the conventional process technology. Where is the model of the forecast if you consider sponge iron production just after the war and since? The capacity of sponge iron production may touch 50 million tonnes and in another five to ten years time maybe 100 million tonnes out of the global projection estimated at 900 million tonnes. Models apply insofar as the conventional approach is concerned, but there is much more than that, there is a pragmatic and dynamic model and that is the basis of developing countries.

My second comment relates to productivity. Productivity is a three-dimensional model: cost of steel per tonne; the man hours per tonne of steel; and what the workers in turn get out of their salaries and wages. That is the third dimension, and when you consider this productivity in statistical terms you have to consider the third dimension.

My document like many others is somewhat incomplete because 1979 figures are not covered. During 1979 global steel production touched an all-time record of 745.3 million tonnes with the Western world achieving a remarkable growth rate of only 0.3 per cent between 1974 and 1979. This is something to ponder on; what is the reason?

The industrialised countries of Western Europe, North America, Japan, South Africa and Oceania produced 442 million tonnes in 1979 representing an increase of 5.1 per cent over 1978. The developing world during 1979 according to the estimates which we have prepared had a crude steel output of approximately 101 million tonnes. But here I agree with Dr. Korf that when we talk of crude steel we are not really projecting the actual figures in terms of finished steel because much of the production has taken place via the continuous casting route and naturally finished steel production will be higher in comparison with the previous figures.

I would not like to go into each of the individual tables which are with you, but I would mention that compared to 1967 the developing world increased their production of crude steel from 36.6 to approximately 76 million tonnes in 1977, 91 in 1978 and 101 in 1979, representing a growth rate of 7 per cent. This comprises 7.4 per cent of the global share in 1967, 11.3 per cent in 1977, 12.8 in 1978 and 13.5 per cent in 1979. If this trend continues we hope that we should be in the neighbourhood of, if not close to, the Lima Declaration figure of 25 per cent of world steel output in 1990.

In projecting the steel making capacity in the developing countries over the years 1985 and 1990 I made a differentiation between those figures which are adjusted by the project realisation ratio and those which are not so adjusted. In this connection I would like to point out that the data contained in Document 5 are not our estimates but the estimates of the developing countries as submitted by them to us through a very detailed questionnaire which we circulated. These are their figures based on their announced planned expansion and growth potential which have been summarised and put in various tables. You will find in different tables various groups of the developing countries classified in different areas of steel processes. We have included also in these tables the centrally planned economy countries including China and North Korea, and we have said that although the capacity might be of the order of 180 million tonnes in 1985, the production may well be below 140 million tonnes because of the difference between capacity and production and I would say also consumption.

There is a clear distinction between steel production, steel consumption and steel capacity. For example in India the steel capacity may be of the order of 14 million tonnes but the production is about 10 million tonnes for various reasons. However, consumption in many of these developing countries will tend to increase as production goes up and as capacity goes up. Figure 1 in Document 5 compares steel consumption with production in the case of three countries — Brazil, Mexico and Turkey. You will see that up to 1974 consumption rises more dynamically than production, after 1974 there is a drop, due to overall economic recession. In figure 2 you will note that in the year 1974 consumption reached approximately 80 million tonnes excluding China and North Korea; production was much lower, and imports were rising consistently up to 1974. After 1974 for various reasons the curve does not follow the trend it followed up to 1974, but I agree with the speaker this morning that the establishment of growth of the steel industry in the developing countries should not cause any misunderstandings because the consumption and imports from the developed world will also increase.

With these premises, I have given certain tabulations where you find the process route comparison of direct reduction with blast furnace and conventional oxygen steelmaking. We have also given the capacity utilisation ratio for various regions of the developing world. The last few tables concern the announced steel capacity in the developing countries of around 200 million tonnes by 1985; but that does not necessarily mean achievement of that capacity. Tables 12, 13 and 14 show the capacity utilisation ratio, production and consumption. From these you will see that imports over the years to come or over the decades to follow will increase proportionately with the growth of the steel capacity and steel production in the developing world.

In our opinion the developing countries will increase their steel capacity more dynamically and more progressively, despite the many difficulties we have witnessed in the developed countries during the last ten years.

CHAIRMAN:

I have asked Mr. Anderson to comment briefly on the two presentations that we have had on capacity and then we will open the floor for comments and questions.

Mr. ANDERSON:

The papers presented this morning by Messrs. Kono and Wienert as well as those presented now by Mr. Miles and Dr. Nijhawan all suffer from the fact that there is indeed great confusion regarding the term «capacity». We have heard this already this morning, terms like theoretical capacity, effective capacity, nominal capacity, rated capacity, maximum possible production and finally the American term production capability. The papers suffer from this fact because in these tabulations — and we all have the same problem — we have to add up all these things that are called capacity. Another thing from which they suffer is, as Mr. Korf mentioned this morning, that they do not speak about all the other things that go with crude steel capacity. The really important thing is rolling capacity and all the ancillary steps including availability of raw materials and coke.

I think that there is a clear case that international action should be taken in order to get some order into this so that we can get a clearer view of what actual capacity might be in a year like 1985. I single out 1985 because it is obviously a year for which most of the larger projects are either now under construction or in a very advanced stage of engineering and planning, so that one can be almost sure that they will be there. However, there is a source of error even on that because if one rounds out capacity by adding electric furnaces or builds a scrap-based or direct reduction-based electric furnace plant the capacity figure might change very quickly in either direction; it could be higher or it could be lower, so even the 1985 capacity figures that we have here can really not be taken for granted.

But I must say that the figures that have been mentioned here do not really vary very much because we all use roughly the same information; we use announcements in the press, official announcements, Dr. Nijhawan is using his questionnaire he sent out to the governments of developing countries, and we are also using the reports of companies where one can clearly see whether a decision to undertake a project or not has really been taken.

Now this goes for 1985 but when we come to capacity figures which were presented here by Mr. Wienert and by Dr. Nijhawan for the period beyond 1985, that is to the year 1990 and then even further to the year 2000, I am afraid that if one uses the already shaky basis I have described for 1985 and extrapolates capacity figures to 1990 and further on, even if one supplements this by annonced projects the result is likely to be very uncertain. So one wants to be careful about the figures as soon as they go beyond four to five years forward. My institute had a bitter experience in the year 1974 when in the summer of that year we were convinced that between 1974 and 1985 about 240 million tonnes of additional capacity would be built, and in fact some people considered us to be on the conservative side in those days. Yet when we made this assessment again using published information and eliminating what seemed to us to be vague hopes and vague plans, after three years recession about 100 million tonnes of this capacity or over 40 per cent had definitely disappeared; nobody talked about it.

I want to warn the users of these papers that when we speak about capacity anything that goes beyond 1985 becomes very uncertain. If one uses project realisation ratios and even if one speculates about utilisation rates and if one then derives from that a figure on supply I am sorry to say that among the better fraternity of forecasters this is a figure that we cannot accept because it is so uncertain that it should not be used. It should particularly not be used in any comparison with figures that are clicked out by a global model on future consumption. If one compares that to these uncertain capacity figures then one gets a net trade figure which is really uncertain.

You will have noticed perhaps that Mr. Kono, who is an old hand at forecasting, does not walk into the trap of making these comparisons; when he speaks about future trade he just says that this is going to develop in the climate of future production and future consumption which he has described in his paper. He does not go as far as Mr. Wienert and make an actual calculation.

I want to stress once again the necessity perhaps in this body or somewhere else that we get some order into the methods of evaluating capacity according to the probable degree of realisation. Perhaps we could follow ore where one has measured, inferred and possible reserves and also get together to say from a technical point of view what is actual crude steel capacity and get order into the terminology.

CHAIRMAN:

Now I would like to turn to the audience for any questions to our speakers and our panellists. Moreover I do not mean to inhibit any general observations or any perspective that anyone wants to advance independently of the particular points raised by the speakers.

Mr. J. DOYEN:

I would like to respond to your invitation and voice a few thoughts. We feel a great deal of respect and admiration for the people who made the forecasts for the coming five or ten years which we have before us. We note from what we have heard today that there is a great deal of uncertainty and that many questions are still open, but I come back to what has already been said by my trade union friends: we feel justifies in raising a number of questions.

First, it would seem that for the steel industry a fairly moderate growth rate is forecast for the next ten years. This growth rate for the industry is linked with the overall growth rate of our economies. We get the impression that a social choice is being made. We feel that we are becoming established in a situation of underemployment. The documents which we have seen so far would not seem to indicate that there is a political will to swim against the stream with regard to this moderate growth rate. How are we going to solve in the coming decade the problems of underemployment and of unemployment for millions of workers throughout the world?

A second thought: given this situation, is it possible to envisage some kind of co-ordination, some degree of consultation at the investment level or will each country consider its own problems and try to deal with them at a purely national level? Or could we imagine the possibility of consultation in the area of investment, rationalisation and modernisation at a level higher than the pure national level? It seems to me that this dimension should be considered.

Another problem: I come back to the trade union viewpoint; in all the approaches to the problems and in all the analyses that have been made, we do not have either a quantitative or a qualitative balance sheet with regard to employment. All that is going to happen in the next ten years will have an impact on the overall level of employment and I do not subscribe to the conclusion of one report which considers that the steel industry will have higher employment levels in the future in industrialised countries. The steel industry will have fewer jobs overall. As workers therefore we must be aware, together with the economic forecasts, what will be the repercussions on employment and their extent.

A final point on the subject of developing countries: in the horizon year 2000 these countries will produce 30 per cent or 25 per cent of the world's steel production — the precise figure does not matter. The fact which we take into account is that the onward march of these countries is going to take place. This being so it is possible to envisage some degree of harmonization and consultation both at this level and with the developing countries.

Mr. STEFFEN MØLLER:

I would like to follow up what we just heard. From the papers, I cannot see how the production capacities will develop in the coming years from the point of view of increased productivity. I still think it is a very good idea to have some dimensions on the future so that you can see — if you have certain percentages in productivity increases — how many people might then be employed, to give a chance to us who are going to monitor employment in the future steel industry. It looks like that. It is the trade unions who are going to take the people out of the companies — not the management — as we might have heard this morning from the employment bureau mentioned.

Dr. NIJHAWAN:

I think it is a valid point to which however there is more than one angle. For example, during my visits to China, at least 30 per cent of the workers in the steel plant were engaged only in being trained. Which means that 70 per cent of the workers in the steel plant at Anshan, were operating the plant and the rest were being trained; but it is very difficult to make a distinction between the two. So the developing countries normally employ many more than would be deemed necessary for the purpose of producing 75 per cent of the rated capacity output.

In the developed countries the situation may be different, because they do not have to train their manpower to the same extent as in the developing countries. In the developing countries when you are starting with a raw base on a green field site with no trained workers — for the steel industry is not only highly capital-intensive but also highly capitalised in its maintenance and repair operations to keep it up to peak maintenance — you require a much bigger skilled base. The employment in the developing countries covers not only those who operate the plant but also those who are being trained for it.

The manpower in the developing countries per tonne of steel over a specific period is much higher and the productivity is lower. Still the cost per tonne will not differ to the same extent as the input of the man-hours or man-months, because of the lower wages in the developing countries. But as I said earlier, it is not the cost of the labour considered in terms of per ton of steel output, but what the workers can buy out of it which is the third dimension which plays an all-important part.

I would like to mention that productivity, which perhaps will come up tomorrow, may be analysed in somewhat greater depth. The cost of production, the man-hours or man-months required to produce a tonne of steel and the quality of steel, these are interrelated factors which have to be taken into account when assessing the price of steel. The price of steel for the domestic home market is different in quite a few countries from the price of exported steel. Export prices are in many cases subsidised by governments and that brings in a lot of other factors which affect the trade. I believe that is an important subject which requires separate treatment, if I may suggest this.

41

Mr. ALEKSANDER ČAVIĆ:

We have heard many interesting and sometimes controversial statements about the development of steel demand and capacities in the world steel industry in the 80s. A few general remarks about the Yugoslav position in these respects might be of some interest.

The steel production in Yugoslavia is below domestic market demand and the economic development level of the country. In 1979 Yugoslavia produced 3.5 million tonnes of crude steel. The consumption of steel products in the same year in Yugoslavia amounted to over 4.5 million tonnes in terms of specific net steel imports; Yugoslavia is among the significant steel importers in the world. The present specific steel consumption in Yugoslavia amounting to about 250 kilos per capita it is low compared with gross domestic product per capita of about $2,000. It should be twice as high. The policy of steel industry development in Yugoslavia is based on the principle of meeting domestic requirements. About 25 per cent of necessary scrap quantities and about 15 per cent of requirements of iron ore are imported and will be imported.

With these development plans, whose implementation is slowed down by the shortage of financial funds, Yugoslavia intends to install by 1985 a total steel production capacity of about 8 million tonnes in order to meet approximately 80 per cent of total domestic market requirements. It is estimated that 20 per cent of total requirements are to be imported and exported. We think that such a concept of the development of the steel industry in a developing country will help to improve general development of the trade between developed ans developing countries.

We are interested to know whether somebody from the panel could comment on the prospects of the trade in and availability of steel semis and the future availability of steel scrap.

Mr. WIENERT:

The availability of semi-finished steel is a subject which has been discussed at great length for some time. The concept behind it was originally that it was a good thing to leave your stocks in places where costs were not high and then you could transport the semis to the mills. This is a fairly convincing concept at first sight, but it proved in practice to give rise to considerable obstacles. In order to build up the mill capacity you have to be sure that your stocks of semis will not be interrupted so that you can use them at any time. This question of deliveries, of guaranteed supplies is of great importance.

The stocking places may be favourable from the cost point of view but they are situated far from the consumption centres. For example, this sort of thing was discussed in Australia for a long time and also in Sweden and the Soviet Union. But it is always difficult to guarantee supplies.

There are also the same sort of questions on the part of people who offer semi-products. Why should they in the long run content themselves with dealing only with semi-products when they themselves could build up finishing mills and of course increase their income? The interests on both sides of the question have up till now prevented people from creating large-scale projects in these fields and this seems likely to continue to be the situation in the 1980s. Because of overcapacity it seems unlikely that anyone would set up a big installation to deal with semis. This would have to have fairly big dimensions in order to be profitable at all.

As to the availability of scrap, I cannot tell you very much. But I think that the situation is likely to become rather more favourable. There should be more scrap through the use of certain techniques. Directly reduced ore could be used — certain modifications would have to be made to existing equipment. But these questions of structure have always been difficult to solve and it has always been difficult to answer questions in this connection. From the point of view of processes, there is a considerable difference in the production of scrap.

I could perhaps add a remark on the subject of co-ordination of investment. This is connected with the question of certainty of supply of semis. To my mind, this leads us onto the next point. At the present time, what do we mean by investment co-ordination? In the situation that we have at present it can only mean preventing the building of new steelworks. It means maintaining capacity in highly industrialised countries at the expense of the expansion plans of the developing countries. So if you speak about investment co-ordination you have to make yourself perfectly clear. This leads us into a field where Dr. Crandall has put forward a very interesting report and where Dr. Peco also has something to contribute. Perhaps we should bear this point in mind because we are likely to come back to it later.

d) The Future Supply/Demand Balance and its Implications Including Developments in International Trade and Steel

"The Economics of the Current Steel Crisis in OECD Member Countries" (Document No. 7)

Dr. CRANDALL:

I might say that, given the scepticism I have heard around this room today about the validity and value of economic forecasts, I wonder whether this conference might better have been held in Monte Carlo than in Paris.

I think we have to go on the basis of what little we know. As you may be aware, in my country you can learn all you need to know about economics from reading the Wall Street Journal. About three years ago at a time in which the United States economy was just beginning to come out of the deep 1975 recession — which everybody in this room well remembers followed a frantic period in 1973/74 — the Wall Street Journal was very much concerned about economic exercises somewhat of the sort that we have heard here today projecting demand, projecting capacity and projecting supply without any regard to relative prices. They referred to them as inventory clerk economics, because the inventory clerk doesn't really know or care or have responsibility for setting prices. All he must be sure of is that as his shelves run down he restocks them to make sure that he doesn't run out and have a shortage. He is totally immune to relative prices, he has no responsibility in that area.

I think as we look at the future of this industry we have to be concerned about relative prices, because in the absence of government intervention I think relative prices will determine the distribution of production and consumption around the world.

The essence of my paper is to look at how relative prices impact upon investment decisions around the world. It is quite clear that it is very difficult to generalise in this area because it is difficult to break the world down into neat little packages. We have heard much discussion today about how we cannot characterise the developing and developed world in any neat little way and that the number of sub-categories increases as time goes by and particularly as energy implications grow and multiply.

I am talking basically in my paper about the carbon steel industry. I have analysed principally the blast furnace — BOF process of production. When we get into other technologies one draws slightly different conclusions, some of which are not necessarily that apparent, by the way.

The thrust of what I have to say in this paper was expressed surprisingly enough in a doctoral dissertation written at the Massachusetts Institute of Technology in the year 1968. So for those of you that think economists never predict correctly one might go back to look at the dissertation by a young man — young at the time — by the name of Richard Mank who saw the world steel industry changing very drastically from the perspective of 1968. Looking backward to the late 1950s and early 1960s, with the beginning of the opening up of the world iron ore market, and the sharp decline in the real cost of shipping the relative advantage of the United States Industry, to which he was addressing himself, began to decline. I think that exactly the same deduction would be true for the European Economic Community. Unfortunately, none of us could recognise that at the time and perhaps even from the perspective of 1968 would not be able to agree with Mr. Mank's thesis. I think as time goes on the thesis gains credibility.

The decline in real delivered raw material prices and the decline in real shipping costs over the past twenty years has reduced the comparative advantage of the developed world and increased the comparative advantage of the developing world. In my analysis I look at various parts of the world. I try to characterise those areas of the world where I think we are likely to get the most growth in the developing world and that is Latin America and Eastern Asia other than Japan. In the case of non-Japanese Eastern Asia and Japan I see the total cost of production of flat rolled carbon steel products being substantially below the cost of production for similar new mills in the Western European and North American industries. As a result, if economics determines capacity decisions, it will surely drive capacity towards that part of the world. Japan now, with her current exchange rate — when I was analysing it it was 240 to the dollar — is back in a very commanding position for expansion if in fact she feels that she could get access to the world market. By the same token, Eastern Asia is in a similar position. I am more pessimistic about Latin America only because of reports on construction costs.

One of the things which I must disagree with in the discussion earlier by Dr. Nijhawan on his analysis of the developing world — which I generally agree with and which I think was very well presented — is his assumption that steel is necessarily a capital-intensive process. There tends to be a confusion of terminology here. In the popular domain, capital intensity often refers to having large agglomerations of capital in one location. To economists capital intensity means a large amount of capital relative to labour. In fact, I think if you were to examine other types of basic materials industries you would find that steel is relatively labour-intensive and not relatively capital-intensive for carbon steel flat rolled products. It may not be true for some of the simpler products, particularly some of the simpler bar mills. Moreover, in the construction of a steel mill an enormous amount of labour is required over a very long period of time, as you all know, and that adds to the competitive disadvantage of the developed world. So the competitive disadvantage of the developed world stems from its high labour costs not only in the direct and indirect costs of producing steel but also in the cost of producing new mills.

Despite this projection, which may sound very gloomy for those in this room from the European and North American steel industries, I hardly foresee a major shift from the developed world to the developing world in a short period of time. There is an enormous amount of inertia in the system for several reasons. First of all, Dr. Nijhawan has pointed out that as the developing countries grow they increase their consumption of steel more rapidly than their production of steel over some period of time. I do not think that is necessarily a fundamental economic law; it happens because of a variety of institutional constraints. Secondly, the greatest advantage for the Western European and North American industries is the enormous amount of capacity they have in place today. I think that the North American industry, which was restrained by capital markets who saw these trends developing, has very little capacity to shed. It may have 5 or 10 million tonnes to shed over the next ten years but you rarely hear any estimates higher than that. It is not a large share of the United States capacity and with a growth of mini-mills the net reduction in United States capacity is bound to be rather small despite these trends.

In the Western European case the bitter irony is that they expanded at a time when the economics were turning against them to the water in order to take advantage of the declining real shipping costs. Unfortunately, they are not able to pay the wage rates of the developing countries and as a result they found themselves at a competitive disadvantage.

The United States industry pulled back away from the coastal sites towards the Great Lakes and therefore away from their import competition and are accordingly in a much better position today to compete. So I see the future holding a shift towards the developing world — a slow shift — probably at the expense — this is the pure economics of it — of the European industry and less at the expense of the United States industry. I would also see a substantial possibility for renewed growth in Japan if they were able to work out the politics of their trade relations. I don't think their failure to grow to the 160 million tonne level is any way solely a reflection of the economics of the situation. It has, I believe, probably as much to do with the politics of the situation. I think they are in an excellent position to continue to renew their growth. Whether they will or not depends entirely on politics.

That is basically the picture as I see the future of the industry. I am equally evasive about timing. Since there is a great scepticism about economic forecasts I think perhaps I should close by explaining to you that as I was taking my graduate training in economics many years ago a professor explained to me that one ought to have a great deal of modesty about what one was doing because the principal new development in the field was in the field of econometrics which he defined as simply an attempt to perfect the prediction of the past.

"How Steelmaking Enterprises can Become Internationally Competitive" (Document No. 6)

Dr. PECO:

As a logical follow-up to today's discussion covering the future development of the world economy, forecasts of steel consumption and the possible or probable development of capacity, we now come on to the question what companies should actually do. I am the one to present the last report. I am glad about this both because it can simplify what I have to say and also because I can take advantage of the discussion which has taken place up till now.

If we ask ourselves what should be the logical behaviour of companies given the rather grim horizons which have been examined today, I think that there is only one solution, and that is a continuous search for greater international competitiveness. This search for competitiveness has already been mentioned in Mr. Kono's paper and it is on this particular subject that I would like to try to say something. I am referring to any company in any country which is or should be viable and should try to make profits each year. Mr. de la Palisse would perhaps say that the obvious solution is simply to try to improve revenue and keep down costs. It is simple to say this but rather more difficult actually to do it in our market.

As regards revenue we come up against developments in steel markets. I am not thinking so much in terms of each national market, these markets can be closed or open, protected or not, I am thinking about above all of the world market — in other words of the sum of all exports which give rise to international trade in steel products. As you know, these are markets of the classical marginal type. We heard this morning and Mr. Wienert said in his paper that there are two elements in world trade which are split 50/50 today. Fifty per cent is trade in surplus products sent from one country to cover the necessary imports of another country. The other half

45

consists of purely competitive trade in steel products between countries which do not need to balance out their own production or who add to their balancing-out imports more competitive forms of import. Because of its marginal position this world market suffers much more than the national markets from cyclical fluctuations.

To meet this demand which of its nature is rather rigid there is a potential supply which consists of the sum of the potential supplies of all the countries which export either from necessity or for competitive reasons. Given the existence of excess capacity today and given the desire to export on international markets, the result is that the world steel market is in a state of disequilibrium in which the search for an economically fair price, as the Secretary-General of the OECD recalled this morning, is becoming more and more difficult. Mr. Kono in his paper spoke of marginal capacity and it is in fact the existence of the marginal capacity which comes into play on the world market which gives rise to the difference in prices between the world and national markets, the world price being lower because of competition rather than, as Dr. Nijhawan said, because of state aid for exports.

In fact, national output searchs out markets everywhere so that it can be used. Competition on the world market is therefore fierce. Over the years the number of countries operating on this market has continually increased. Only twelve countries were exporters before the First World War and even between the two wars but today at least 50 countries export on the world market. This shows the very competitive — perhaps too competitive — character of this market and is perhaps one of the reasons for the establishment of the OECD Steel Committee and the fact that we are here at this Symposium. Price policies have become difficult and will go on being difficult in the 80s because of the disequilibrium in the world market, and it is not easy for companies to devise revenue policies which are consistent with the objective of international competitiveness.

As regards reduction of costs companies have to make continuous efforts and perhaps should make even greater efforts than at present. Technical innovation has to be continued and stepped up, investments have to be made, not, in present circumstances, to increase capacity, but to re-structure, to modernise and to streamline operations. At present we are unfortunately living in conditions of overcapacity. This was examined by Mr. Kono, Mr. Miles and others.

I think that we have several different kinds of excess capacity throughout the world. In some markets and in some national situations there is temporary excess capacity caused not by too large an expansion of equipment but because of the weakness of demand. On the other hand, there are other forms of excess capacity which are caused either by obsolete equipment which must unfortunately be scrapped or by structural problems or because of the necessity to export has led to a state of disequilibrium when compared with the possibilities of domestic consumption; although here we find also well established historical positions which we could discuss.

To state simply that excess capacity should be re-absorbed is to over-simplify matters; you have to see where the capacity can be re-absorbed and where existing capacity cannot be reduced. We heard, this morning, that although we are rather pessimistic, there is nonetheless a gradual development of consumption, so the search for a balance between supply and demand by re-absorption of capacity will differ according to the region of the world in question, and according to the products under consideration. I think that excess capacity should be treated not exactly on a case-by-case basis, but in different ways depending on the region concerned.

Reduction of costs involves not only the technical efficiency of installations but other cost factors, including labour. Unfortunately there are employment problems. Mr. Doyen was right to remind us of the need to draw up a manpower balance sheet. But from the point of view of costs I think that we are faced not only with the problem of wages but also with the problems of productivity and industrial relations. Productivity in terms of production per man differs from country to country as do industrial relations. In some countries people strike, but work goes on. In other countries strikes involve bringing about as much loss of production as possible compared with the loss of wages. This diversity affects costs and competitiveness in the world market.

The net effect of these factors leads to the question of profitability. I would like to make a distinction between two elements in profitability. The first depends directly on the management of a company and gives us the gross producer's margin which is the relationship between the company's net costs and its losses. I consider this to be one of the most useful indices to show which industries should seek to become more competitive on the market.

But there are also within the concept of profitability factors which are outside the company's control. These may be shareholders or owners or completely external factors such as the capital market. Financial charges can vary considerably. Sometimes the capitalisation is well related to the sum invested. But sometimes it is inadequate and does not allow self-financing; then it is necessary to increase the capitalisation and temporarily to borrow, which, depending on the country and the market, affects the rate of inflation and the rate of interest.

That brings me to some kind of conclusion. This search for competitiveness which each company should undertake takes place within the framework, which can be defined by the present Symposium. It is a sensitive framework but one to which the companies themselves could give a rational form; but it is also something which they could and have a right to ask for.

As regards the contribution which companies should make they should aim to ensure that they are competitive. They ought to make sure that there are loyal forms of competition on the market. I think this is one of the aims of the Steel Committee of OECD. From the point of view of investment there must, of course, be co-ordination, there must be some form of consultation. I would be satisfied if we could get greater transparency of investments, so that everyone knows who is going to invest what and where and so that those who need to invest know what the consequences of their investment decisions will be.

On the other hand, the steel industry is also entitled to make certain claims. The steel industry asks for an open market but at the same time this market has to eliminate disloyal forms of competition which appear very easily in very competitive markets.

The companies should ask for a normal financial situation within companies. They would like to see greater harmonization at the international level in the matter of industrial relations. They would like to see a comparison of investments so as to be clear as to the decisions which should be taken. In the matter of international co-operation the companies would like to see as wide co-operation as possible at the world level.

Finally, I believe that our work and the work of the Steel Committee of the OECD should not be considered as a point of arrival, as a conclusion to studies that have been made. We should try to arrive at a starting point so that we can try to do something more in the field of international co-ordination. This will work to the benefit not only of companies as a whole including their social elements but in the interests of international trade throughout the world.

PROFESSOR RUIST:

We have now come to the real long-term projections and we have got two views which are in a way quite opposed to each other. According to Dr. Crandall the United States and particularly the European Steel Industry have, in Schumpeter's terms, really embarked on negative economic development, whereas Dr. Peco, who may perhaps be taken as a representative of this poor European steel industry is convinced that if governments provide a decent climate for business then the European steel industry will be able to become competitive. My only comment on Dr. Peco's paper is a small question: if everybody is going to be more competitive, what will be the result? To me it is more or less that if everybody tries to be above average, then the average will certainly also change. But I will concentrate more on Dr. Crandall's paper and raise two questions on it.

Dr. Crandall stresses the importance of marginal costs on prices. I think that is certainly something which has to be stressed. I certainly agree with him that there are many false conclusions drawn from calculations of average prices for countries. But using these marginal costs, he finds that the European and United States steel industries will have such high marginal costs that they will not in the long run be able to compete with new plants in certain less developed countries. He finds that the present crisis started in the early 1960s when transport costs came down rapidly. Today transport costs have gone up quite a lot. I must confess that I have not tried to find out whether they are again relatively speaking back in the same position as they were in the 50s, but I am afraid that we cannot expect that transport costs will get much lower. So probably one of the reasons for the very high competitiveness of the less developed countries has to a certain extent vanished. But granted that they have an advantage in their lower labour costs it is quite true that at least they should be able to produce steel more cheaply than American and European plants. Whether they will be able to ship it or not is a different question.

But even though it would be possible for the less developed countries to produce steel at lower cost than steel industries in other countries, would not that be true for most other industries too? Their low wages certainly apply in most industries and they would be able to compete with European and American industries in the same way. The less developed countries cannot, however, take over all industrial production from Europe and America. So my question is: on which industries will they concentrate? Is it necessary or is it probable that steel is the best industry for them to concentrate on? It is quite true that steel is more or less a prestige industry and that may be one reason for them to concentrate on steel. But otherwise I think there really is a big question, will they take over the steel industry, will they take over the engineering industry or will they take over the electronics industry? I think it is rather ridiculous to think that it will be a one-sided development. There will certainly be a mix of industries and there is a limitation after all in the capacity of these countries whichever industry they concentrate on.

So I would not be so sure as Dr. Crandall that the steel industry is going to move altogether to the less developed countries. It may very well be that we shall get in these countries as well as in Europe and America a rather standard mix of industry. The mix may be a little different in different areas of course because of different advantages. But I do not think there will be a complete takeover.

Dr. MARSHALL:

Let me try and focus mainly on Dr. Crandall's paper. He said two interesting things: one is that most Americans get their economics out of the Wall Street Journal and in fact I think there may be more truth in that than he believes. I think it is one of the basic problems we face in dealing with the steel problem. The Wall Street Journal believes, as many Americans do, that the market — free market — economy will determine and should determine almost all economic activity. He then went on to say a second thing which I found most telling — he said it very quietly so you may not have heard it — so let me emphasize it: he said that in the absence of government intervention relative prices will determine investment decisions. But to claim that governments are not intervening in the world steel market is to ignore reality. To use the concept of a free competitive market to describe the steel industry also ignores reality.

There are at least three assumptions which I want to talk about. One I have already mentioned — that there is no government involvement; that is ludicrous — there is government involvement. There is government ownership in large parts of Europe. My own government is actively involved in maintaining a trigger price mechanism. I would even argue — and I will later — that the governments have to be involved — but they are.

My second point: a competitive market theory assumes that no one can isolate a market and then effectively price discriminate. To me one of the tremendous strengths of the Japanese steel industry is their ability to isolate the export market from their domestic market in their pricing mechanisms. By doing that, they are tremendously competitive and successful; market theory would argue that they cannot do it and yet we know they have.

My third point it that there are in competitive market theory many small producers, no one of which is able to determine the terms of a market. That is crazy too. There are maybe ten significant steel companies that could determine the price of steel in any given regional or product type market. The sum and substance of what I am saying is that the theory does not work because the reality is that the world steel market is an oligopoly. Most economists, I think, would argue that an oligopoly must be regulated if the public interest is to be upheld and the form of that regulation is usually a government. Now what does all that mean?

I think Dr. Crandall's prescription is probably correct in some world that does not exist. In the world that we have I think there must be some form of co-ordination and co-operation other than a world cartel. In the past that co-ordination and co-operation has ignored important groups, one being consumers. It has also ignored those people who are not in the group — in this case the less developed countries; and it has probably ignored the worker. The form of worker adjustment that will come about in some kind of solution of the steel problem will, I think, be different from what we have seen in the past because the unions are so very strong that they are not going to allow the solutions that we have seen in the past, namely massive displacements of people.

So my response is that somehow we have to get the presence of governments in solving this problem in a responsible way. I might say again that I admire the ability of the Japanese to do long-range planning of their industry and the government's role of encouraging and supporting an industry which is without match in the world today. I do not think that my own government has at the current time — the Chairman may cut off my microphone but I will say it anyway — the mechanism to look at the long run future as to where our steel industry should be. Our solutions have been short run and reactive. That is partly because the American steel industry does not encourage our government to have an ongoing consistent policy towards the industry. To me one of the great weaknesses of a lot of us is that we would like to use some excuse for ignoring the very difficult and real problem we now face. Some of us fall back on old theories that we learnt at school — they are hard to get rid of — while some of us predict that the problem will go away. I submit that neither of those things will happen and that we are going to have to solve the problem somehow collectively. I hope that this forum is a first step in that direction.

Mr. BRUNNER:

Most of what I might have wanted to say about the two papers in question has in fact been said by earlier panellists. I too would have put a lot of emphasis on the freight rate point in Dr. Crandall's paper — freight rates on both raw materials and finished products. I think it is not impossible that we are going to see a fall in real wages in industrialised countries vis-à-vis the developing countries and that too could make quite a difference to Dr. Crandall's calculation. So what I want to do is react to these papers and what has been said this morning in a more general way. I am conscious that the Chairman is using me as an *agent provocateur,* and having alienated my fellow economists this morning, I am well aware that what I am going to say now is going to alienate almost all my fellow steelmakers.

I do think that there is a danger in international gatherings like this that we put altogether too much emphasis on the international dimension, by which I mean foreign trade. It is understandable; governments don't get together to discuss their internal business with each other; they get together to discuss their relations with each other. And industry — I am not only getting at the steel industry here — all industry tends to ask for protection from foreigners rather than from its compatriots because the foreigners don't have votes and politicians are fair touches as far as the foreigners are concerned. But it is no good going whining to your government and saying that some fellow countryman of yours is making life difficult for you, because governments don't tend to be too responsive to that.

The burden of what I am trying to say is that for the most part the enemy is within the gates. The enemies I am referring to are, on the one hand, the aluminium, concrete, glass and plastics industries and on the other — as far as the traditional steelmakers are concerned — the electric steelmakers.

We all know perfectly well that inter-material competition has made major inroads into our market. Just to mention the case of aluminium: we have lost out considerably in the container business, in transport, in building. I do not think that we should kid ourselves that this is not the case by putting all this emphasis on other people's steel. Moreover, the more successful we are in keeping other people's steel out, the higher we maintain domestic prices and the more we are inviting inter-material competition.

As for the electric furnace, let me say I am not just getting at the Americans: we are all in this together. You can see from Mr. Kono's figures that the growth of electric steelmaking in the United States has been three or four times that of steel imports. To some extent the American integrated steel industry has adopted the very sensible principle of "if you can't lick 'em, join 'em". But a lot is still coming from the mini-mills, and I think in part it is a reaction to what I was getting at this morning: that there is a desperate need to get down the break-even point. This of course is one of the great advantages of electric steelmaking: for every dollar you lose in revenue because of the market downturn you probably save about fifty cents in the bought-in cost of scrap. You can also turn the thing off at the week-end. It is more flexible — it has all sorts of attractions.

I am not in the business of selling electric steelmaking. Australia is almost unique in having virtually no electric steelmaking. But from where I sit it does seem to me that the problems of the integrated steel industry are as much provided by other materials and by the electric steelmakers as from outside. I know full well the argument that the electric steel industry is making a different product line. That for the most part is true; but it is undeniable that it has robbed the integrated steelmakers of a lot of the economies of scale which they would have got up to the steelmaking stage. I think probably, I have already said too much.

Mr. MICHAEL MARSHALL:

I am representing the United Kingdom Government which, despite what my namesake said earlier, in one important respect is strictly not intervening at the present time.

I think that the two papers we heard and some of the comments which have followed have been very useful in pulling together many of the themes which we have heard throughout the day. I think Dr. Peco's paper in talking about the balance of supply and demand was in effect perhaps reflecting much of the conventional approach to the problem. Certainly I was impressed with one or two of the differences which he brought out: for example, differences in industrial relations between various countries in the world in steel-making terms. But I think it was Dr. Crandall's paper which — as it were developed a whole new area in my thinking because what he was asserting has been said a number of times: that the newly-industrialised countries are in fact making inroads into the traditional steelmaking balance in the world. But I admired his willingness to attempt to try to spell this out; and although he was quite modest in saying that like everybody else he was reluctant to give a lot of numbers I think the fact is that in his paper he refers to a possible 10 per cent of reduction in steelmaking capacity in Europe over the next decade. If that were correct and taking Dr. Nijhawan's estimate for that capacity in 1990 it would mean a reduction of something like 24 million tonnes of capacity. That is pretty significant because if what he says holds good, it has major implications for the European steelmakers and indeed for European governments. I wonder though, when he argues that this will come about because of the ability of the newly industrialised countries to benefit from low wage cost to offset relatively lower productivity, what evidence he has? Would he not recognise that there are examples the other way where higher productivity has been justifying higher wages? I would be very interested to hear his more detailed thoughts on that aspect.

Dr. CRANDALL:

The productivity experience across these countries varies enormously. You heard Dr. Nijhawan talk earlier about the number of man-hours used essentially in training new personnel as opposed to producing steel; as well as, of course, the number of personnel present simply to shake down new mills or to shake down the expansion of new mills in these emerging countries. It is certainly true that there are countries in which the productivity experience is very bad; one would expect, however, that, with the rather abundant technical assistance in the world through the sale of technology and people to help implement the technology, there will be some countries in which productivity will begin to be comparable to some of the Western countries. The assumptions in my paper are of rather substantial productivity differentials. As a matter of fact, our own United States Industry is beginning increasingly to look to engineering from its rivals to help it improve the productivity in its own facilities.

Secondly, it is probably true that in most of these countries wages will not follow productivity improvements in particular industries dollar for dollar or increment for increment. But as these countries become more developed then of course the distinction will go away. If we are all developed countries with the same factor endowments my distinctions would go away. Let me refer back to that because Professor Ruist was commenting that other industries also face higher labour costs in developed countries. It is surprising to hear that from a Swedish economist. The Swedish school was in the foreground in the early part of this century developing the theory of comparative advantage and factor price equalisation and the like. In fact, I pointed out quite clearly that it is my contention that steel is not among the most capital-intensive of new industries. There are many more capital-intensive industries than steel. It is indeed labour intensity, especially in some of the more sophisticated flat rolled products which drives the comparative advantage away from the developed towards the developing world.

Dr. VONDRAN:

In order to come to Paris we had to read a great amount of documentation. We had to consider many figures. I would like to have clarification on one point to take home with me. This is a question for Dr. Crandall. It was with great interest that I saw the costs he calculated for the American steel industry as compared with the European steel industry. Briefly, his conclusion is that costs are more or less the same on both sides of the Atlantic. But I would refer to another paper that has probably had much wider circulation — the Orange Book of the American steel industry — and here the figures look quite different: the German steel industry has a forty-five dollar higher cost than the American industry and our British colleagues are in a still worse situation, eighty dollars worse than the American situation. I realise that we are not necessarily comparing like with like. There is the question of the rate of capacity use assumed in the two calculations. I have two comments: first, the range is far too wide to be fully explained and secondly the use of capacity levels shown for the European steel industry are far too low. Capacities which cannot always be utilised are included. The question for Dr. Crandall then is, why are the two American sources so very different from each other? I think there are some difficult policy questions involved here. I think we should ensure before tomorrow's discussion that we are speaking the same language and using the same terms.

Dr. CRANDALL:

You may be surprised to hear that I have read that document.

We are talking about different things here. In the first place there is, at least at the back of the minds of some people, the United States trade loss and it is not clear what cost means. Mr. Vanik is in the room; perhaps he could explain it to us in terms of the trade loss.

What I was looking at was not the average unit cost at existing capacity utilisation for all plants regardless of whether they are at the efficient margin, are approaching a zero net asset value or have a negative net asset value but continue in operation. I was looking only at the efficient margin for existing plants as I could see it and the cost of producing from new facilities.

I will grant you that there is a great deal of difference of opinion about some of these figures, although nowhere near so wide a range as you spoke of for the precise concept which I was using. For instance I would be interested in talking with some people from your country about one study by Jarvis in 1978 which argues that German capacity costs are substantially below United States costs. It is my speculation that that is because the figure is based upon an ECSC product definition and my flat rolled mill figures do not coincide with that — they are based on a broader product mix.

But I would be very interested to hear whether my assumptions regarding the cost of building and operating a new plant in the United States and particularly in Northern Europe are accurate; whether indeed they are very close. It is my information, and I can assure you that I have talked to any number of steel engineers and steel companies, that they are approximately correct.

Dr. KORF:

I think I can contribute something to the cost differences between the United States and Germany, because we operate similar plants of the same age and with the same capacity on both sides. The total inclusive cost difference in favour of our United States plant last year was about 15 per cent. This is simply due to the currency exchange rate between the dollar and the Deutschmark. If you look back two years, when the dollar was much more expensive, the situation was completely different. The cost changes today are mainly due to the change of currency rates.

I also want to come back to Mr. Brunner's statement about electric arc furnace steelmakers. If I remember, there are two kinds of steelmakers — the integrated and the electric. The first are the baddies and the others are the goodies — or the other way round. I think that Australia, where there is no electric arc steelmaking, is really a unique case; and I think that we should state here that in the United States more or less all large steelmakers also operate large electric arc furnaces and are going to build new ones. What is the reason for that? It is very simple, I think:

  i) electric arc furnace technology has very much improved in recent years: one can convert in this furnace much more cheaply than before; and
  ii) it is the cheapest way to add new steel capacity quickly.

There is admittedly a limitation on the scrap market. Some day if electric arc furnaces continue to be built, scrap will run out — but for this there is now the alternative of direct reduction. This technology is not limited to small plants; it is also feasible for big plants. There are plants at the moment under construction in the world not only for half a million or one million tonnes but for two, three or four million tonnes.

The initial investment costs on a green field site plant are about 60 per cent of those for an integrated plant based on blast furnaces. The published figures from the United States say that a new green field plant costs today $1,200 to $1,400 per tonne of annual capacity. Sixty per cent of this is very adequate to build a plant on the direct reduction/electric arc furnace route. So I think that most of the new capacity in future years will be based on this route, because this is the only way to finance it.

There are a lot of places in the world where we have plenty of natural gas to operate these plants; these are mainly underdeveloped countries. But there will be the alternatives of using coal for direct reduction, by gasification; this gives a completely new direction.

I would summarise by saying that in the future the electric arc furnace route will not be a side route. I am convinced that, together with direct reduction, it will be the main route for steelmaking.

CHAIRMAN:

You are totally objective about that of course?

Mr. SCHUBERT:

I think that my comments probably bear on what Dr. Korf has been talking about. In the dialogue which occurred earlier Dr. Crandall painted an extraordinarily pessimistic view because he simply did not take sufficient account of the possibilities of roundout which exist in brown field facilities. Dr. Crandall reiterated that he basically looked at existing facilities in green field sites. Our conclusion is that whether or not one accepts what Dr. Korf has said the possibilities of rounding out in existing brown field facilities with large electrics or rounding out with existing steelmaking facilities, mean that as a practical matter, if we get the kind of government climate which is not hostile and perhaps not even indifferent over the longer term then we clearly are in a position to secure significant cost reductions and to maintain our competitive position, including that with the LDCs, at least in our own market. We do not share the pessimism of Dr. Crandall in that regard. We do not think that the facts bear that out.

CHAIRMAN:

And surely you do not think that your government is hostile or indifferent?

Dr. CRANDALL:

There are two issues I should respond to here. One is Dr. Korf's. I would not dream of getting between Dr. Korf and the rest of the industry about where coal-based direct reduction is going; what he says about the exchange rate is certainly quite true. If the United States dollar continues to depreciate against other currencies such as the German currency, other than that against the Japanese currency, certainly many of my conclusions are changed. That is, as we approach the developing countries through the devaluation of the dollar in terms of our real economic welfare we begin to look better as a steel producer.

Secondly, I must take some issue with Mr. Schubert. I think that this indicates some difference of opinion within his own industry. There has been a decided movement away from the brown field roundout theory, that is the theory that there is a substantial amount of roundout and brown field capacity available in the United States at a cost substantially lower than green field capacity — and it is, I think, reflected in the Orange Book. There apparently was considerable disagreement within the industry on this issue. But I must tell him — and I have these figures in confidence from some people who work for him so I dare not offer them publicly — that my analysis of green field/brown field possibilities is more optimistic than the analysis from within his own firm. I therefore find it very surprising that he would tell us that the brown field roundout offers substantial capacity expansion possibilities at dollar values substantially below the green field route.

Mr. SCHUBERT:

I was referring not so much to the expansion of capacity as to modernisation and the replacement of facilities and rounding out in that fashion with the result of improved cost. I am sure that if he listened carefully to my associates within my own company he would find that they are extraordinarily optimistic about cost improvements that can be effected in existing facilities through rounding out. That is what I was referring to.

Mr. ACLE:

Most of the time which we have spent this morning and a part of the afternoon has been related to predictions. In a certain sense that reminds me of something which a professor used to tell me at the university: that an economist spends the first half of his life predicting and the second half trying to understand why his predictions did not work. I think that were are in that position at the moment.

One problem which I think is very important to consider is that predictions do not normally take changes in structure into account. For example, in the developed countries now there is a move towards more sophisticated industries, while the developing countries are moving towards the base or basic industries. I think that one fact which has to be recognised is that this is a pattern and that it has to be taken into account. Developed countries need the steel industry because it is strategic. I cannot remember any case of a developed country being developed without the steel industry — so that has also to be taken into account.

Reference has been made to the problem of rationalisation. I think that it is very clear that that will not be accepted by the developing countries as a reason for stopping their development of the basic industries.

On the other hand I am a little surprised that the steel industry has been defined as labour-intensive. I think that all this has to be looked at in relative terms. Compared with the capital goods industry the basic steel industry is labour-intensive; but in the developing countries the capital goods industries do not exist, so the comparison is not fair.

On the other hand the comparison does not take into account the cost of money, which in the developing countries is quite important because it is the scarce resource. In the case of Mexico, for example, the cost of money is about 60 per cent more than the cost of money in the United States. So it is very important to take this factor into account.

CHAIRMAN:

We are rapidly making up that cost of money differential.

Mr. MORGUES:

With regard to the basic problem of maintaining economic equilibrium in the development of a basic industry like the steel industry, by extrapolating the different points put forward today by the various distinguished economists one could draw the conclusion that in the long term we may be concurring in a general transfer of steel production capacity from industrialised countries to developing countries; always bearing in mind that we should qualify the term developing according to the stage of development reached by the country considered.

Basically I think that the problem is one of profitability. Not much has been said about it although it has in fact underlain all our discussions. What appears to us to be major risk is the following: the estimates we have been given of the increase in production capacity in developing countries would not seem to be envisaged as much related to the requirements of the developing countries so that their development can be economic and as smooth and harmonious as possible. It would seem, on the contrary, that part of this increase in production capacity is largely the result of assumptions by producers coming from developed countries thinking entirely or almost entirely in terms of present costs of labour and raw materials in their own, developed, countries. Investment will not necessarily be cheaper in the long term in developing countries: there will be fairly heavy interest charges to be borne; but it is important that we say that as trade union representatives we refuse to consider that the present and immediate future position regarding the use of manpower in developing countries will not change. This is a factor which we feel, has not so far been considered. It is also untrue that investment charges added to prime costs will remain at the same level as today. We are very interested in the preparation of medium-term and long-term forecasts regarding the estimated needs of all countries for all types of steel. It is particularly important to us for it is to be accepted that effective consultation can be established, especially within OECD, so that balance can be maintained and production capacities can be developed without destroying the equilibrium which is necessary within every country, whether industrialised or developing.

Another important point (and I would refer back to Dr. Marshall's comment on it) is that the manpower problem cannot be considered simply as one of shedding manpower in the same proportion as it is considered necessary to close down facilities regarded as obsolete. Having regard to the work done in the Community and in our own country voluntarily, under trade union pressure and as a result of discussions, worldwide efforts should be made in the light of the existing situation to ensure regional balance in producing countries. It was said that consistent rules would have to be established in the OECD to allow the market to be transparent. This will also be necessary as regards the development of production capacities. Finally, we should also consider that it is imperative that we consider all three aspects of the problem: the increase and adjustment of production capacity should be undertaken at the same time as efforts to ensure that economic replacements are established in areas dominated by steelmaking so that employment levels can be maintained and changes are assured. Once again, manpower cannot conceivably be considered as an accessory factor. So far as we are concerned it is the main, the fundamental factor.

It appears to us necessary too that we should not allow an increase in protectionist attitudes or behaviour of any sort. What is important to us as a result of the conclusions drawn from the Symposium is that appropriate tasks should be defined for OECD so that the development of the steel industry in the best possible conditions is envisaged.

Mr. KONO:

I would like to give my reply to three points which have been raised during discussion.

The first point was that the low level of fuel and raw material costs during the 1960s contributed largely to the competitiveness of large steel mills in particular. The situation changed, of course, after the oil crisis in view of the large cost increases that we experienced. This made it imperative for the Japanese steel mills to resort to other action, to keep our steel mill operation competitive, namely our extraordinary efforts in the field of oil conservation and technological progress. To be more specific, although there was a considerable decrease in crude steel production between 1973 and 1979 from 120 to 112 million tonnes, our finished steel output in 1979 was far larger than 1973 thanks to our efforts in energy saving and for technical improvement such as higher production yield.

I think all this is proof of what Mr Schubert said, that rounding out rationalisation of existing facilities plays a far more crucial part in enhancing the competitiveness of our steel mills today.

One way in which I think that developing countries can better develop their steel industries in the future is to find optimal combinations of electric furnace mills and direct reduction mills. The reason that the electric furnace route in developing countries is more advisable is because per tonne construction cost for integral steel mills in developed countries runs to something like $1,200, whereas if you build a similar large integral steel mill in developing countries, including the costs of infrastructure, in our calculation the cost runs to something like $1500, per tonne. This is one reason why I say that the electric furnace cum direct reduction route may be preferable in the case of developing countries.

One more point wich I would like to make is in reference to Dr. Paul Marshall's earlier comments. While we admit that it is true that we wish to maintain certain levels of export activities because of our need to import most of the requirements of raw materials which are deficient domestically, I would like to emphasize at the same time that there is a consensus among us that it would be extremely dangerous to raise the ratio of our exports to total steel production. To sum up my comment about our export activity, I would like to say with the current depreciated exchange rate of something like 240 or 250 yen to a dollar our export business to the United States market is highly lucrative business for our steel mills.

Mr. WIENER:

I can understand that costs give rise to a highly interesting discussion, but it is a pity that we have not considered Dr. Peco's contribution. Reference has been made to effective co-ordination and harmonization of investment. It has been said that market is not as it should be and we should try to alter this. This will certainly be a topic discussed tomorrow but we could perhaps put the following question at the end of today: what criteria exists which allow us to have effective co-ordination of structures or whatever we call it? The result of our dicussion on forecasting was that it was practically impossible to give figures as to which country would produce how much and to ensure that we have the right level of capacity. Efficient co-ordination sounds very good but what does it look like in fact?

To come back to the matter of costs I found the argument very interesting. In relation to the contribution from Dr. Crandall it was said that we are competitive on our domestic markets; this corresponds with the argument about rising transport costs. There could be a change in general trade movements and steel, but if we adopt a positive approach young industrialised countries whose domestic market is large enough for the minimum economic capacity, which have enough capital to undertake its construction and which have reached a certain treshold with regard to the training of labour could, independently of the cost calculation, still develop their capacity. I am afraid that co-ordination of investments does not take account of certain aspects. It was said, why should we build new capacity when we already have capacity? I think we must look for other ways of acting. The representatives of the trade unions have made points which are important. The consequences for workers are very considerable but I think we should consider whether we will not make the consequences still worse. What do we mean when we say that we should maintain a high level of employment in the steel industry? This would mean that steel prices for domestic industry would be higher than steel from countries where labour costs are lower. This could lead to further protectionist measures. I do not wish to insult our United Kingdom colleagues, but this may be an element in the British malaise. I think there are some problems in this general discussion area to which we can perhaps return tomorrow.

CHAIRMAN:

I have a feeling that members of the panel don't trust my summary and want to continue talking. But I do think it is totally understandable and almost appropriate that we end on some unanswered questions and on some highly significant issues because surely this is not the last time we will gather on such a subject and you have tomorrow to focus further. Today's discussion has been a rather broad one ranging all over the place, but today its role was to be informative, instructive if you will, and I think not as focused or issue-oriented as tomorrow, and I think you will continue to probe in these very interesting areas. I want to express my appreciation to the speakers and to the panellists and to you as an audience for being so very understanding and helpful.

CHAIRMAN'S STATEMENT ON CONCLUSION OF THE FIRST DAY SESSION

I think we can all agree that the panellists have made excellent presentations and the comments from the participants have added significantly to our collective judgment and understanding of the industry's future.

To summarise the main points of our discussion today I will organise my comments on the following topic areas:

1. The macro-forecast;
2. Problems in forecasting generally;
3. World steel demand;
4. Industrialised country steel demand and capacity;
5. LDC demand and capacity;
6. Implications:
   a) Shortage/surplus
   b) Adjustment issues
   c) North/South issues
   d) Energy factors.

Of course, there are overlapping topics but I will proceed even with this difficulty in mind.

1. The Macro-Forecast

We began as a starting point with a forecast of world GDP growth in the eighties of about 4 per cent a year in real terms. This figure has been questioned extensively as being optimistic due to the energy, inflation, and other political uncertainties.

With regard to the longer term, perhaps we can summarise the sense of this session as a view of a moderate growth for the world in the 3.0 to 4.0 per cent range, but one with considerable downside risk.

2. Problems in Forecasting Generally

Much of our morning session addressed the grave difficulties and risks inherent in any forecasting effort. Forecasts can't change uncertainty into certainty, and today's economic and political risks are especially difficult to assess.

Personnally, on balance, I think that forecasting represents a necessary, if difficult, exercise. While it does not reduce uncertainty, it allows us to understand the basis for our differences. The forecasts represent a useful take-off point for policy discussions since they highlight assumptions and provide the basis for a systematic evaluation.

3. World Steel Demand

It appears that overall economic activity will continue to be restrained by the rising real cost of energy and the negative effects of continued inflation. The rapid growth of steel consumption during the 1960s and early 1970s in the developed countries is unlikely therefore to recur in the foreseeable future. On balance, I think, the sense of the session is that the best we can hope for is an average increase of 3 per cent per annum in steel consumption, yet this may be viewed by many in this group as optimistic.

4. Industrialised Country Demand and Capacity

The demand in industrialised countries by one forecast in the morning was at about 2.5 per cent annually. Like the main forecast, the sense of this group is probably that this represents an optimistic view, yet one, of course, that we all hope comes true!

On the capacity side there is greater agreement, I think: capacity in industrialised nations over the next few years, on balance, will remain stable or rise only slightly.

5. LDC Demand and Capacity

On the demand side there was some agreement, but in a range, for LDC demand, of 6 to 8 per cent annually.

On the capacity side there is greater variation and certainly on the balance between capacity and demand there seems to be wide disagreement.

Some think that LDCs will increase as net importers, others think they will remain stable and, of course, others see that they will become net exporters.

These differences have important implications for policy, which we will explore tomorrow.

6. (a) Implications: Surplus or Shortage

On balance, regarding capacity there was general agreement that even after taking into account the re-structuring plans already underway, there will be a continuing and formidable excess supply problem in at least some of the industrialised countries, in relation to demand estimates. One question continues to bother some steel people: how much effective capacity really exists today? In other words, how accurate are capacity figures?

On the optimistic estimate of demand (at least 3 per cent annually) we can expect the current capacity surplus to decline gradually over the next five years.

This remains, however, the less likely outcome in the view of this group; there is the distinct possibility that as the capacity surplus is reduced, there may be regional or product specific shortages.

6. (b) Adjustment Issues

Against this background of macro-economic uncertainty and low growth in steel demand in the industrial world, the world industry will be engaged in fundamental re-structuring efforts. "Re-structuring" encompasses a variety of actions or programmes which can differ from country to country, but which all involve the closing of obsolete facilities, modernisation, and achieving a better balance between supply and demand.

Trade unionists have now indicated the understanding that with re-structuring, industry employment will decline.

This represents a serious social problem, and one that I hope will be discussed fully tomorrow.

These uncertainties, however, make clear that, in the future, steel companies must adopt and live with the fact that, if necessary, they must be able to produce at well below this maximum capability and still make a reasonable profit.

6. (c) North/South Issues

The question of LDCs as net importers or progressively stronger exporters is likely to be one of the more important issues in the next decade.

It is an issue that has real implications for adjustment in industrialised countries.

6. (d) Energy

The energy issue has arisen repeatedly in our discussions. The obvious first effect is on our own macro-forecasts, but there is also the hidden effect on capacity. Energy factors may make much capacity uneconomic and it is not clear that this has been fully appreciated yet.

Final Remarks

Since steel is so important to every nation's economy, and restructuring involves considerable social and economic costs to the affected communities, governments have become more involved than ever before in the industry's affairs.

As the steel industry's adjustment process evolves globally in the coming years, I would urge strongly that government policies be directed toward positive adjustment. National policies must be designed to facilitate, rather than resist, changes that are indicated by market and price mechanisms. Internationally government activity should be aimed at co-operative efforts to ensure that national policies do not lead to distortion in trade flows.

To put it bluntly, the uncertain and unpromising conditions we have recently experienced is the environment we are likely to have to live with in the first half of the 1980s and we are going to have to learn somehow to survive and prosper in these conditons.

We must not underestimate the difficulties we face in dealing with the world capacity questions we have discussed today. The problems are not going to be alleviated in the short term by a rise in demand, nor is any one country or group of countries going to be willing to bear the brunt of the adjustment process for all others. All of us face severe trade and domestic policy problems over the next several years, which will tax our ability to adjust without yielding to rigid arrangements.

THE WORLD ECONOMY
IN THE 1970s AND 1980s

by
Dr. Sylvia OSTRY
Head Economics and Statistics Department, OECD

The purpose of this paper is to provide a macro-economic background, or at least some elements of it, for the Symposium on "The Steel Industry in the 1980s". The paper is in two parts. The first reviews developments during the past decade focussing on the apparent constraints to growth that countries seem to have encountered. The second, after presenting the Secretariat's short-term outlook for the current year, examines, on the basis of some quantitative projections, the prospects of restoring faster and more stable growth patterns over the medium-term.

I. THE PAST RECORD

Developments during the past ten to fifteen years have been characterised by increasing instability in the world economy. Signs of social and political tensions became evident in the late 1960s associated with the Vietnam conflict, the emergence of the post-war student generation and the ratchetting up of inflationary pressure. The earlier seventies were dominated by such events as the collapse of the fixed exchange rate system, the unparalleled peace-time commodity and food price boom and the outbreak of the oil crisis. The deflationary demand and policy response to these events culminated in the unprecedented experience of "slump-flation" in 1974-75 from which the recovery has been hesitant and rather uneven as between countries. When the upswing finally began to show some self-sustainable features, a second "oil shock" hit the oil-importing countries in 1979 probably submitting the world economy to a renewed period of stagflation.

A. *The 1970s in retrospect: initial expectations and outcome*

The dramatic change in the pattern of development since the early 1970s is clearly revealed by the behaviour of traditional macro-economic performance indicators. As can be seen from Table 1, the level of real output in the OECD area rose over the ten years to 1979 1.4 percentage points less per annum than during the preceding decade. The average annual rate of inflation almost tripled as compared to the 1960s and was running about four times as high by the end of 1979[1]. At the same time, the area's traditional and rising current surplus with the rest of the world was replaced by a deficit.

---

1. In November, consumer prices (weighted average) exceeded their corresponding year's level by 11.4 per cent.

It is interesting to compare the actuel outcome of the past decade with the initial projections of the growth of output, employment and productivity (Table 2). Except for a few countries, growth was expected to be even higher than, or at least the same as, during the 1960s. Between 1969 and 1979 the combined gross domestic product of the OECD area as a whole was expected to increase more than 60 per cent, or at an average annual rate of about 5 per cent, after a rise of 4.9 per cent per annum during the preceding decade. In the event, the area's total output probably rose by no more than 40 per cent — or 3.5 per cent per annum — with below average growth recorded for the post 1973 period (2.7 per cent per annum). As can be seen from Table 2, the biggest discrepancies between actual and predicted growth among the major OECD countries occured for Japan, Italy and France, and among the smaller Member countries for Greece and Portugal. Differences of less than 1 percentage point per year were recorded for six countries with growth in Norway coming closest to initial expectations.

Before considering the major factors which seemed to have been responsible for the unexpected weakening of growth, it may be useful to recall some of the basic assumptions explicitly, or implicitly, made on which the long-term projections were based:

i)   On available historical evidence, the traditional presumptions that rates of productivity increase would decline, as levels of output per head rise, and existing technological gaps close, was rejected.[2]

ii)   Inflation was recognised to be a problem, and a growing one in some countries, but it was felt that it was manageable and that it would not negatively affect potential growth. Likewise, it was thought that actual demand growth in line with potential supply could be assured through appropriate fiscal and monetary policies.

iii) There were no signs of demand saturation and there would be no supply constraints to growth resulting from shortages of raw materials (including fossil energy sources).

iv)   The world economic order would continue to be dominated by the industrialised market economies and the United States dollar as reserve currency. Pending changes in the international monetary system were expected to improve the efficiency of markets and policies and hence help growth and stability.

B.   *Main reasons for slower growth relative to the 1960s*

As noted above, the past decade has been strongly influenced by a series of major disturbances. The sequence of "shocks" which struck the world economy has not only dampened the rate of economic activity and added to inflation, but for a time also nourished an excessive growth pessimism ("Club-of-Rome mentality") and has heightened scepticism, or even hostility, to demand management and traditional tools of stabilization policies. There were also fears that liberal democracies had lost their remarkable capacity to adapt to change and that increasing stress and tension placed intolerable strains on political leadership in a world economy which had become so integrated as an economic system and so polycentric as a political system, that the two were intolerably inconsistent.[3]

2.   "…. there is little doubt that rates of growth in productivity are positively associated with the rate of productive investment, technical progress, and labour skills and training. There are no reasons to suppose that any of these factors will become weaker, or have less influence, in the 1970s than in the 1960s. On the contrary…." See page 84, *The Growth of Output* 1960–1980, OECD, December 1970.

3.   See Otto Eckstein, *The Great Recession,* The Data Resources Series 3, 1978, pages 3–5.

The new energy situation with its disruptive effects on prices, real national income and the balance-of-payments, has probably been the most important single cause of increased instability, but neither the oil shock nor some of the other events (e. g. the agricultural price explosion in 1972/74, the speculative commodity price boom in 1973/74 and the collapse of the fixed parity system) should be seen as exogenous to the system even though their sequence and timing were partly accidental. In fact, most of these incidences may be seen as manifestations of an intensified income distribution struggle at a national and international level, reflecting changes in power balances and serious inadequacies in the timing and design of macro-economic and structural policies which became increasingly inappropriate to the changed economic, social and political circumstances. Consequently, the initial shocks have not been smoothly absorbed and neutralised, but have entailed a series of secondary (i. e. domestically generated) shocks which cumulatively often exceeded the first shock by considerable margins. The resulting unusually large imbalances, emerging mainly around the middle of the decade, constituted serious impediments to buoyant and sustained growth during the recovery from the post-oil recession. While varying in relative importance over time and between countries, the imbalances have become manifest in four different areas: inflation, current external account, private investment-saving gap, and the financial position of the public sector[4].

While the greater part of the actual slowdown of growth between the two past decades may be explicable in terms of constraints which have been operating on the demand side of the economies, there is little doubt that there have also been longer-run negative effects on the course of potential output. Endogenous technical progress may have slowed as research and development activities suffered, and the quality and quantity of investment was impaired. Changes in relative energy prices are very likely to have affected actual and potential labour productivity via a number of channels, both on the supply and the demand side of industrialised economies[5]. Furthermore, gains in overall productivity resulting from structural shifts have tended to decline, mainly because of reduced labour outflows from agriculture.

II. GROWTH PROSPECTS FOR THE 1980s

As witnessed by the huge forecasting errors recorded in Table 2 predicting world economic trends, ten years ahead is indeed a hazardous task! This is particularly true when major economies are out of balance and when fundamental changes in behavioural patterns and attitudes are required to remove these imbalances and to create the necessary conditions for returning to a more stable and steeper growth path. It is for this reason that the section following the discussion of the Secretariat's short-term outlook will consider the scope for progress rather than presenting detailed projections based on necessarily uncertain assumptions and hypotheses[6].

4. For a brief discussion of the medium-term constraints on growth see *Economic Outlook* *N⁰ 25*, OECD, July 1979, pages 9–11.

5. They may have caused an acceleration in technical absolescence, factor substitution in the production process, shifts in the pattern of demand, and changes in the level of demand affecting capital accumulation.

6. For a discussion of alternative growth scenarios up to the end of this century see *Facing the Future: Mastering the Probable and Managing the Unpredictable*, Interfutures Report, OECD, Paris 1979.

A. The 1980 outlook

The latest Secretariat estimates of likely 1980 developments are summarised in Table 3. As can be seen, the OECD area's GNP/GDP growth was strong during 1978, reflecting a renewed policy stimulus and favourable terms of trade developments. Growth decelerated somewhat in 1979 as oil and commodity prices rose and policies (particularly monetary policy) became more restrictive. Although demand remained surprisingly strong in the closing months of the year, output growth is expected to remain virtually flat through 1980 for the area as a whole, perhaps averaging ½ a per cent in the two half-years. The sharp rise in energy prices and continuing strength in some primary commodity prices is expected to result in an acceleration in private consumption deflators to around 10½ per cent in the first half of 1980, possibly decelerating to around 9½ in the second half of the year. Despite stagnating real output, the "oil bill" will rise sharply and probably result in more than a doubling of the OECD area's 1979 deficit of around $ 30 billion to 60-65 billion. Correspondingly, the OPEC surplus may attain a new record of $ 95-100 billion, while the payments position of other developing countries is likely to deteriorate quite dramatically.

If these predicted results are not too far off the mark, the starting position for the 1980s does not look particularly encouraging, even though some countries are expected to do much better than others. Despite generally sluggish demand conditions and rising unemployment, inflation will continue to exceed tolerable rates in most countries, and the current balance of payments deficit may, at least in some countries, be perceived as an additional constraint on growth[7]. Moreover, most governments are likely to experience a cyclically-induced weakening of their financial positions making them even more reluctant than otherwise to take expansionary fiscal measures.

As the industrialised countries slide into more general growth recession, it can be expected that the economies of the developing world will be set back in a cyclical sense, because many of their best export markets will be sluggish. On the other hand, it is to be noted that the developing countries now trade more among themselves and have more stability than if they were to adjust in a passive way to every set-back in the industrial areas. The Far East nations exemplify this internal trading propensity and can sustain their gowth at a pace far above the developing country average, even though the industrial nations are expected to sag.

B. The World Economy in the 1980s

On these considerations 1981 will probably be another year of below capacity growth. However, with some prospective easing of inflation and some "technical rebound" of aggregate demand, prospects for faster growth may begin to improve. Recent Wharton projections (Table 4) suggest a small net addition to real GDP in the OECD area as a whole of some 1½ per cent between 1980 and 1981 and average annual rates of growth of more than 4 per cent between 1981 and 1985. The projected marked pick-up of growth after 1981 is, however, based on a number of rather optimistic assumption:

    i)   absence of major world-scale disrupting factors like bad harvests, cut-backs in oil production and shortages of other raw materials;

---

7.   In the likely event of continued pressure from OPEC, there will also be a continued shift of world income shares away from oil-importing countries. Within these countries emphasis on investment over consumption plus slower growth implies pressure on real income growth of workers. Both factors suggest possibility (probability) of increasing social tensions.

ii) annual real crude oil price increases (OPEC price) of no more than 4 per cent;

iii) a successful winding down of average OECD inflation rates to 5 per cent by 1985 and,

iv) a relatively strong growth performance of non-OECD countries[8].

While the above model projections may not fall outside the range of "realistic" or "feasible" outcomes, they would nevertheless seem optimistic. Even so, the underlying proposition that the world economy is not "condemned" to continue to grow slowly for ever is not unreasonable. Growth rates not far short of those recorded up to the oil crisis could indeed become physically attainable before the end of this decade provided that effective energy policies are being put in place without any further delay[9]; and rates significantly above those likely to be recorded in the first half-decade would certainly be sustainable once real wages and other income claims have been brought into line with the present relatively low trend increase of productivity and adverse movements in the terms of trade. However, such income moderation and easing of domestic cost pressure will have to be followed by a strong pick-up of productive investment and a consequent rise in the investment/GDP ratio.

While the prospective energy situation and insufficient capital formation will rule out a quick return to past growth rates, there are at present no constraints on future growth to be seen on the side of demand. Some countries have seen a rise in private savings propensities, but this seems to reflect job insecurity, inflation expectations and income uncertainty rather than a general saturation of consumer demand. In fact, current claims for higher incomes are clearly based on excessive expectations about the availability of both private and public goods and may be seen as an important root of present problems. Moreover, an increased preference for leisure rather than material goods could be expected to show up in less buoyant supply of labour, which would hardly be consistent with present rising trends in participation rates.

From the above considerations, it may be concluded that success or failure of current stabilization and "positive" adjustment policies will largely determine the growth prospects for the 1980s and beyond. Given the backlog of inrealised technical progress, the unused potential for improved international division of labour and the present depressed rates of capacity widening investment, the scope for accelerating potential output growth (and productivity) is certainly quite considerable[10]. However, whether such a desired improvement in the expansion of world output will actually occur will depend on the will and wisdom of policy- and decision-makers, both private and public, to face up to and solve the structural problems which are the legacy of the 1980s.

8. Following rather modest growth predicted for the near term, developing countries are expected to advance in the LINK projections to advance at a 6 per cent rate with OPEC and the countries of South and Sout East Asia above average and Africa (except for the net oil exporters) and Latin America (except for Mexico, Venezuela and Brazil) below average. The centrally planned economies of Eastern Europe and Asia are not foreseen to grow much faster than developed market economies with China (assumed growth of 8 per cent) being the only major exception.

9. If the OECD area is to grow at anything like past performance standards in the medium-term, huge investments will have to be made to push up indigenous energy production besides necessary substantial economies in energy use in relation to GDP which also require more investments.

10. A successful stepping-up of capacity growth would also help to remove potential supply bottlenecks, and hence assist in the fight against inflation.

Table 1   MACRO-ECONOMIC PERFORMANCE INDICATORS OF THE OECD AREA

	1959-1969	1969-1979	1963-1973	1973-1979
Gross domestic product (annual average rate of growth)	4.9	3.5	4.7	2.7
Consumer prices (weighted annual average increase)	2.9	8.5	5.9	10.2
Current external balance (cumulative, $ bill.)	30.1	− 56.8	34.7	− 91.5

Source: Economic outlook, OECD.

Table 2   GROWTH PROJECTIONS AND OUTCOME, 1969-1979
Average annual rate of change
A.   SEVEN MAJOR COUNTRIES

	Projected			Outcome		
	GDP	Employment	Productivity	GDP	Employment	Productivity
Canada	5.1	2.5	2.5	4.2	2.9	1.3
United States	4.2	1.6	2.6	2.8	2.2	0.7
Japan	10.1	1.0	9.1	6.1	0.9	5.2
France	6.0	0.7	5.3	4.0	0.5	3.5
Germany	4.8	0.3	4.5	3.2	− 0.4	3.6
Italy	5.5	0.7	4.7	3.1	0.5	2.6
United Kingdom[1]	3.0	0.1	2.9	1.6	0.1	1.5
Total Big 7	4.9			3.5		

[1] Excluding North sea oil.

B.   OTHER COUNTRIES

	Projected GDP	Outcome GDP
Austria	5.2	4.3
Belgium	4.9	3.8
Denmark	3.7	2.7
Finland	4.9	3.9
Ireland	4.6	3.9
Netherlands	4.8	3.6
Norway	4.2	4.1
Sweden	3.8	2.1
Switzerland	3.6	1.2
Greece	7.7	5.3
Portugal	7.3	5.0
Spain	5.6	4.3
Turkey	6.6	6.1
Total OECD	5.1	3.5

Source: The growth of output 1960 to 1980, OECD 1970; *Economic outlook,* Nº 26, December 1979, OECD.

Table 3 SHORT TERM OUTLOOK FOR THE OECD AREA
Seasonally adjusted annual rates

	1978 I	1978 II	1979 I	1979 II	1980 I	1980 II
A. Domestic Indicators						
GDP Growth	3.4	4.3	3.0	3¼	¾	½
Prices						
GDP Deflator	7.4	7.7	7.4	9	9	9¼
Consumer prices	6.9	6.8	7.6	9¾	10¼	9¼
Unemployment rate	5.2	5.2	5.1	5	5½	6
B. Foreign Trade Growth (Volume)						
Total exports	5.7	7.4	5.3	12½	3½	3½
To: OPEC	6.0	− 4.4	− 29.2	28	17½	20
Non-oil developing countries	7.0	8.0	6.0	14	0	0
Other non-OECD countries	6.0	9.0	4.0	13	4	2
Total imports	6.3	8.1	9.0	6	¾	½
From: OPEC	− 7.8	12.1	− 2.7	− 5	− 10	− 6½
Non-oil developing countries	7.5	8.7	10.0	12	3	0
Other non-OECD countries	6.0	8.5	9.5	5	4	4
Balance of payments						
Current account						
Deficit ($ billion)	2.6	16.5	− 13.6	− 47½	− 74½	− 52

Source: OECD Secretariat.

Table 4 GLOBAL ECONOMIC TRENDS

	1960-65	1965-70	1970-75	1975-80	1980-85
GDP: World	5.0	5.5	3.8	3.9	4.5
GDP: Developed countries[1]	5.1	5.5	3.3	3.3	(4)
Of which:					
Canada	5.7	4.8	5.0	3.9	7.0
United States	4.6	3.1	2.4	2.7	4.1
Japan	10.1	11.6	5.5	5.0	5.9
France	5.8	5.4	4.0	2.8	4.4
Germany	5.0	4.5	2.0	3.4	3.7
Italy	5.1	6.0	2.5	3.4	4.8
United Kingdom	3.1	2.5	2.1	1.4	3.4
Trade: World[2]	6.8	9.2	5.7	5.6	5.8
Inflation: World[2]	4.0	5.0	10.0	11.4	7.5
Current Account (OECD)					
(End of period — $ Bill)	3.8	6.7	− 0.3	− 24.4	− 90.0

[1] 13 major industrialised OECD countries.
[2] Excluding Sino-Soviet countries.
Source: Direct communication to OECD by Lawrence Klein, Wharton School.

OUTLOOK FOR WORLD STEEL INDUSTRY UP TO 1985 DEMAND, TRADE AND SUPPLY CAPACITY

by
Tsutomu KONO
General Manager
Corporate & Economic Research Department
Nippon Steel Corporation

1.   INTRODUCTION

I deeply appreciate the opportunitiy that this OECD Symposium on *The Steel Industry in the 1980s* affords me to express my opinion on the outlook for steel demand and the development of international trade in steel for the decade of 1980s. I am fully aware of the extreme difficulty involved in making such a forecast for the 1980s, particularly in view of the multitude of uncertainties surrounding the political and economic situation of the world.

As we all are well aware, any forecast of world steel demand in the 1980s depends upon one's predictions of the world economic growth in that decade. In this paper, I will explain my views on steel demand by referring to the economic forecasts which were published last year by two authoritative international organizations.

The first forecast I will refer to is the one worked out by the OECD in its *Interfutures* project. The OECD forecast runs up to the year 2000 and contains several different scenarios for economic growth depending on such factors as the extent of cooperation among the industrialized countries themselves and between industrialized and developing countries and the formation of protectionistic regional economic blocs. Depending on the different scenarios, the OECD pictured medium world GDP growth rate of 4.4 per cent, a high GDP growth rate of 5.0 per cent and a low rate of 3.5 per cent as shown in Table 1.

The second forecast I wish to refer to is one made for the GDP growth rates of the Western World for the 1980s and found in the 1979 World Development Report of The World Bank (August 1979). Using three different scenarios also, The World Bank predicted a basic, a high and a low growth rate of 4.5 per cent, 5.2 per cent and 3.8 per cent, respectively (Table 2).

Both the OECD and The World Bank forecasts stress the over-riding importance of international cooperation for future world economic growth. However, considering the unstable political and economic situation of the world it is difficult at the present time to have an optimistic view of future economic growth, and, in accordance, any projected growth rates for the period extending to the full range of 1980s would be unrealistic. Therefore, I would like to limit my present discussion on the world economy and steel supply and demand to the first half of the 1980s.

2. WORLD ECONOMIC OUTLOOK UP TO 1985

As shown in Table 3, I predict that world economic growth up to 1985 will continue at an annual rate of 4 per cent. I am assuming also that, during the period up to 1985, there will be no sharp down turns in economic growth as we experienced in 1974–1975. My predicted 4 per cent economic growht rate is between the lowest and the medium growth rates in the forecasts made by the OECD and The World Bank. My figure is also lower than the 5.4 per cent growth rate which the world enjoyed before the oil crisis (1960–1973) and higher than the post-oil crisis rate of 3.3 per cent (1973–1978).

The economies of industrialized countries, which suffered through four years of hardship after the 1973 oil crisis and which only recently returned to moderate gowth, is now faced again with the possibility of stagflation emerging as a result of a recent series of oil price hikes. In the 1980s, energy supplies will probably tighten, thus leading to more upward pressure on the price of energy, and the price of oil in particular. Under such circumstances, the governments of various countries will tend to continue their anti-inflationary economic measures, and the private sector will probably keep investment at a low level except in the electronic and energy-related industries. No consumer consumption boom can be expected to lift demand as consumer spending will most likely be restrained by the slowdown in real income increases.

Looking at the economic growth of developing countries, I predict their growth rate, which has been relatively high since the 1973 oil crisis, will decelerate somewhat in the first half of the 1980s. The reason for this is that I foresee a slowdown in the economic growth of two groups of developing countries which heretofore have enjoyed above-average growth rates.

One group is the oil-producing countries. Partly because the oil-producing countries will not increase their oil production to a large extent and partly because they, especially those known as "high-absorbers", will experience a slowdown of their pace of industrialization.

The other group is the newly industrializing countries (hereinafter referred to as NICs). Their economic growth will be curbed by mounting inflation and the growing deficit in their international balance of payments.

In this respect, the economic liability of many developing countries will depend on whether the enormous amount of dollars flowing into the oil-producing countries can be as effectively recycled as it was after the first oil crisis.

Judging from the factors mentioned above, the world economy will still be in an adjustement stage during the first half of the 1980s, and international cooperation will not be sufficiently developed during the first half of the 1980s.

3. WORLD STEEL DEMAND IN 1985

What I am going to present today is my personal view on this subject. I expect that under the projected economic conditions I mentioned earlier, the world's apparent crude steel consumption in 1985 will probably reach the 900 million ton level. As shown in Table 4, the apparent steel consumption in the world will increase at an average annual growth rate of approximately 3 per cent during the period 1978 to 1985. This rate is about half the rate that prevailed before the oil crisis (1960–1973), but higher than the rate as shown in a period from 1973 to 1978. A breakdown of the 900 million-ton figure shows the industrialized countries as a group will consume 445 million tons and developing countries 135 million tons. This amount to 580 million tons at an annual growth rate of 3.2 per cent for the Western World. The corresponding consumption figure for centrally planned economies is 320 million tons, bringing world consumption to 900 million tons. However, if the world steel demand remains stagnant, the total figure may decline to 850 million tons.

When we forecast the world steel demand in the first half of 1980s it is important to take a closer look at the steel demand in developing countries and centrally planned economies. As we can see, steel demand in developing countries in 1978 was at a remarkably high 153, as against 100 for 1973, and the demand in centrally planned economies recorded a high 124, as compared to a low level of 86 for industralized countries (Figure 1). This is because high steel demand in developing countries has been primarily caused by high growth of steel demand in both the oil producing countries and the NICs due to their high level of the economic growth.

However, the slowdown of the economic growth rate and the review of development plans by oil producing countries and NICs will bring about a decrease in the growth of steel demand for developing countries to around 5.8 per cent in the first half of 1980s (Figure 2).

Although it is difficult to examine the factors influencing steel demand in centrally planned economies, it is said that their steel demand growth rate is influenced largely by demand trends of Eastern European countries and China. The growth rate of the steel demand in this region is estimated to be about 3 per cent annually through 1985. On the other hand, the steel demand growth in industrialized countries will remain at around 2.4 per cent during the first half of the 1980s if their annual economic growth rate is 3.8 per cent. This is because the growth of steel demand heavily depends on the level of fixed capital formation, and for the foreseeable future we cannot expect any high growth in fixed capital formation in the industrialized countries (Figure 3).

The growth rate of private capital investments will be low in this period due to investment constraints caused by uncertainty about future market developments, the lower return on investment, and the high cost of capital.

4.   WORLD STEEL SUPPLY CAPACITIES

Now, let us look at world steel supply capacity through 1985. Here I see decelerating investment the world over — whether in industrialized, developing, or centrally planned economies — in response to decelerating demand. I therefore predict that the steel industry investment will be sluggish, with most existing expansion plans being either scaled down or postponed.

Under these circumstances, it becomes a question of whether world steel demand and supply will be sufficiently balanced or not in the year 1985.

A number of economists and steel analysts and international research organizations have published forecasts on world steel supply capacity up to 1985[1]. However, their forecasts are limited mostly to supply capacity in the Western World. Therefore, I will also limit my discussion to capacity in the Western World.

1.   Peter F. Marcus, Karlis M. Kirsis *A Western World Steel Supply/Demand Scenario for the 1980's*, October 23, 1979. Peter F. Marcus *Selected Exhibits for Presentation in Tokyo*, November 15, 1979.
William T. Hogan *Steel Supply and Demand in the Mid-1980's*, May 1979.
UNIDO *Background Working Document on Global Scenario of World Steel Industry Growth Particularly up to 1985*, October 1979.
CIA *The Burgeoning LDC Steel Industry, More Problems for Major Steel Producers*, August 24, 1979.

The most important point of disagreement among the various published predictions concerns whether or not there will occur a tight steel demand-supply situation in the Western World before 1985.

Their outlooks seem divided into two views. One view is that a tight world steel demand and supply situation will exist by 1985, with a supply shortage likely in the United States by 1981–1982. The other view, exemplified in a speech by Mr. Lenhard J. Holschuh[2], Secretary General of IISI at its 13th General Meeting in Sydney in last October, asserts: "the Western World as a whole could have by 1985 a total crude steelmaking capacity sufficient to produce 675 million metric tons. ...., it would now seem that this year 1979, Western World crude steel output is likely to be at 490 million metric tons; if we compare this with 1985 effective capacity of 675 million metric tons, demand for steel would have to grow between 1979 and 1985 at the rather high rate of around five percent a year before a global shortage would occur." Also, I would like suggest that 5 per cent is an unrealistically high rate of demand growth for this period. In this sense, I personally believe that world steel supply capacity will stay well ahead of demand, and that there is little possibility of a steel shortage before 1985. Before taking a further look at the steel supply capacity, and point out that there is no uniform formula for calculating steel capacity in all steel producing countries in the Western World. At present, the steel capacity in each country is calculated by either nominal capacity, effective capacity, or production capability, as shown in Figure 4. The results of these supply capacity assessments significantly differ from one another.

We need in any case to establish an internationally acceptable method of assessing capacity. In this paper, the steel supply capacity of the Western World has been calculated on the basis of effective capacity, with adjustments made for the equipment put out of service due to restructuring in the USA and the EC and for the postponement of existing expansion plans in the developing countries.

Now, let us look at steel capacity by group, as shown in Table 5. The capacity of the industrialized countries for 1985 is predicted to be 560 million tons, a level which shows little increase over present capacity. In contrast, the capacity of developing countries will be 120 million tons, a large increase commensurate with the magnitude of their growth in demand and the earnest desire of these countries to increase their degree of self-sufficiency.

The total capacity of the Western World, a combination of effective capacities of both groups, is 680 million tons. On a basis of an effective capacity of 680 million tons and our past production experience, the figure of possible production in 1985 can be calculated as 600 million tons.

Another issue involved in assessing supply capacity is the question of how can we calculate the steel capacity of the Western World counting only the competitive and effective capacity of equipment and facilities.

In this respect, Dr. D. F. Barnett [3], Assistant Vice President of American Iron and Steel Institute, has made an informative analysis of the "vintage" of steel plants in the USA.

Also, the concept of marginal capacity adopted by Mr. P. Marcus[4], The First Vice President of Paine Webber Mitchell Hutchins Inc., affects measurements of capacity due to the loss of competitiveness of equipment and facilities.

Dr. Barnett's analysis and Mr. Marcus' concept could prove useful in measuring capacity and should be discussed more in the future. Looking at the

    2. Lenhard J. Holschuh *Annual Report of the Secretary General*, October 15-17, 1979 IISI.
    3. D.F. Barnett *The American Steel Industry in the 1980's: Capital Requirements for Modernization*, October 12, 1979.
    4. Marginal capacity is interpreted here, though subject to further study, as the capacity of equipment which has lost competitiveness.

present situation, however, we cannot accurately calculate steel capacity because there is no uniform method for measuring competitive and effective capacity.

Now, I would like to compare the projected steel demand of the Western World for 1985 with the production capacity for the same year. As I mentioned earlier, steel demand will be about 580 million tons. Adding to this approximately 10–15 million tons for net exports to the centrally planned economies, we find that the steel demand and supply for 1985 will be tight. This means that the present surplus capacity in the Western World will diminish through the year of 1985 (Figure 5). Furthermore, due to the structural change that has taken place in steel demand since the 1973 oil crisis, the production capacity of some specific steel products may fail to meet demand because of a lack of rolling capacity.

I cannot agree, however, with the view that the world steel supply capacity on a global basis will be such that a steel shortage can be expected in the period 1981–1982. However, if we look at the steel capacity by country, there is a small possibility that a supply shortage may occur in certain industrialized countries. For example, in the case of the United States, the steel production capability utilization rate exceeded 94 per cent in the first half of 1979, leaving little apparent surplus capacity. This situation is created by a shortage in coke oven capacities, as seen in the large quantity of coke imports by the USA. Another reason is the number of plant shutdowns experienced recently. Therefore, Americans naturally feel there will be some supply shortages in the coming years.

There are several ways to overcome this potential shortage however. The United States has recently increased its electric furnace capacity, as shown in Table 6, and it may add further capacity in this way in the future. In addition, there appears to be still more possibilities for expanding the actual steel supply capacity, such as improving coke and fuel ratios in blast furnace operations and increasing the continuous casting ratio. Therefore, I feel there is little likelihood of any steel shortage in the US before 1985.

Concluding my discussion of steel demand-supply problems up to the year 1985, I would like to emphasize the following:

The steel industry in industrialized countries should be strengthening their international competitiveness through rationalization and modernization. On the other hand, the steel industry in developing countries will face ever steeper construction cost for large scale greenfield integrated steelworks. I believe, therefore, that it is important for the steel industry in the industrialized countries to develop "Appropriate Technologies"[5] and cooperate with developing countries in the construction of efficient steelworks best suited to local conditions in developing countries (Figure 6).

5.   TREND OF WORLD STEEL TRADE

We shall now look at the prospects of world steel trade under the kind of the climate of world steel supply and demand as I have just discussed. World steel trade in the period preceding the 1973 oil crisis developed in parallel with the steady growth of world economy and world commodity trade supported by the framework of IMF and GATT. During the period from 1960 to 1973, world commodity trade had grown 8.3 per cent annually in volume, while world steel trade enjoyed an annual growth rate of 8.5 per cent as shown in Table 7. From 1973 to 1978 since the

5.   Shigeo Hosoki and Tsutomu Kono *Japanese Steel Industry and its rate of Development*, September 1979.

oil crisis, the growth of world commodity trade came down to 4.5 per cent per year and world steel trade grew at a reduced 5.3 per cent annually.

During the period from 1973 to 1978, steel imports by industrialized countries decreased appreciably, while the imports by developing countries and centrally planned economies did not decrease as much. Here I would like to take a closer look at this trend.

Steel imports by industrialized countries, developing countries and centrally planned economies grew at annual rates of 9.1 per cent, 7.7 per cent and 7.9 per cent, respectively, before the oil crisis (1960–1973), but changed to 2.8 per cent, 7.5 per cent and 10.5 per cent after the oil crisis (1973–1978). Accordingly steel imports by developing and centrally planned economies together accounted for 54 per cent of total world steel trade volume in 1978 as compared to 38 per cent in 1970 (Figure 7). It is, therefore, important to fully understand the import trend of these two groups of countries when projecting the future world steel trade.

As far as steel exports are concerned, we find that the EC and Japan have continued to be the two major steel exporters in the world through the year of 1970s. Steel exports from the EC and Japan accounted for 61 per cent of all exported steel in 1970 and 65 per cent in 1978 (Figure 8). In developing countries, mainly NICs in Asia and Latin America, have grown steadily since 1973. Though small in quantity now, they have a great potential and merit attention.

I foresee the future growth of world commodity trade up to 1985 staying at about 5 per cent, as compared to the 4 per cent level during the 1973–1978 period. However, the growth of the world steel trade over the period from 1978 to 1985 will be relatively slower than the level attained the period from 1973 to 1978 because the growth of steel demand in centrally planned economies and in developing countries, especially, oil producing countries and NICs, as I mentioned earlier, will fall below previous levels.

In particular, NICs in Asia and Latin America will be reducing their steel imports as a result of having increased their level of self-sufficiency in steel supply. Viewed from export side, the EC and Japan will continue to stay as the major steel exporters in the world without any drastic change even if the increase of steel exports from NICs is considered.

A very informative study entitled "The analysis of the evolution of trade flows 1970–1978 and the assessment of the economic factors influencing them" was made last September by the Steel Committee of OECD. I strongly hope that such analyses of OECD member countries trade will continue to be made together with a special emphasis on essential points such as the import and export behavior of developing countries and centrally planned economies. Although the forecasting of world steel trade is very difficult, it is one of the most important tasks of the Steel Committee of OECD.

6. CONCLUSION

Under changing economic environs in the 1980's characterized by an intensifying energy crunch and continuing inflation, the foremost task facing the world steel industry and the steel industry of industrialized countries in particular is not just to resolve demand-supply problems but more importantly to carry out basic structural counter-measures. First, it is imperative for the steel industry of industrialized countries to regain its earning capacity by improving its labor productivity and international competitiveness which can be done through actively investing in the restructuring and rationalization of obsolete facilities. Second, it is equally important that the industrialized countries act to meet the strong demands of developing countries that desire to be more self-sufficient in steel supply. We believe that a reasonable approach to this problem is for industrialized countries to provide suitable economic and technical cooperation for such steel plant construction projects as deemed appropriate in the light of the various economic conditions prevailing in the particular country desiring to receive such cooperation and the degree of economic development achieved in that country.

It is essential that the governments of industrialized countries act rationally vis-a-vis the steel industry, fully respecting the principles of fair competition and market economy. We are concerned that the unstable economic conditions predicted for the 1980's might tempt some governments to resort to such negative policy action as trade protectionism. What is needed is a package of "positive adjustment policies" that restructure and strengthen the industry, promote R and D activity, and create new job opportunities rather than protective policies which would only serve to safeguard outmoded industries and preserve inefficient economic activities. Any aid measure taken by government to benefit an individual industry must be provisional, dispensed in full recognition of the preservation of market principle and entrepreneurial viability.

In order to create an economic environment conductive to the adoption of these positive adjustments, each government should pursue adequate macroscopic economic policies geared to establishment of a system between the North and South, fight against inflation, sustenance of proper levels of economic growth and employment and effective energy conservation.

I am convinced that the making of efforts to sustain the growth of the world economy while upholding the workings of international cooperation lend itself, in the long run, to better profitability and employment within the steel industry through the expansion of steel demand.

REFERENCES

OECD, "Interfutures" — Facing the Future —, July, 1979.

The World Bank, "The World Development Report, 1979", August, 1979.

OECD, *The Impact of the Newly Industrializing Countries on Production and Trade in Manufactures*, 1979.

Tsutomu Kono, *The Steel Industry in Developing Countries* — Steel Demand and Supply Condition —, October, 1978.

Hans Muller, Conference Papers Series: 41 Business and Economic Research Center, Middle Tennessee State University, February, 1979.

Tsutomu Kono, *Study of the World Steel Trade 1960-75* — A Structural Analysis —, January 1, 1978.

Tsutomu Kono, *Economic and Management Implications of Technological Advances in the World Steel Industry*, June 20, 1979.

C.A. Bradford, *Steel Industry* Quarterly Review, August, 1979.

Industrial Economics Research Institute Fordham University *Analysis of the US Metallurgical Coke Industry*, October, 1979.

C.E.C., *General Objectives for Steel 1980, 1985 and 1990*, July, 1978.

V. Davignon, *Restructuring the 80s* The Address to the Conference held by Metals Society, The Steel Industry in the Eighties, September 12, 1979.

R.L. Deily, *IISS Commentary January-February, 1977*, Institute for Iron and Steel Studies.

Table 1   GROWTH OF GROSS DOMESTIC PRODUCT UNDER
INTERFUTURES SCENARIOS

Average annual growth rates, 1975-2000 (percent, at 1970 prices)

	Scenario A	Scenario B2	Scenario C	Scenario D
Industrialized countries	4.4	3.5	2.3	3.4
Developing countries	6.7	6.0	5.3	5.9
Western world	4.9	4.0	2.9	3.9
Centrally planned economies	5.3	5.1	4.5	5.1
World	5.0	4.4	3.5	4.3

Scenario A: High Growth, *Scenario B2:* Traditional Moderate Growth;
   Persistent Disequilibria.
Scenario C: A Breakdown in North/South Relations; The South Opts Out.
Scenario D: A Fragmented World.
   Source: Interfutures, Facing the Future (August 1979), OECD.

Table 2   GROWTH RATES OF GDP FORECASTS OF THE WORLD BANK

Percent, at 1975 prices

	Average annual growth rates, 1980-90		
	Base	High	Low
Industrialized countries	4.2	4.9	3.5
Developing countries	5.6	6.6	4.8
Western world	4.5	5.2	3.8

Source: World Development Report 1979 (Aug. 1979) World Bank, Washington, D.C.

Table 3   AVERAGE ANNUAL GROWTH RATES OF GDP UP TO 1985

Percent. at 1975 prices

	1960-73	1973-78	1978-85
Industrialized countries	4.9	2.4	3.8
Developing countries	5.8	5.0	4.8
Western world	5.1	2.9	4.0
Centrally planned economies	6.7	4.9	4.0
World	5.4	3.3	4.0

In this paper classification by type of area is based upon that of IISI.
Industrialized
   United States. Canada. Oceania. EC (7). Other Western Europe. Japan. South Africa.
Developing
   Latin America. India. Other Asia. Other Africa. Middle East.
Centrally Planned
   USSR. Eastern Europe. China and DPR Korea.
   Source: Nippon Steel Corporation.

Figure 1
APPARENT STEEL CONSUMPTION INDEX BY REGION
1973 – 100

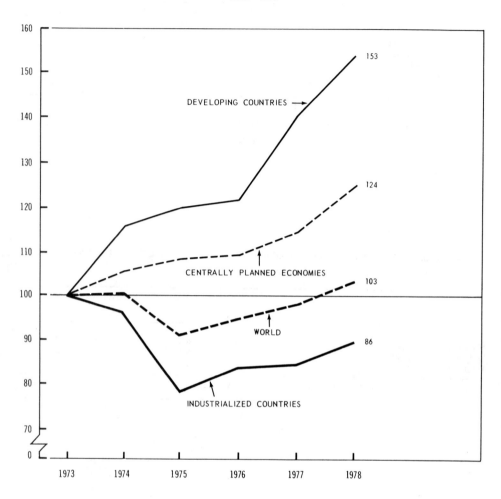

Source : IISI.

Figure 2

APPARENT STEEL CONSUMPTION IN DEVELOPING COUNTRIES
Average annual growth rates by country group 1960-1978

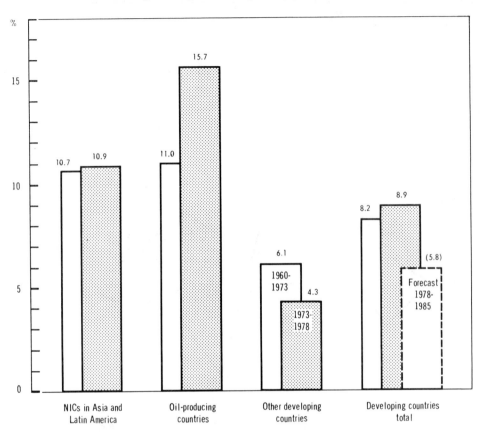

Sources : IISI and *Nippon Steel Corporation.*

Table 4   WORLD APPARENT STEEL CONSUMPTION ESTIMATE FOR 1985

	(million M/T) Apparent steel consumption		(%) Average annual growth rates of ASC		
	1978	1985	1960-73	1973-78	1978-85
Industrialized countries	376	445	5.5	▲ 2.9	2.4
Developing countries	91	135	8.2	8.9	5.8
Western world	467	580	5.7	▲ 1.1	3.2
Centrally planned economies	262	320	5.4	4.5	2.9
World	729	900	5.6	0.6	3.1

Source: Nippon Steel Corporation.

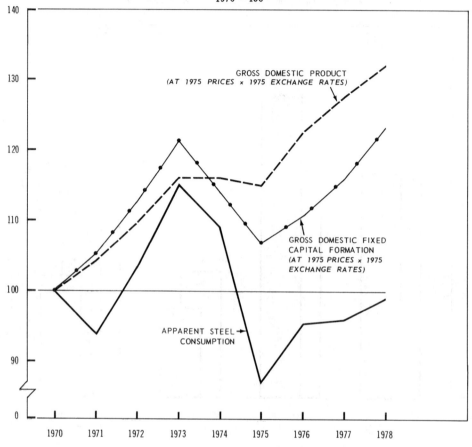

Figure 3

RELATION BETWEEN ASC AND GDCF IN OECD MAJOR 7

1970 = 100

GROSS DOMESTIC PRODUCT
(AT 1975 PRICES x 1975 EXCHANGE RATES)

GROSS DOMESTIC FIXED
CAPITAL FORMATION
(AT 1975 PRICES x 1975
EXCHANGE RATES)

APPARENT STEEL →
CONSUMPTION

Sources : OECD and IISI.

Figure 4

CONCEPTS OF STEELMAKING CAPACITY

Nominal Capacity

Production capacity signifies that of main equipment i.e. converter, electric furnace etc. without regard to factors concerned with raw materials supplies, restraints imposed by the preceding and/or following process, and excludes that of equipment not in use long or intended abandoned.

Effective Capacity

Maximum possible production is the maximum production which it is possible to attain during the year under normal working conditions, with due regard for repairs, maintenance and normal holidays, employing the plant available at the beginning of the year but also taking into account both additional production from any new plant installed and any existing plant to be finally taken off production in the course of the year. Production estimates must be based on the probable composition of the charge in each plant concerned, on the assumption that the raw materials will be available.

Production Capability

Capability is defined as the tonnage capability to produce raw steel for a full order book on the current availability of raw materials, fuels and supplies and of the industry's coke, iron, steelmaking, rolling and finishing facilities, recognizing current environmental and safety requirements.

Sources: ECSC, AISI.

78

Figure 5
WESTERN WORLD STEEL SUPPLY / DEMAND OUTLOOK FOR 1985

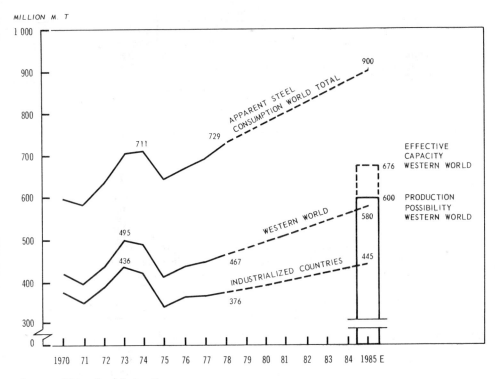

Source : Nippon Steel Corporation

Table 5 STEEL PRODUCTION CAPACITY ESTIMATE FOR 1985
Effective capacity: million M/T

	1978	1985
Japan		144
EC (7)		177
U.S.A.		150
Others		91
Industrialized countries	560	562
Developing countries	68	114
Total western world	628	676

Source: Nippon Steel Corporation.

79

Table 6   UNITED STATES STEELMAKING CAPACITIES BY PROCESS

	(million S/T) Steelmaking capacities			(percent) Process composition		
	1957	1973	1977	1957	1973	1977
Basic oxygen processes	1	92	105	0.8	50.8	57.7
Electric arc furnaces	12	33	45	9.3	18.2	24.7
Open hearths and others	116	56	32	89.9	30.9	17.6
Total	129	181	182	100.0	100.0	100.0

Source: Institute for Iron & Steel Studies.

Figure 6

COMPARISON OF MASS-PRODUCTION TECHNOLOGY,
ADVANCED TECHNOLOGY AND APPROPRIATE TECHNOLOGY

	Mass-production technology	Advanced technology	Appropriate technology
Definition	Integrated steelworks technical system aimed at mass production by large capital and equipment investments. The system has been completed and its results already established.	Technology for meeting future new supply environment and market requirements and for supporting the management of steelmakers as material manufacturers.	Technology aimed at capital, production scale and other levels most appropriate to local conditions under given environment.
Background	○ Expanded growth of the supply of steel materials. ○ Stable supply of raw materials and energy.	○ Uncertainty of the supply of raw materials and energy. ○ Diversification of market requirements.	Local special conditions of energy, market, transportation, capital, etc.
Necessary technical basis	○ Accumulated production techniques.	○ Systematic R & D activities. ○ Transfer of advanced technology.	No particularly high technical basis is needed.
Examples	1. Large BF operation. 2. BOF, CC. 3. High-speed rolling (hot, cold). 4. Production control. 5. Preventive maintenance.	1. Formed coke. 2. Continuous steelmaking. 3. CC-DR. 4. CAPL. 5. High-strength sheet, rail. 6. Tin-free steel. 7. Line pipe material for frigid regions.	1. Charcoal BF. 2. Direct reduction using low-cost natural gas. 3. Minimill utilizing scrap.

Source: Nippon Steel Corporation.

80

Table 7 STRUCTURE AND GROWTH OF WORLD STEEL TRADE[1]

Percent

	Country composition			Average annual growth rates	
	1960	1973	1978	1960-73	1973-78
Exports					
Industrialized countries	90.9	86.5	87.2	8.1	5.5
(Japan)	(8.5)	(32.5)	(31.2)	(20.3)	(4.5)
(EC (7)	(57.6)	(34.7)	(34.3)	(4.3)	(5.0)
Developing countries	0.8	4.0	5.6	22.9	13.0
(Other Asia)	(—)	(1.9)	(3.1)	(—)	(16.9)
(Latin America)	(0.5)	(1.9)	(2.1)	(19.2)	(7.2)
Centrally planned economies	8.3	9.5	7.2	9.6	(▲0.5)
World total	100.0	100.0	100.0	8.5	5.3
Imports					
Industrialized countries	48.2	52.3	45.7	9.1	2.8
(U.S.A.)	(10.8)	(18.5)	(19.7)	(13.0)	(7.0)
Developing countries	36.9	33.7	36.8	7.7	7.5
(Other Asia)	(7.8)	(11.6)	(13.3)	(11.8)	(8.6)
(Latin America)	(11.9)	(9.2)	(6.0)	(6.2)	(3.0)
(Middle East)	(6.1)	(7.7)	(11.1)	(10.4)	(13.5)
Centrally planned economies	14.9	14.0	17.5	7.9	10.5
World total	100.0	100.0	100.0	8.5	5.3

[1] Excluding intra EC and intra COMECON trade.
   Source: Nippon Steel Corporation.

Figure 7

TRENDS IN WORLD STEEL IMPORTS

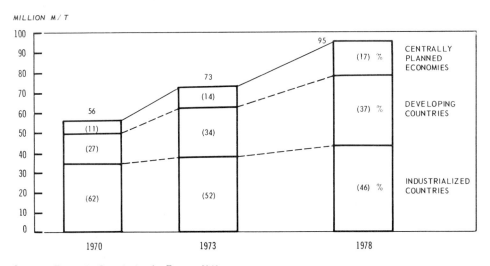

Source : Economic Commission for Europe, U.N.

81

Figure 8
TRENDS IN WORLD STEEL EXPORTS
Area composition

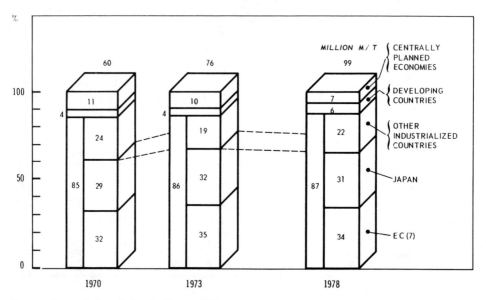

Source : Economic Commission for Europe, U.N.

82

WORLD TRENDS IN STEEL
CONSUMPTION AND PRODUCTION
TO 1990

by
Helmut WIENERT
Research Economist
Rheinisch-Westfälisches Institut
für Wirtschaftsforschung

FOREWORD

In 1975 the upward trend in steel consumption and production faltered and although world production of crude steel has admittedly risen steeply again since then (750 million t in 1979, which is at least 100 million t more than in 1975) it is not yet back again on its earlier growth path. Consequently all medium-term forecasts for the years after 1975, made before the collapse of world steel consumption in that year, are to varying extents off target, which is why — unlike earlier years — there is so much uncertainty regarding future growth in steel consumption in the medium term.

In all logic, it would therefore seem sensible to deal with the margins of error in earlier forecasts before attempting new forecasts of world steel consumption and crude steel production to 1990.

In 1959 the United Nations published an estimate which put world steel consumption in 1972–1975 at about 630 million t [1]. The actual trend figures [2] for world steel consumption were about 640 million t in 1972 and about 690 million t [3] in 1975. This degree of accuracy [4] was possible because the growth rate for steel consumption in the fifties was maintained through the sixties.

In 1971 the Rheinisch-Westfälisches Institut für Wirtschaftsforschung [5] expected world steel consumption in 1975 to be 760 million t and in 1972 the International Iron and Steel Institute (IISI) [6] expected it to be 750 million t. Subsequently both estimates proved to be about 65 million t, or almost 10 per cent,

---

1. United Nations, *Long-Term Trends and Problems of the European Steel Industry,* Geneva 1959, page 132
2. Three-year moving average.
3. All figures in traditional crude steel equivalents.
4. For some countries, however, like Japan or India, this is not true, as some forecasts and actual values are far apart (in Japan's case, for instance, the forecast consumption figure was 38 million t whereas actual consumption was 75 million t).
5. Spiegul, U., Rohstahlkapazität und Rohstahlverbrauch in der Welt im Jahre 1975 in "Mitteilungen des Rheinisch-Westfälischen Instituts für Wirtschaftsforschung (RWI)", 22, year 1971, page 21 ff.
6. International Iron and Steel Institute (IISI), Projection 85, World Steel Demand, Brussels 1972, page 26.

too high. The change of direction in the growth trend for consumption since 1975 comes out still clearer in the forecasts for 1980; in 1972 the International Iron and Steel Institute put consumption then at 950 million t and as late as 1976 the Commission of the European Communities [7] put it at about 900 millon t, whereas it will in fact be well below 800 million t [8].

Thus experience with medium-term forecasting argues for caution, because deviations in growth trends are difficult to foresee.

Forecasting errors are almost always due to the fact that assumptions and/or growth postulate prove wrong. Assumptions [9] are made when other indicators are not available. Growth postulates [10] are usually deduced from past trends, but they also incorporate subjective expectations. They oversimplify interrelationships and therefore reflect only their general tendency.

THE FORECASTING METHOD

In the medium term world steel production is determined by consumption.

The trend postulate in this forecast is that, apart from structural factors, steel consumption in a region depends mainly on the level of development reached by the regional economy. As the various regions of the world are at different stages of development, steel consumption needs to be calculated separately for each of them [11].

In an ideal model, each region — although at different times and speeds — would go through an evolutionary process from simple agricultural economy to highly developed industrial economy [12].

At first, steel consumption hardly grows at all, because economic growth depends on the agricultural sector. The steel intensity of real GDP, i.e. the average amount of steel used per unit of GDP, is low.

During the industrialisation process, steel consumption rises faster than GDP, so steel intensity increases. Sooner or later, however, steel consumption will rise at only the same speed as GDP and then even more slowly that the latter, because industrial (i.e. steel-intensive) production will no longer grow faster than GDP, less steel-intensive and more "intelligent" products will come to the fore in industrial production and the services sector will gain more importance. Steel intensity reaches its zenith and then declines.

Thus the growth rate for steel consumption in a region depends not only on GDP growth, i.e. the rate of economic development, but also on the current level of that development [13].

Even if GDP was growing at the same rate in all regions, the growth rate for world steel consumption would vary because of the considerable difference between the development level in different regions.

An example will make this clear. Let us take two regions, A and B, which start with the same GDP and steel consumption — 100. We shall then suppose that

7.   Commission of the European Communities, General Objectives Steel 1980 to 1985, in the Official Journal of the European Communities, Vol. 19, No. C 232 of 4th October, 1976.
8.   All figures are trend values in traditional crude steel equivalents.
9.   e.g. about future GDP trends.
10.   e.g. that steel intensity of GDP depends on per capita GDP.
11.   For the purposes of the forecast the countries of the world have been grouped in 12 regions.
12.   This assumes the historically "regular" upward development of an economy by steps or economic stages.
13.   The influence of technological progress is not taken into account here.

industrialisation starts in region A at time t = 1, but in region B not until time t = 10, that this leads to a steady growth in GDP of 10 per cent per unit of time and that, at the same time, steel consumption increases at first faster and then more slowly than GDP. The result is shown in Table 1.

Although the relationship between GDP and steel consumption depending on the stage of economic development is reliable in this very general form over long periods of time, it is by no means sufficiently stable over periods of five or ten years. Deviations are always to be expected when the regularity of economic development is disturberd or interrupted by prolonged investment crises, sudden structural adjustments or similar events. In North America[14], for example, steel intensity fell from about 0.20 kg per dollar of GDP[15] in the first half of the fifties to 0.17 or 0.16 kg in the first half of the seventies. This decline was fairly steady except from 1958 to 1962, when a pronounced drop in investment drove steel intensity down to the low level it reached in the first half of the seventies (see diagram 1).

However, even if such disruptions[16] fail to occur in future, there remains a considerable area of uncertainty. Admittedly, steel intensity in highly industrialised countries may be expected to decrease further in future, but the question is by how much.

For North America past trends would suggest that steel intensity will drop to about 0.13 kg per dollar of GDP by 1990, but consumption per head would then have risen to well above 1,000 kg. However, previous experience points to a saturation limit of 700 or 800 kg, so that a steeper decrease in steel intensity to 0.115 kg per dollar is more likely. Figures of 0.11 or 0.12 could also be defended, but given the high GDP of this region they would involve a forecasting margin of 16 million t[17] or almost 10 per cent of expected steel consumption.

As there is no historical model for future steel intensity trends in the regions with the highest per capita income, deduction and speculation are the only guides. One possibility is that steel intensity is decelerating faster than per capita income is accelerating, which means a decline in steel consumption in absolute terms as well[18]. It could be, for example, that industries with a high indirect steel export content[19], still located in highly industrialised countries, are moving increasingly to recently industrialised countries or that there is faster progress in steel-saving technologies.

Consequently estimates of future steel consumption based on GDP and steel intensity trends are to be treated with considerable caution[20]. The search for forecasting methods which promise "more valid" results will and must not cease, although it often merely means exchanging new risks for old.

To allow for uncertainties, the consumption forecast described here for the various regions is in the form of a bracket with high and low limits[21].

14. United States and Canada.
15. in 1963 prices.
16. Cyclical deviations from the trend are not meant here, but rather "growth pauses" in the general trend, visible in the five-year moving averages for instance. However, the borderline between cyclical downturns and growth pauses is ill-defined.
17. 177 million t for 0.11 and 193 million t for 0.12.
18. In North America steel consumption in 1990 would be about as high as in 1974 (trend value), if steel intensity fell to about 0.10.
19. Vehicle manufacture, shipbuilding and unsophisticated engineering.
20. On top of the risk of miscalculating steel intensity there is the risk of miscalculating GDP trends.
21. These limits only indicate the area of uncertainty accepted in the forecast and arising from the given data and the trend suppositions. Other postulates and hypotheses would give another range of uncertainty. Consequently this range of forecasts is narrower than the range which can be plausibly deduced.

Table 1   EFFECT OF LEVEL ECONOMIC DEVELOPMENT ON GROWTH RATE FOR TOTAL STEEL CONSUMPTION

Numerical example

Period	Region A GDP		Region A Steel consumption		Region B GDP		Region B Steel consumption		Total GDP		Total Steel consumption	
	Volume	Percentage increase	Volume	Percentage increase	Volume	Percentage increase	Volume	Percentage increase	Volume	Percentage increase	Volume	Percentage increase
0	100	0	100	0	100	0	100	0	200	0	200	0
1	110	10	106	6	100	0	100	0	210	5,0	206	3,0
2	121	10	114	8	100	0	100	0	221	5,2	214	4,1
3	133	10	126	10	100	0	100	0	233	5,5	226	5,3
4	146	10	141	12	100	0	100	0	246	5,7	241	6,7
5	161	10	161	14	100	0	100	0	261	6,0	261	8,2
6	177	10	187	16	100	0	100	0	277	6,2	287	9,9 max.
7	195	10	213	14	100	0	100	0	295	6,4	313	9,1
8	214	10	238	12	100	0	100	0	314	6,6	338	8,2
9	236	10	262	10	100	0	100	0	336	6,8	362	7,0 min.
10	259	10	283	8	110	10	106	6	369	10,0	389	7,4 max.
11	285	10	300	6	121	10	114	8	406	10,0	414	6,6
12	314	10	312	4	133	10	126	10	447	10,0	438	5,6
13	345	10	318	2	146	10	141	12	491	10,0	459	4,9 min.
14	380	10	324	2	161	10	161	14	541	10,0	485	5,7
15	418	10	331	2	177	10	187	16	595	10,0	517	6,7 max.

Author's own calculations

Diagram 1
STEEL CONSUMPTION TRENDS IN NORTH AMERICA

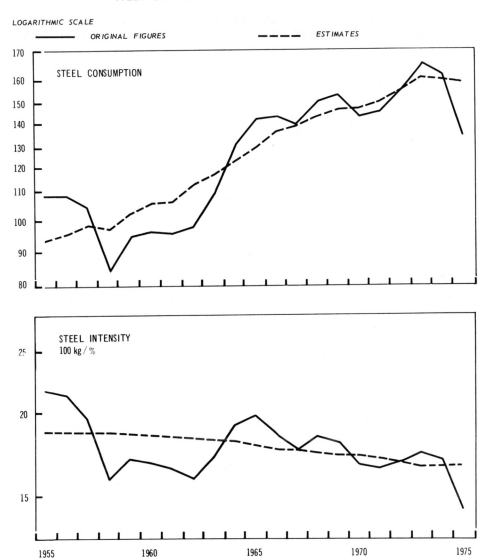

Author's own calculations and estimates based on U.N. data.

1. $\ln \left(\ln \dfrac{220}{ASC} \right) = 0.8816 - 0.00208 \cdot GNP$
    $\qquad\qquad\qquad (5.5) \qquad (9.5)$

    $\qquad R^2 = 0.83 \qquad DW = 0.96$

87

Once consumption is determined in this way, production can be deduced from it. Here the forecasting problems are not less but rather more difficult than in the case of consumption.

It is not sufficient to know the consumption in the different regions in order to calculate regional production, because there are inter-regional exchanges of steeel products.

Firstly, import and export flows are a result of regional differences between demand and the possibilities of satisfying it. Such differences may be due to lack of raw materials or manufacturing know-how, lack of capital or regional specialisation in the economy's production facilities. This kind of foreign trade is called "deficit trade", whereby demand which cannot be covered by the region's production capacity is met by imports.

Secondly, import and export flows are also a result of competition between suppliers for customers. For this reason there are exchanges of similar goods between regions whose production capacity is sufficient to satisfy local demand. Unlike "deficit trade" this trade is "pure exchange" and is often in balance.

In practice both these kinds of trade co-exist, so that the trade balance is only a small part of total trade.

However, as a "residual quantity" resulting from the comparison of consumption with production [22], the foreign trade balance is very sensitive, so that despite its relatively small volume it may fluctuate considerably from year to year

In 1978 the Federal Republic of Germany's balance was equivalent to about 4 million t crude steel, or at least 10 per cent of steel consumption. It was the difference between imports of almost 14 million t and exports of almost 18 million t. A third of the market's supplies was imported, while two-fifths of production was exported. When countries are grouped together in larger regions, the relative importance of the latter's foreign trade balance is still less, if countries with export or import surpluses are included in the same region.

To calculate a region's production one must know its foreign trade balance as well as its consumption, and to calculate its foreign trade balance one must know its production.

The first clue is the argument that a region's foreign trade balance or production will follow the same trend in future as in the past, which would suggest making a forecast based on a time series. For a specified consumption in a region, however, the figures given by such trend forecasts for the foreign trade balance and for production will not as a rule agree.

Moreover, the total of the trade balances of all regions will not be nil and total production will not agree with the figure given for world consumption. While these discrepancies can be resolved by stage-by-stage adjustments, forecasting entirely on the basis of trends would still warrant objections.

It therefore seems preferable to calculate production from the trend of production capacity in the region.

Owing to the long time required to build steelworks this is already known to some extent several years in advance. Existing plans for extending capacity provide rough pointers to the trend up to 1990.

Capacity utilisation rates in the various regions are decided by competition between all the suppliers for customers.

---

22. Changes in stocks are disregarded, so that production equals regional consumption minus imports plus exports.

Meanwhile whether suppliers in the different regions are successful or not depends not only on market conditions, but also on power relationships and administrative conditions. At all events, steel markets cannot be adequately described by a model based on free competition between numerous small suppliers. Technical and operational factors would lead instead to a suppliers' oligopoly in which the market behaviour of a single supplier could influence market results decisively. Moreover, a considerable proportion of world steel capacity is State-owned and in many countries imports and exports of steel products are influenced by government regulations.

Consequently cartel-like divisions of the market, dumping practices, import restrictions, export subsidies and government compensation for losses determine success in the market just as do advantages in production and transport.

Owing to the merging of market factors with power relationships that is found in many steel markets it is extraordinarily difficult to forecast the results of competition in different regions.

Two extreme cases may be postulated:

— Competition leading to similar rates of capacity utilisation throughout the world. Production would then be distributed between regions in accordance with the capacity of each.
 This case is supported by the argument that every steel industry tries to run its production plant at maximum capacity so as to reduce overhead costs per unit of output. Consequently in their export business steel-makers often content themselves with earnings which help to cover their costs, but do not cover all of them. This produces a tendency towards reciprocal market penetration and similar rates of capacity utilisation.
— Competition limited to essential imports, so that a region's consumption is covered by its own production until its capacity is fully utilised.
 This case is supported by the argument that every region's steel industry is the most competitive in its own markets because, unlike its competitors, its transport costs in them are low, it knows them very well and often enjoys government or industry support.

Although neither of these cases alone is a true reflection of reality, the first could be said to apply more to capacity in exporting regions and the second to capacity in importing regions. A point in favour of this is the empirical finding that production in net exporting regions is usually subject to wide cyclical fluctuations, whereas in net importing regions it fluctuates only slightly and increases in parallel with production capacity. Production in the different regions is then calculated as follows.

First production potential is compared with consumption in the various regions. This shows that there are some regions whose consumption exceeds their production potential so that they have to import, and other regions which can produce more than they consume and have therefore an export potential.

It is assumed that production facilities in regions which need to import work at full capacity. If a region has an export potential it is ascertained whether, and if so to what extent, it may nevertheless be a net importer. This will give the volume of world imports, which will then be distributed roughly over the remaining exporting regions.

Lower and upper limits are set for production capacity and consumption trend; when they are compared, four production forecasts can be made.

The countries of the world were grouped together in five regions[23].

Region 1: EEC countries and Japan. This region taken as a whole is the only net exporting region.

Region 2: North America. From the point of view of its historical development this is an old steel region which has become a net steel importer, in spite of its adequate production capacity.

Region 3: Eastern Europe. This region consists of the centrally planned economies on the Soviet model and is largely self-sufficient in steel.

Region 4: Other West European countries, Oceania and South Africa. This very heterogeneous group is in the nature of a residual quantity; it includes both old and young industrialised countries.

Region 5: Developing countries including the People's Republic of China and North Korea.

World steel consumption[24] grew:

from 336 million t in 1960
to 457 million t in 1965
to 588 million t in 1970
and to 691 million t in 1975.

According to our estimate it will be:
  794 million t in 1980
  from 889 to 953 million t in 1985
  from 1,011 to 1,101 million t in 1990.

The corresponding annual average rates of increase are:

6.3 per cent from 1960 to 1965
5.2 per cent from 1965 to 1970
3.3 per cent from 1970 to 1975
and then
2.8 per cent from 1975 to 1980
2.3 to 3.7 per cent from 1980 to 1985
2.6 to 2.9 per cent from 1985 to 1990.

23. Forecasted values were calculated for twelve regions, namely North America, the EEC, the rest of Western Europe (including Turkey and Yugoslavia), Eastern Europe (including the Soviet Union), Japan, Oceania, South Africa, the rest of Africa (including Egypt), Latin America, the Middle East, South East Asia and the People's Republic of China (including North Korea).
24. Expressed in traditional crude steel equivalents. If RS is crude steel production, X is the proportion of continuously cast steel and RÄ is the crude steel equivalent, then:
$$RÄ = RS (1 + X \times 0.14).$$
The figures for 1960, 1965, 1970 and 1975 are three-year moving averages for visible consumption, which is the same as production, taking the world as a whole. Visible consumption was calculated from United Nations figures and correlated with world production.

The shares of the regions in total world consumption changed as follows:

	EEC and Japan	North America	Eastern Europe	Western Europe, Oceania and South Africa	Developing countries
	Percentage shares				
	until now				
1960	30	28	27	6	9
1965	28	30	26	7	9
1970	31	25	26	7	11
1975	28	21	29	8	14
	expected				
1980	26	20	29	8	17
1985	25	18	29	8	20
1990	23	17	28	8	24

The steep increase in world steel consumption from 1960 to 1965 was mainly due to North America, which raised its share of world consumption from 28 to 30 per cent. The main reason for this was the exceptionally sharp rise in steel consumption in the United States, which after a long period of low investment rose from almost 100 million t in 1962 to over 140 million t in 1966. Without this leap forward the average growth rate of world consumption during these five years would not have been 6.3 per cent, but only 4.4 per cent.

From 1965 to 1970 the increase in consumption occurred mainly in the EEC and Japan, when this region increased its share from 28 per cent to 31 per cent. Consumption in the EEC rose by 25 million t and in Japan by as much as 30 million t. Thus in these five years Japan's steel consumption rose by an annual average of 15 per cent. Without this increase, world consumption would have risen by only 4.1 per cent per year instead of 5.2 per cent.

Since the seventies the growth of world steel consumption has been marked more and more by increases in the developing countries. Their share of world consumption is assumed to increase from 11 per cent in 1970 to 17 per cent in 1980 and 24 per cent in 1990. This is the result of accelerating growth in developing countries and decelerating growth in highly developed countries.

If one attributes the accelerating increases in consumption in developing countries to higher steel intensity resulting from their low level of development and the decelerating increases in highly developed countries to lower steel intensity resulting from their high level of development, the falling rate of increase in world steel consumption can be explained entirely by differences between the levels of economic development in the different regions.

For decades the main European and North American countries developed through partnerships with highly developed industrialised countries. In the industrialisation process some European countries were at first in the lead, but were later overtaken by North America, after which development in Europe speeded up on a broader basis. In addition there was Japan's rapid industrial progress in the sixties.

These countries left a "development gap" behind them. When at the end of the sixties steel intensity had passed its maximum in the main European countries and Japan — as it had already done in North America — steel consumption in the other countries, although their steel intensity was increasing, was still not sufficient to act as a counterweight.

If steel consumption in the various regions is considered in relation to GDP with the aid of a function showing at first a slower growth of steel consumption than GDP, then faster growth and finally slower growth again, it can be shown for the period from 1955 to 1990, assuming constant growth rates for GDP in the different regions, that the annual growth rate for world steel consumption — due entirely to differences in levels of development — was 5 per cent from 1955 to the end of the sixties, has dropped to 3.5 per cent in the seventies and will remain at that level in the eighties [25]. The growth rate will remain steady in this way, because by then the developing countries will have a large enough share of total world steel consumption not to allow the average growth rate to fall further. Indeed this context will provide a renewed increase in the growth rate of world steel consumption in the nineties.

The drop in the growth rate of world steel consumption in the seventies already started by the difference between development levels in different regions was aggravated by the worldwide recession in 1975.

If the trends of consumption and production capacity are compared, it is found that in 1980 there will be excess in world capacity of 103 million t and an average utilisation rate of 89 per cent [26]. As in this forecast production capacity is so defined that economic optimum utilisation is 100 per cent, the steel market will consequently remain burdened with surplus capacity at world level.

A comparison between the lower and upper limits for consumption and production capacity in 1985 and 1990 gives four possible combinations:

| | | Production potential | |
		Lower limit	Upper limit
Consumption	Lower limit	I	II
	Upper limit	III	IV

Combinations I and IV are plausible straight away; if consumption increases slightly or steeply, production capacity should expand slightly or steeply, because demand acts through capacity utilisation on the profitability of steelworks, thereby influencing how much more capacity is constructed. However, this relationship is not always rigid, because increases in capacity are determined not only by present demand, but more by expected future demand. Governments' economic objectives often play an important part, too. This is why in some years low consumption may coincide with high capacity (combination II) or conversely high consumption with low capacity (combination III). However, these incongruous combinations will not last long.

If in all regions
— the lower limit of production capacity and
— the lower limit of consumption

are reached at the same time (combination I), there will be surplus capacity of 63 million t in 1985 and of 48 million t in 1990. The average utilisation rate will then be about 93 per cent in 1985 and about 95 per cent in 1990.

25. Wienert, H., Zum Einfluss der wirtschaftlichen Entwicklung in verschiedenen Regionen der Welt auf das Wachstum des Stahlverbrauchs bis 1990 (the influence of economic development in different regions of the world on the growth of steel consumption up to 1990), in: Rheinisch-Westfälisches Institut für Wirtschaftsforschung, Mitteilungen, 30, year 1975, p. 101 ff.
26. Trend value.

If in all regions
— the upper limit of capacity and
— the lower limit of consumption

are reached at the same time (combination II), there will be a surplus capacity at world level of 135 million t in 1985 and of as much as 165 million t in 1990. The average utilisation rate will then be about 87 per cent in 1985 and about 86 per cent in 1990.

If in all regions
— the lower limit of capacity and
— the upper limit of consumption

are reached at the same time (combination III), demand will exceed production potential by 1 million t in 1985 and by 42 million t in 1990.

If in all regions
— the upper limit of capacity and
— the upper limit of consumption

are reached at the same time (combination IV), there will be a world capacity surplus of 67 million t in 1985 and 75 million t in 1990. The average utilisation rate will then be about 93 per cent in 1985 and about 94 per cent in 1990 (see Tables 2 and 3).

Table 2 FUTURE TRENDS IN WORLD PRODUCTION OF CRUDE STEEL 1985

Million t traditional crude steel equivalents

	EEC and Japan	North America	Eastern Europe	Western Europe, Oceania and South Africa	Developing countries	World
Combination I Capacity lower limit & Consumption lower limit						
Capacity/consumption	+ 105	+ 2	− 9	− 2	− 33	+ 63
Foreign trade balance	+ 59	− 15	− 9	− 2	− 33	0
Consumption	220	163	264	70	172	889
Production	279	148	255	68	139	889
Combination II Capacity upper limit & Consumption lower limit						
Capacity/consumption	+ 135	+ 13	0	+ 5	− 18	+ 135
Foreign trade balance	+ 33	− 15	0	0	− 18	0
Consumption	220	163	264	70	172	889
Production	253	148	264	70	154	889
Combination III Capacity lower limit & Consumption upper limit						
Capacity/consumption	+ 89	− 6	− 17	− 7	− 60	− 1
Foreign trade balance	.	.	.	.	.	.
Consumption	236	171	272	75	199	953
Production	325	165	255	68	139	952
Combination IV Capacity upper limit & Consumption upper limit						
Capacity/consumption	+ 119	+ 5	− 8	0	− 45	+ 67
Foreign trade balance	+ 68	− 15	− 8	0	− 45	0
Consumption	236	171	272	75	199	953
Production	304	156	264	75	154	953

Author's own estimates.

The following will be the picture in the various regions in 1990.

The developing countries will have raised their capacity to 201–231 million t and their consumption to 228–278 million t [27]. Only if their consumption is low and if they greatly increase their capacity (combination II) will they be able to meet their consumption fully from their own production. In the other three cases they will have to continue importing considerable amounts of steel, although imports will decline as a proportion of consumption.

The region comprising Western Europe, Oceania and South Africa will produce slightly more steel than it consumes, if it increases production capacity to the upper limit (combinations II and IV). If capacity is only increased to the lower limit (combinations I and III), this region will import rather more steel than it exports, as it has done hitherto.

Only if Eastern Europe increases production capacity to the upper limit while keeping consumption down to the lower limit (combination II) will it be able to meet its consumption fully from its own production; in all other cases it will have to import.

Table 3   FUTURE TREND OF WORLD PRODUCTION OF CRUDE STEEL 1990
Million t traditional crude steel equivalents

	EEC and Japan	North America	Eastern Europe	Western Europe, Oceania and South Africa	Developing countries	World
Combination I Capacity lower limit & Consumption lower limit						
Production potential/Consumption	+ 94	− 6	− 11	− 2	− 27	+ 48
Foreign trade balance	+ 57	− 17	− 11	− 2	− 27	0
Consumption	231	171	296	85	228	1011
Production	288	154	285	83	201	1011
Combination II Capacity upper limit & Consumption lower limit						
Production potential/Consumption	+139	+ 9	+ 5	+ 9	+ 3	+ 165
Foreign trade balance	+ 21	− 17	+ 2	+ 3	− 9	0
Consumption	231	171	296	85	228	1011
Production	252	154	298	88	219	1011
Combination III Capacity lower limit & Consumption upper limit						
Production potential/Consumption	+ 73	− 12	− 19	− 7	− 77	− 42
Foreign trade balance	.	.	.	.	.	.
Consumption	252	177	304	90	278	1101
Production	325	165	285	83	201	1059
Combination IV Capacity upper limit & Consumption upper limit						
Production potential/Consumption	+118	+ 3	− 3	+ 3	− 47	+ 75
Foreign trade balance	+ 64	− 17	− 3	+ 3	− 47	0
Consumption	252	177	304	90	278	1101
Production	316	160	301	93	231	1101

Author's own estimates.

27.   Figures in traditional crude steel equivalents.

Although North America has sufficient capacity, it will probably meet about 10 per cent of its consumption from net imports as hitherto. Only if there is excess consumption at world level (combination III) will North American capacity be fully utilised [28].

The risks of capacity working at full load are greatest in the exporting region comprising the EEC and Japan. According to the forecast described here, the utilisation rate of its production facilities in 1990 will range from 68 per cent to 100 per cent. Consequently this region will be in the awkward position of reserving considerable capacity [29] to meet fluctuations in demand in the other regions. Its capacity will be all the more stretched, the higher steel consumption rises and the slower production capacity increases in the other regions.

There will be more combinations, if all the regions do not always have the same combinations but sometimes have different ones. In 1990, for example, the combination of capacity and consumption at their lower limits in the EEC, Japan, North America, Eastern Europe and the residual group together with the combination of capacity and consumption at their upper limits in the developing countries would put world consumption at 1,061 million t with a production potential of 1,089 million t. The increased import requirements would then benefit the EEC-Japan region, so that production there would reach 308 million t.

It is hardly possible, however, to attribute different degrees of probability to individual cases, because the influence of subjective expectations is too great. If we were to venture to do so from today's standpoint, we would rate the combination of capacity and consumption at their lower limit in all regions as more probable than the others.

In this case the average utilisation rate of world steel capacity would rise from 89 per cent in 1980 to 93 per cent in 1985 and 95 per cent in 1990 [30].

The existing imbalance between production potential and consumption would then be reduced.

In order to eliminate the influence of the continuous casting process which economises crude steel we have expressed consumption, capacity and production in traditional crude steel equivalents. From 1978 to 1990 continuously cast steel should increase as a proportion of total production, as follows:

in the EEC and Japan	from 36 to 64 per cent
in North America	from 15 to 38 per cent
in Eastern Europe	from 8 to 30 per cent
in Western Europe, Oceania and South Africa	from 32 to 56 per cent
in the developing countries	from 13 to 50 per cent

Owing to advances in continuous casting techniques, world production of crude steel in 1990 will be 6 per cent or about 65 million t lower than with traditional ingot casting. With the combination "capacity lower limit and consumption lower limit" world production in 1990 would then be 949 million t of crude steel instead of 1,011 million t traditional crude steel equivalent, including

---

28. For this reason North America might reduce its capacity in the eighties more than is assumed in this report.

29. The region has already been fulfilling this function, although mainly in order to meet cyclical fluctuations in consumption.

30. The utilisation rate of 89 per cent forecast as a trend value for 1980 corresponds to the actual rate in 1978, and the trend value forecast for 1985 corresponds to the actual rate in 1979.

the EEC and Japan [31]   264 million t
North America   146 million t
Eastern Europe   274 million t
Western Europe, Oceania
and South Africa   77 million t
and the developing countries [32]   188 million t

The shares of the different regions in world production will change considerably, as follows:

	EEC and Japan	North America	Eastern Europe	Western Europe, Oceania and South Africa	Developing countries
	Percentage shares				
	past				
1960	35	28	26	5	6
1965	34	28	26	5	7
1970	38	22	27	6	7
1975	37	19	28	7	9
	future				
1980	32	18	29	8	13
1985	30	17	28	8	17
1990	28	15	29	8	20

Thus the distribution of production will follow that of consumption, but with a time lag. A striking feature is the steep fall in the share of the EEC and Japan in the eighties, the counterpart of the steep rise in the share of developing countries. In 1990 the trend value of production in the EEC and Japan should approximate to the cyclical production peak reached in 1974.

CONCLUSION

The forecast of world demand for and supply of steel up to 1990 pointed to

— slower growth in consumption and production, and
— a considerable shift of emphasis from the highly industrialised to the less developed regions, with production following consumption in both space and time.

This shift will probably occur in a stage marked again by surplus capacity, when the imbalances will decrease only slowly. In the EEC and more so in Japan the main problem will be how to adjust capacity to the unexpectedly slow increase in demand, whereas in the developing countries there will be financing problems and considerable difficulties in starting up new production capacity.

31. Japan 114 million t and the EEC 150 million t.
32. Including Latin America 62 million t, China and North Korea 55 million t and South East Asia 44 million t.

THE GROWTH OF STEEL-MAKING CAPACITY IN THE 1980s

by
Kimiro SUZUKI
and
Tudor MILES
OECD Steel Secretariat

SUMMARY

The paper provides estimates of steel-making capacity which aim to be realistic, i.e. they indicate what steel production is likely to be when demand is strong. These new estimates are known as "effective capacity".

In the OECD area (the main industrialised countries) there was a round of capacity expansion in the 1970s which continued long after demand had turned down and these two factors resulted in under-utilisation and widespread financial losses. The excess capacity is estimated to have been 80 million tonnes, or 16 per cent of effective capacity.

The rate of capital expenditure in the EEC and Japan fell by over 50 per cent in two years after 1976 and investment is now directed mainly to energy saving and other cost reduction, product improvement and environmental protection. At this lower level it is probably insufficient to maintain existing capacity intact in the EEC, Japan and the USA taken together. However, there will be significant additions to capacity in some of the smaller countries. For 1985 capacity in the whole OECD area is forecast to be 515–535 million tonnes compared with 513 million tonnes in 1979.

Capacity in the non-OECD[1] countries expanded consistently fast in the 1970s at an annual average rate of 9 ½ per cent in response to an even faster growth of steel consumption. There is ample evidence that a rapid capacity growth will continue up to 1985. Capacity in 1978 is estimated to have been 56 million tonnes and it should grow to approximately 96 million tonnes by 1985.

However, it is more likely that OECD steel producers' net exports to these countries will expand in the years up to 1985 than that they will diminish. The first call on their new production will be their domestic economies and as a group they will not have surpluses.

Utilisation of capacity in the OECD countries should gradually improve brought about by:

  a) a gradual growth in consumption and net exports to non-OECD areas;

---

1. Throughout this report "non-OECD countries" excludes the Centrally Planned economies.

b) an initial fall in capacity from restructuring schemes and the low investment rates, particularly in the EEC area.

The key factor in determining capacity is how decision makers perceive the trends in the market, namely steel demand and profitability, and how they respond to the perceived situation given the restraints on their managerial freedom. The present prospects in OECD countries seem most unfavourable for capacity expansion.

INTRODUCTION

There have been many estimates of world and individual country's capacities for producing crude steel, but there has been no generally accepted definition of capacity. The estimates of capacity reported to OECD by Member countries are theoretical capacity and actual production has never reached the stated capacity levels when strong market conditions required maximum output.

As a contribution to the examination of the prospects for the steel industry in the 1980s we aim in this paper to apply stricter criteria to the information to give new estimates of capacity which correspond with maximum annual crude steel production actually available, and this definition we call "effective capacity".

The paper provides estimates of current effective capacity for OECD and non-OECD areas. We use known information on investment developments, planned plant closures and some judgement to attempt a tentative estimate of likely capacity levels in 1985. The centrally planned economies are excluded from this survey because of information gaps, but this is not a serious omission as these countries are almost self-sufficient in steel and their net trade with the rest of the world is likely to remain quite small.

PROBLEMS OF ASSESSING CAPACITY

OECD has for many years collected estimates from Member countries of their crude steel making capacity. Definitions of nominal capacity are known not to be strictly comparable from country to country and experience has shown that actual production has never reached the level of nominal capacity. However, study of the relationship between actual production and nominal capacity over a number of business cycles has enabled us to modify the nominal capacity figures to produce fairly satisfactory estimates of obtainable production, i.e. effective capacity.

Projecting forward, the development of capacity follows with a certain time lag the decisions taken on investment, and investment decisions by and large are a function of perceived rather than actual trends in steel demand. Static interpretation of given figures might miss the intrinsically dynamic character of steel capacity relative to steel demand.

Also involved are the difficulties in assessing the degree of ultimate realisation and the precise timing for each future capacity expansion or closure programme.

It is already clear that the following discussion on capacity cannot avoid a significant element of subjective judgement. The judgement element is even larger when the developing countries' capacity has to be estimated for there does not exist in non-OECD areas any regular reporting system for capacity or investment, and much of the information is fragmentary.

RECENT CAPACITY DEVELOPMENTS AND PROBLEMS
IN THE OECD AREA

The nominal crude steel-making capacity in 1978 estimated from figures reported to OECD by Member countries was 567 million tonnes. On this basis, effective capacity, which is empirically measured maximum production possible is assessed at approximately 513 million tonnes (or 9 ½ per cent below the nominal capacity). Crude steel production in 1978 was 412 million tonnes and on the basis of these estimates of nominal and effective capacity the operating rates were 73 and 80 per cent respectively. Although in 1979 output increased to some extent, underutilisation was still substantial, the production level being 80 million tonnes, or 16 per cent, below the effective capacity (see Tables 1 and 2).

During the period 1970–79 capacity development proved to be inflexible and insensitive to the change in trend in steel demand. Nominal capacity increased during 1970–77 from 451 million tonnes in 1970 to 566 in 1977 with a remarkably consistent pace of 3.3 per cent annually (Figure 1). The growth rate even accelerated in 1975, 1976 and 1977 when steel demand had been undergoing a sharp fall and the extent had been widely perceived by the industry. Figure 2 well illustrates the contradictory movements of capacity and demand (production) at the critical turning point in 1975 for the EEC, Japan and the United States. The figure shows that by the time the changed market circumstances had been clearly recognised by the decision makers, many of the investment projects commenced during the years 1970–74 were too deeply committed to be stopped.

This rather expensive lesson tells us that it is much more difficult than we usually imagine to respond effectively to a changed demand-supply situation. Accordingly, the operating rate (to effective capacity) dropped disastrously from 99–97 per cent in 1973 and 1974 to 76 per cent in 1977 (which is equivalent to 70 per cent on nominal capacity). Of the OECD countries, the ones which suffered most were the EEC and Japan.

It was not until 1978 that the growth in capacity levelled off, and it even declined slightly in 1979 reflecting the reduced rate of capital expenditure and the restructuring efforts made by the industry. With some recovery of demand in OECD countries in 1978 and 1979 there was a moderate improvement in operating rates. However, among the OECD countries there was a wide variety of utilisation rates. The EEC and Japan were still operating 20 per cent or more below their effective capacities whereas the USA, Canada and Oceania achieved utilisation rates well over 90 per cent of their effective capacities. The trade regulating measures introduced by some countries simultaneously functioned to some extent to bring about this patchy picture.

CURRENT TRENDS IN INVESTMENT AND IMPLICATIONS
FOR THE 1980s

Capital expenditures in current as well as constant price terms began to fall substantially in 1977, although the downward trend was concealed to some extent by inflation which was particularly severe for the construction cost of a steel plant.

Seen in terms of 1970 constant prices, the extent of the decline during 1977–78 was so severe that the level of investment for the EEC and Japan in 1978 was only 46–47 per cent of the 1976 expenditure (see Figure 2 and Table 4).

The rate of investment since 1977 in the OECD area may well have been insufficient to maintain the industry at its present size.

It has been argued that the rate at which plant could be operated may well be lower now than in the last peak demand period of 1973–74 because of:

a) the current low rate of capital expenditure mentioned above;
b) the accelerated obsolescence of old facilities caused by the widening gap in the variable cost disadvantage between inefficient and average facilities following the large energy and labour costs increases;
c) the seemingly increased proportion of facilities permanently out of commission but still included in nominal capacity;
d) possible non-availability of energy and coke on a large incremental basis.

In the circumstances, the industry's current investment to meet the changed environment in steel demand and supply are being directed to the following areas:

a) variable cost trimming by energy conservation and operation improvements such as:
— increased Continuous Casting
— replacement of Bessemer and Open Hearth processes by Basic Oxygen and Electric Furnaces
— utilisation of wasted heat
— maximum utilisation of scrap;
b) the intensified need to keep capital costs down illustrated by the preference for Electric Furnaces and mini-mills to the Blast Furnace-Basic Oxygen Furnace process;
c) adjustment to changing market requirements such as meeting the relatively high demand for special steels as well as meeting high quality standards for flat products, and
d) non-productive environmental protection investment.

An important factor in determining future capacity levels is also the extent of disinvestment through closures.

The above list indicates that only a minor proportion of total investment is directed to capacity expansion and that the industry's efforts are aimed at slimming down capacity further. Each steel enterprise is trying to create the foundation of cost competitiveness which is essential for renewed growth in the 1980s.

LIKELY CAPACITY DEVELOPMENTS IN THE NON-OECD AREAS

The picture is, however, quite different for the non-OECD countries. Crude steel production, which is proximate to effective capacity in this region, increased between 1970 and 1978 at a remarkably consistent annual rate of 9.5 per cent to reach 56 million tonnes in 1978.

The rate of increase in production and capacity was hardly interrupted in the difficult years since 1975, when apparent steel consumption paused at around 80 million tonnes in 1975 and 1976 after the rapid growth of 1970-1974.

Consumption growth has since resumed, rising at around 9 per cent per annum between 1977 and 1979. This area import approximately 40 per cent of its steel requirements from the OECD countries (Figure 3).

Nominal steel-making capacity of the non-OECD area is estimated to have been around 76 million tonnes in 1978. However, the dispersion among the countries is noteworthy (Table 3).

The top 5 countries (Brazil, India, Mexico, South Africa and Korea) account for 71 per cent of the total and the top 10 countries (additionally Argentina, Taiwan, Venezuela, Iran and Egypt) for 88 per cent.

100

Then to what extent will the non-OECD countries expand their steel supply capabilities in the 1980s? To answer this question we examined the prospects in all developing countries which had an annual apparent steel consumption above 500,000 tonnes in 1978.

From experience it is well known that steel production in developing countries tends to grow steadily once the production reaches the annual level of 300,000–500,000 tonnes. Furthermore, 500,000 tonnes could now be the minimum size of a domestic market to allow a viable steel establishment.

The 32 countries in this category are listed in Table 5 and they account for more than 99 per cent of the area's total production. In addition, Trinidad/Tobago, Qatar and the United Arab Emirates were taken into account because of their fair-sized steel projects.

For the above 35 countries, each steel plant project was individually scrutinised using published, and in addition unpublished, data and was evaluated as to whether the particular project is likely to be in operation by 1985.

For 1985, the criteria of the assessment were:

a) current stage of each project — under planning, feasibility study or construction, and

b) availability of funds for each project — the availability of export credits, international agency loan of the World Bank Group or regional lending agencies, and local sources.

For the period after 1985 steel development projects will not yet have been formulated so it is important to assess a country's potential, particularly in relation to the development of its domestic market, the following factors playing the determining role:

c) domestic steel market — GDP, population and apparent steel consumption in terms of current size and likely growth rate;

d) stage of national economic development — per capita GDP, share of net material product of manufacturing industries in GDP;

e) intention of government to establish and expand the industry;

f) availability of iron ore, metallurgical coal and energy.

In order to take into account the factors (b) to (f) comprehensively, (although normatively), the 32 countries were classified into the following categories, the idea of which was originated by the IISI, which recognises the different economic potential of the non-OECD countries in the new economic environment existing since 1974:

a) industrial countries

b) oil producing countries

c) primary goods producing countries (high income and low income).

(See Table 5 for the full classification and criteria.)

In the assessment, high dynamism was assumed for industrial countries and oil producing countries.

FINDINGS

1. The magnitude of the increase in nominal capacity is estimated to be 40–52 million tonnes during the 1978–85 period, which would raise the total nominal capacity of the region to 122 million tonnes in 1985 from 76 million tonnes in 1978. Accordingly, effective capacity, conceived equivalent to production, will increase to 96 million tonnes in 1985 from 56 million tonnes in 1978, an annual rate of 8.1 per cent (Table 3).

2. Out of the 46 million tonnes increase in nominal capacity, approximately 80 per cent comes from the following 10 countries:

4 industrial countries (Brazil, Korea, Taiwan, Mexico)

5 oil producing countries (Venezuela, Iran, Nigeria, Saudi Arabia, Algeria) and India.

The distribution of capacity among the non-OECD countries is likely to some extent to be further dispersed.

3. A not insignificant change, however, is seen in the process route pattern:

The DR-EF route (Direct Reduction-Electric Furnace) accounts for more than one quarter of the total increase during 1978–85. The trend to the DR-EF is remarkable particularly in the case of the oil-producing countries, which have plentiful natural gas supplies, and they will depend upon the DR-EF process for 70 per cent of their capacities increments. Industrial countries and primary goods-producing countries are mainly keeping to the traditional process.

Million tonnes	Conventional Routes	DR-EF Route	Total Capacity Increase by 1985	Percentage of DR-EF Route
Industrial countries	20.6	1.2	21.8	5.5%
Oil-producing countries	4.2	10.2	14.4	70.8%
Primary goods-producing countries	9.1	0.8	9.9	8.1%
Total	33.9	12.2	46.1	26.5%

4. Is it likely then that the probable supply increments will match the increase in steel demand of the non-OECD countries in the years to 1985?

The probability seems to be that they wont, although the outcome will be mainly determined by the pace of development of their domestic steel demand. During the period 1970–1978 this group of countries' economic development called for an increase in apparent steel consumption of 9.6 per cent annually, compared with a global GDP growth rate of 5.5 per cent: the elasticity of apparent steel consumption (ASC) growth to that of GDP was therefore approximately 1.75.

If it is assumed that the current level of net imports, i.e. 38 million tonnes (ingot equivalent) is maintained until 1985 with the elasticity of ASC to GDP also unchanged, (it is more likely to rise in the course of industrialisation) a simple calculation shows that their economic development would have to be limited to less than 3 per cent annually. It is evident that industrialisation really needs steel, and the more likely event is that their steel imports will have to go on rising.

IMPLICATIONS OF LIKELY SUPPLY-DEMAND DEVELOPMENTS IN THE 1980s

Considering the long gestation period between the planning of a project and the commencement of steel production, 1985 is now within the time range when capacity can be forecast on present knowledge with acceptable accuracy.

In the OECD area, as already noted, capital expenditure has already been substantially reduced and some of it is directed to the "mothballing" of plant. The burden of long-term debt is already heavy in the circumstances of inadequate profitability. Therefore, it is apparent that investment levels are unlikely to be raised significantly unless the current margin of available supply over demand is reduced improving the industry's financial capability.

Collectively, the EEC, Japan and the USA are likely to stay at more or less the current level of capacity in 1985. In the case of the EEC, nominal capacity will probably be reduced under restructuring plans.

However, the Other Western European countries have announced the intention to expand their steel industries and these countries account for around 70 per cent of the total estimated capacity increase in the OECD area. Some of the expansion projects are listed below but all of them cannot be in operation by 1985:

Spain: Ensidesa Avites, Ensidesa Veria and AHV Sagunto
Turkey: Eregli-Stage II and III, Ismedir and Karabuk
Yugoslavia: Sisak, Smederevo and Skopje
Portugal: SNS Seixal

They are all rounding out expansions of existing plants.

The OECD area's effective steel-making capacity is likely to be in the range of 516–536 million tonnes in 1985, the median being 526. It is axiomatic, though, that capacity will be determined to some extent by the development of steel demand around 1983 and this cannot be easily predicted.

To sum up the discussion so far: The world nominal steelmaking capacity is estimated by 1985 to increase by about 10 per cent, or 60 million tonnes to reach around 700 million tonnes. The increments during the period will be largely accounted for by the non-OECD countries and we will see a slightly greater proportion of World capacity located in the developing areas. This gradually rising proportion will continue in the 1980s, when the developing countries push ahead with their industrialisation programmes and their domestic steel markets grow faster than the OECD countries' markets.

Now, the final part of this paper will focus on the almost unanswerable question of what will be the likely supply-demand position during the 1980s. The developments in investments and capacity in the second half of the 1980s must be largely dependent upon the demand-supply situation in the first half of the 1980s. By the demand-supply situation we mean not only the quantitative supply-demand tonnage balance, but also the price/cost relationship and resultant financial capability to respond effectively to the perceived situation. Taking into account the degree of uncertainty surrounding these factors, an attempt at estimating the supply capability in 1990 might be too pre-emptive. The following discussion therefore concentrates on the mid-1980s.

Given the estimation on supply capabilities in 1985 already shown, further assumptions need to be made on the following factors to solve the equation on the likely supply-demand balance in 1985:

a) likely net trade balance with the State trading countries;
b) the extent of the increase in continuous casting ratio;
    and as a decisive factor
c) likely developments in steel demand in the OECD and non-OECD areas.

As for the net trade balance with Eastern Europe and USSR, the area is largely self-sufficient and their net trade with the rest of the world is fairly stable, reflecting the autarchy of the economies. It would be reasonable to assume that their net trade balance would remain more or less at the current level until the mid-1980s. However, the growth rate of crude steel production as well as of steel demand will decelerate to some extent in coming years on the basis of historical trends and recent

103

developments in their industries and national economies. On the other hand, China is estimated to need more steel for sustaining sound economic growth of around 5 per cent annually (even if the elasticity of apparent steel consumption growth to GNP growth, currently 1.3, remains unchanged). Taking into account the needs for steel as well as the limit on supply capability, China's requirement for net imports will grow probably to around 15 million tonnes.

A not unimportant factor is the crude steel saving arising from the trend towards increased continuous casting. The magnitude of this can be judged by the fact that as much as a further 15–20 million tonnes of crude steel could be saved in 1985 if the following assumptions are realised:

a) 75 per cent of the capacity expansion in developing countries employs the continuous casting process.

b) a 10 per cent increase in the continuous casting ratio in OECD countries, arising mainly from replacement of existing primary mills.

The development of steel demand in OECD and non-OECD countries in the 1980s is very uncertain and the papers on demand prospects prepared for the Symposium show a number of alternative projections. The view of the authors of this paper is that for the OECD area there will be very limited trend growth in steel consumption in this period and there is a high risk of severe fluctuations in demand while the world economic situation remains unsettled. The expansion of consumption in the non-OECD area however seems more steady and assured, and we expect that the growth rate will continue to be quite fast.

If, as seems likely to us, there is a moderate overall growth of demand up to 1985 (say, 2½ per cent p.a. for OECD areas and 8 per cent p.a. for the non-OECD Countries) the mean capacity case in Table 3 would give a 1985 utilisation rate of 90–95 per cent for the OECD countries (of effective capacity, 80–85 per cent for nominal capacity). This level is similar to the operating rates attained in the first years of the 1970s, but rather below the rates of the 1973–74 period. This scenario implies that capacity utilisation in the OECD countries will be improving gradually in the years up to 1985, but as already noted there will be year to year fluctuations.

To conclude we return to the starting point of this paper. The key factor in determining capacity is how decision makers perceive the trends in the market namely steel demand and profitability; and how they respond to this perceived situation given the financial and other restraints which limit their managerial freedom.

Table 1 A   NOMINAL CRUDE STEEL MAKING CAPACITIES IN THE OECD AREA 1970-79

Million tonnes

	1970	1971	1972	1973	1974	1975	1976	1977	1978	1979
EEC (9)	154.5	168.2	173.8	176.6	176.9	189.9	197.7	200.7	202.1	202.9
U.S.A.	138.7	134.8	142.1	142.5	140.6	138.9	143.6	145.2	143.2	140.0
Japan	110.0^E	117.0^E	121.0^E	132.0^E	140.0^E	141.0^E	142.0^E	151.8	151.3	153.0
Austria	4.1	4.0	4.0	4.7	4.9	5.0	5.5	5.5	4.9	5.0
Spain	9.5	11.7	11.1	10.9	13.5	14.0	14.5	15.1	15.6	15.7
Finland	1.1	1.3	1.5	1.6	1.8	1.9	1.8	2.6	2.5	2.5
Norway	0.9	0.9	0.9	0.9	1.0	1.0	1.0	1.0	1.0	1.0
Portugal	0.4	0.4	0.5	0.5	0.6	0.8	0.5	0.6^E	0.7	0.7^E
Sweden	6.0	6.1	6.1	6.4	6.9	7.3	7.3	7.2	7.1	7.0
Switzerland	0.5	0.5	0.5	0.5	0.9	0.9	1.0	1.0	1.0	1.0
Turkey	1.8	1.8	1.5	2.3	2.4	2.4	2.9	3.2^E	3.6^E	3.6^E
Yugoslavia	3.4^E	3.8^E	4.0^E	4.1^E	4.2^E	4.2^E	4.2^E	4.9^E	5.4^E	5.4^E
Greece	0.8	1.0	1.1	1.3	1.3	1.3	1.5^E	1.5^E	2.5^E	2.5^E
Other W. Europe	28.5	31.5	31.2	33.2	37.5	38.8	40.2	42.1	44.3	44.4
Canada	12.0	12.8	13.6	14.0	16.3	16.7	16.9	17.0	17.5	17.6
Australia	7.5^E	8.0^E	8.0^E	8.8	9.0	9.0	9.0	8.4	8.4	8.4
New Zealand	0.2^E	0.2^E	0.2^E	0.2^E	0.2^E	0.2^E	0.3^E	0.3^E	0.3^E	0.3^E
Total OECD	451.4	472.5	489.9	507.3	520.5	534.5	549.7	565.5	567.1	566.7

Source: (a) OECD Annual Surveys and Steel Committee Continuous Information System.
(b) Figures marked «E» — the Secretariat's estimation.
(c) U.S.A. 1970-74: from a Report by the Council on Wage and Price Stability, published April 1976.
(d) EEC (9) 1978 and 1979: Investment Report by ECSC, published in January 1979.
(e) Japan 1970-76: because of the definitional changes, published figures prior to 1977 are incomparable with those since 1977. The figures listed here are strictly the Secretariat's own estimates.

Table 1 B   EFFECTIVE CAPACITIES IN THE OECD AREA 1970-79

Million tonnes

	1970	1971	1972	1973	1974	1975	1976	1977	1978	1979
North America	143.2	140.3	147.9	148.9	149.1	149.9	152.5	154.1	152.6	149.7
(U.S.A.)	(131.8)	(128.1)	(135.0)	(135.4)	(133.6)	(132.0)	(136.4)	(137.9)	(136.0)	(133.0)
Japan	99.0	105.3	108.9	118.8	126.0	126.9	127.8	136.6	136.2	137.7
EEC (9)	136.0	148.0	152.9	155.4	155.7	167.1	174.0	176.6	177.8	178.6
Other W. Europe	25.1	27.7	27.5	29.2	33.0	34.1	35.4	37.0	39.0	39.1
Oceania	6.9	7.4	7.4	8.1	8.3	8.3	8.4	7.8	7.8	7.8
Total OECD	410.1	428.6	444.6	460.2	472.0	484.3	498.0	512.2	513.5	512.9

Note: 1. Effective capacity: Nominal Capacity x Maximum Operating Rate.
2. Assumptions on maximum operating rate as follows:

U.S.A., Canada	95%
Japan	90%
EEC, Other Western Europe	88%
Oceania	90%

Table 2   DEVELOPMENTS IN OPERATING RATES
Production/effective capacity

Percentage

	1970	1971	1972	1973	1974	1975	1976	1977	1978	1979
North America	91.1	85.7	89.7	101.0	97.8	80.3	84.9	82.6	91.0	92.9
(U.S.A.)	(90.5)	(85.4)	(89.6)	(101.1)	(98.9)	(80.2)	(85.1)	(82.4)	(91.2)	(92.4)
Japan	94.2	84.1	89.0	100.3	92.9	80.7	84.0	75.0	75.0	81.2
EEC	101.3	86.6	91.0	96.6	100.0	74.9	77.2	71.4	74.6	78.4
Other W. Europe	95.3	87.6	98.3	101.0	94.2	86.9	84.8	79.3	79.8	86.0
Oceania	—	—	—	97.6	96.7	97.8	95.6	97.1	99.7	106.0
Total OECD	95.4	85.7	90.4	99.3	97.0	79.3	82.1	76.1	80.4	84.4

Note: Figures of 1979 are partly estimated.

Table 3   ESTIMATED CAPACITIES IN 1978 AND 1985
Million tonnes

	1978 Nominal capacity A	1985 Nominal capacity	1985 Nominal capacity (Mean) B	Nominal capacity increase 1978-85 B-A	B/A	Assumed maximum operating rate %	1978 effective capacity	1985 effective capacity (Mean)
North America	160.7	162.0-168.0	165.0	+ 4.3	1.03	95.0	152.6	156.8
(U.S.A.)	(143.2)	(144-148)	(146.0)	(+ 2.8)	(1.02)	(95.0)	(136.0)	(138.7)
EEC	202.1	196.0-200.0	198.0	− 4.1	0.98	88.0	177.9	174.2
Other W. Europe	44.3	49.2-58.6	53.9	+ 9.6	1.22	88.0	39.0	47.4
Japan	151.3	152.0-155.0	153.5	+ 2.2	1.01	90.0	136.2	138.2
Oceania	8.7	10.6	10.6	+ 1.9	1.22	90.0	7.8	9.5
Total OECD	567.1	569.8-592.2	581.0	+ 13.9	1.03		513.5	526.1
Latin America	31.7	51.0-53.4	52.2	+ 20.5	1.64	90.0 Brazil 70.0 Others	24.3	41.5
South Africa	9.9	10.0	10.0	+ 0.1	1.01	90.0	7.9	9.0
Other Africa	2.0	5.8-7.4	6.6	+ 4.6	3.30	75.0	1.5	4.9
Middle East	3.6	7.9-9.9	8.9	+ 5.3	2.47	50.0	1.5	4.5
India	13.0	18.3	18.3	+ 4.7	1.41	75.0	10.1	13.7
Other Asia	15.5	22.8-28.9	25.9	+ 10.4	1.67	90.0 Rep. Korea, Taiwan 75.0 Others	10.5	22.4
Total non-OECD	75.7	115.8-127.9	121.9	+ 46.2	1.61		55.8	96.0
Total	642.8	685.6-720.1	702.9	+ 60.1	1.10		569.3	622.1

Classification of Regions as in the OECD short-term Outlook Report.

Table 4   CAPITAL EXPENDITURES AND CAPITAL GOODS PRICES INDICES SINCE 1973
EEC, U.S.A. and Japan

	1973	1974	1975	1976	1977	1978	1979	1980
EEC								
Actual expenditures millions of EUA	3028.4	2989.5	3331.5	3293.3	2359.5	2055.3	2017.6	1354.7
Capital goods prices indices (1970 = 100)	121.0	140.6	162.2	182.9	200.9	215.0		
Capital expenditures at 1970 constant prices	2502.8	2126.2	2053.9	1800.6	1174.5	956.0	938.4	630.1
Index (Average 1973-74 = 100)	108.1	91.9	88.7	77.8	50.8	41.3	40.5	27.2
U.S.A.								
Actual expenditures millions of US$	1399.9	2114.7	3179.4	3252.9	2850.3	2538.3		
Capital goods prices indices (1970 = 100)	109.4	124.9	143.9	153.4	163.3	176.5		
Capital expenditures at 1970 constant prices	1279.6	1693.1	2209.5	2120.5	1745.4	1438.1		
Index (Average 1973-74 = 100)	86.1	113.9	148.6	142.7	117.4	96.7		
Japan								
Actual expenditures (billions of yen)	592.8	892.2	1147.4	1264.6	684.1	593.9	E670.0	
Capital goods prices indices (1970 = 100)	110.1	138.6	142.9	144.2	147.2	149.1		
Capital expenditures at 1970 constant prices	538.4	643.7	802.9	864.5	464.7	398.3	449.4	
Index (Average 1973-74 = 100)	91.1	108.9	135.8	146.3	78.6	67.4	76.0	

Note on sources: (a) for Actual Expenditures. EEC: Investment Report by the ECSC, Jan. 1979. U.S.A.: AISI. Japan: Investment Finance Committee, Industrial Structure Council, MITI.
(b) for Capital Goods Prices Indices. EEC: Investment Report by the ECSC, Jan. 1979 pp 54-55. U.S.A.: Survey of Current Business. Japan: Monthly Bulletin of Statistics, U.N.

Table 5.   CLASSIFICATION OF THE NON-OECD COUNTRIES
WITH ANNUAL APPARENT STEEL CONSUMPTION OF MORE
THAN 500 THOUSAND TONNES

Industrial countries:

Argentina, Brazil, Mexico, Chile, South Africa, Israel, Korea, Singapore, Hong Kong and Taiwan	— 10 countries

Oil producing countries:

Venezuela, Algeria, Nigeria, Libya, Iran, Iraq, Kuwait, Saudi Arabia and Indonesia	— 9 countries

Primary goods producing countries:

High income: Colombia, Tunisia, Morocco, Rhodesia, Syria, Malaysia and Peru	— 7 countries
Low income: Cuba, Egypt, India, Pakistan, Philippines and Vietnam	— 6 countries
Total	— 32 countries

Note: a) Criteria of industrial countries:
  i) per capita GDP more than 500 US$ in 1975 constant prices (equivalent to 300 US$ in 1963 constant prices).
  ii) share of net material product by manufacturing industries to GDP larger than that by agriculture plus mining industry.
  Oil producing countries according to the World Bank.
  b) IISI classification different from the above as follows:
  Algeria, Colombia, Tunisia, Rhodesia, Egypt: quasi-industrial countries;
  Peru: industrial country.
  c) "Industrial country" classified here is different from industrialised countries in the OECD area or "newly industrialised countries" analysed in the OECD's report.

Figure 1
OPERATING RATE, PRODUCTION, CAPACITY, CONSUMPTION : OECD AREA

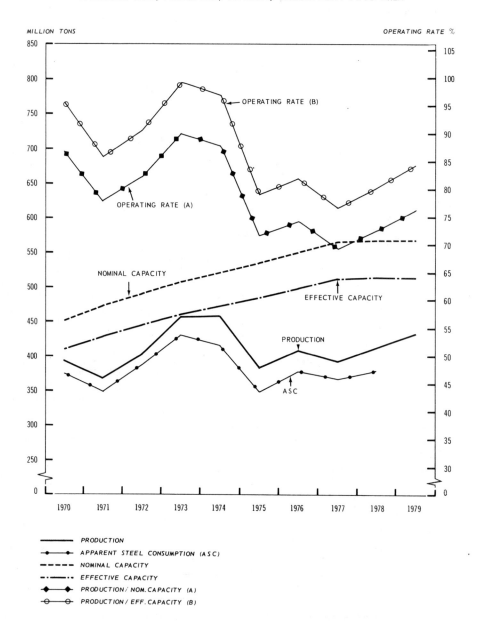

MILLION TONS

OPERATING RATE %

OPERATING RATE (B)

OPERATING RATE (A)

NOMINAL CAPACITY

EFFECTIVE CAPACITY

PRODUCTION

ASC

- ───── PRODUCTION
- ─●─●─ APPARENT STEEL CONSUMPTION (ASC)
- ───── NOMINAL CAPACITY
- ─·──·· EFFECTIVE CAPACITY
- ─◆─◆─ PRODUCTION/NOM.CAPACITY (A)
- ─⊖─⊖─ PRODUCTION/EFF.CAPACITY (B)

Figure 2(a)

PRODUCTION, CAPACITY, CAPITAL EXPENDITURE : EEC (9)

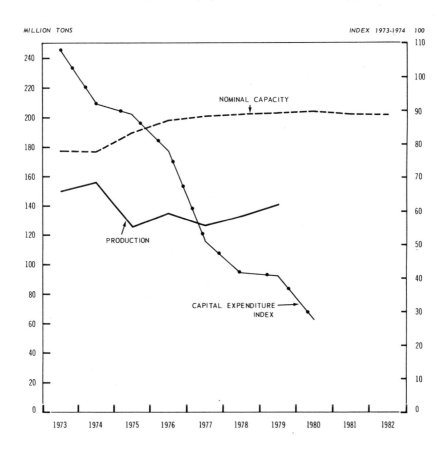

Figure 2(b)
PRODUCTION, CAPACITY, CAPITAL EXPENDITURE : JAPAN

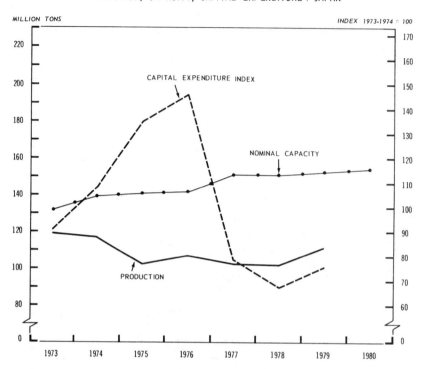

Figure 2(c)
PRODUCTION, CAPACITY, CAPITAL EXPENDITURE : USA

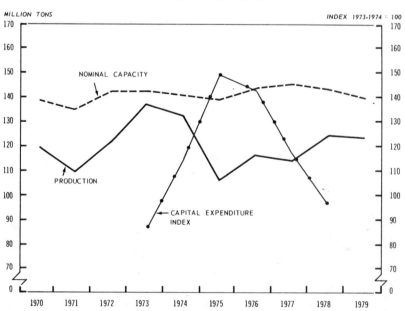

Figure 3

PRODUCTION, CONSUMPTION, IMPORTS, SELF-SUFFICIENCY RATE
IN THE NON-OECD AREA

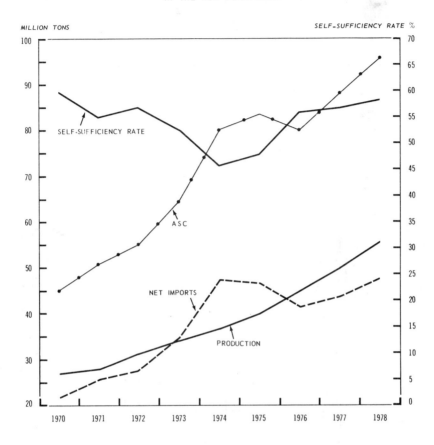

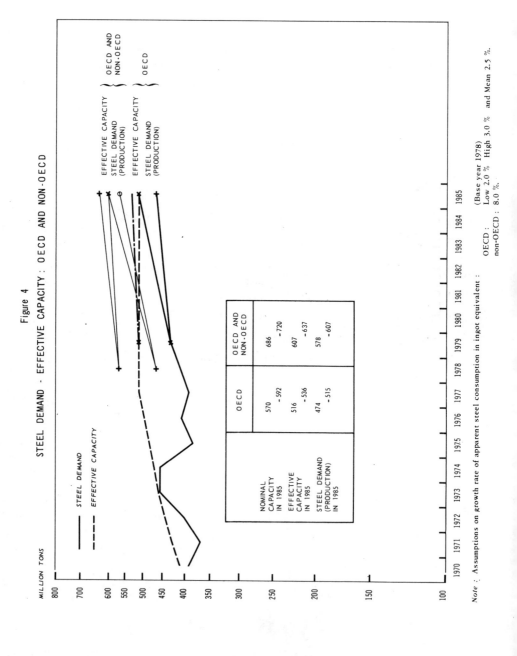

Figure 4

STEEL DEMAND - EFFECTIVE CAPACITY : OECD AND NON-OECD

	OECD	OECD AND NON-OECD
NOMINAL CAPACITY IN 1985	570 - 592	686 - 720
EFFECTIVE CAPACITY IN 1985	516 - 536	607 - 637
STEEL DEMAND (PRODUCTION) IN 1985	474 - 515	578 - 607

Note : Assumptions on growth rate of apparent steel consumption in ingot equivalent :

(Base year 1978)

OECD : Low 2.0 % High 3.0 % and Mean 2.5 %.

non-OECD : 8.0 %.

112

GLOBAL SCENARIO OF WORLD STEEL INDUSTRY GROWTH PARTICULARLY UP TO 1985

by
Dr. B.R. NIJHAWAN
Senior Adviser (Inter-Regional)
United Nations Industrial
Development Organisation

EXPLANATORY NOTES

For the purpose of this document the following definitions have been used:

Developed countries (Dd.C.):

>Eastern developed countries:
>CMEA countries in the eastern Europe
>Western developed countries:
>Australia, Canada, Japan, New Zealand, South Africa, USA and Western European countries (excluding Cyprus, Malta, Turkey and Yugoslavia

Developed countries (Dg.C.): Countries other than developed countries as defined above. They were divided into the following groups for the sake of convenience:

Africa: African countries
Arab: Arabic countries
America: Central and South American countries
ME Asia: Market economy countries in Asia and Pacific other than CPE Asia defined below
CPE Asia: Centrally planned economy countries of Asia; namely China, Korea DPR, Mongolia and Vietnam
Europe: Cyprus, Israel, Malta, Turkey, Yugoslavia
Reference to "tons" (t) are to metric tons.
Reference to "Mt" are to million metric tons.
Reference to "Dollars" ($) are to United States dollars.

Unless otherwise specified, all figures of steelmaking capacity, steel production and consumption refer to crude steel (ingots and continuously cast semis) equivalent.

Process routes are defined as follows:

BF-BOF: Integrated works which produce pig iron in blast furnaces (BF) or electric smelting furnaces (ESF) and process it into steel in basic oxygen furnace (BOF or LD) Bessemer convertors (BC) or open hearth furnace (OHF). Majority of the works are based on BF-BOF or OHF routes. BF include charcoal blast furnaces (CBF).
DR-EF: Integrated works based on direct reduction processes (DR) and electric arc furnaces (EF)
EF: Semi integrated works based on scrap melting in EF or OHF. Majority of the works are based on EF process.

STEELMAKING CAPACITY: PAST, PRESENT AND FUTURE

1.1 *The past evolution of the steel production and consumption in the developing countries*

The past evolution of the steel production in the developing countries is shown in Table 1 by regions, along with that of the developed countries. It is seen from Table 1 that:-

i)   Developing countries as a whole have increased the steel production rapidly during the last ten years and the production reached the level of 76 Mt in 1977 which was 40 Mt higher than that in 1967.

ii) The annual growth rates of the production in the period of 1967–1977 were 7.6 per cent and 2.7 per cent in developing and developed countries respectively, when calculated from the production ratios 1977/1967.

iii) The recent somewhat severe setback in steel production in the Western developed countries has not been observed in the developing countries.

iv) The share of the developing countries in the world steel production has steadily improved from 7.4 per cent in 1967 to 11.3 per cent in 1977.

v)   All the regions have shown the development. Although the steel production in Africa and Arab regions is still low, it is increasing at a high growth rate.

Table 2 shows the increase of the steel production of the groups of countries as classified by the 1977 production level.

The estimated capacity in 1985 and 1990 shown in Table 1a has been mainly calculated on the basis of the announced projects as mentioned above. Announced projects were classified by probable commissioning data (before and after 1985), without critical techno-economic evaluation of each project. Therefore, it is necessary to take into account the project realisation ratio for the estimated capacity increase for 1977–1985 and 1977–1990. For the year 2000 capacity, this is assessed mainly on the potential of country but not on announced project capacity. Therefore it is not necessary to base it on the project realisation ratio. As a rough guess based on past experiences, the project realisation ratio of 75 per cent would be appropriate. Applying this ratio to 1977–1985 and 1977–1990 capacity increase in

Table 1   THE PAST EVOLUTION OF THE STEEL PRODUCTION IN THE DEVELOPING COUNTRIES (MT)

Regions	Years				Growth rate (%/y)	
	1967	1974	1977	1978	1979	77/67
Developing countries	36.60	63.8	75.9	91	101.3	7.57
Africa	0.23	0.62	0.65			10.95
Arab	0.48	0.86	1.20			9.60
America	9.79	17.74	21.82			8.34
ME Asia	7.33	11.05	16.53			8.47
CPE Asia	15.89	29.10	30.95			6.89
Europe	2.91	4.42	4.76			5.04
Developed countries	450.8	644.1	598.4	622	644	2.65
World total (rounded)	497	709	674	713	745.3	3.09
Share of developing countries (%)	7.4	9.0	11.3	12.8	13.5	—

Table 1 a   ESTIMATED STEELMAKING
CAPACITY IN THE DEVELOPING COUNTRIES

Not adjusted by the project realisation ratio

Estimated steelmaking capacity (in Mt)

Region	1977	1985	1990	2000
Africa	0.88	4.5	7	15
Arab	2.37	10.4	20	44
America	31.03	59.2	87	154
ME Asia	21.50	45.3	66	128
CPE Asia	37.0	65.5	96	176
Europe	6.90	17.7	24	43
Total	100	203(a)	300	560

a)   Because of rounding of figures for estimated capacity at country levels, the figure does not exactly coincide with that of Table 1.

Table 1 b   PROBABLE STEELMAKING CAPACITY AND STEEL PRODUCTION
IN THE DEVELOPING COUNTRIES
ADJUSTED BY THE PROJECT REALISATION RATIO, INCLUDING CPE ASIA

	1977	1985	1990	2000	Growth rate % 1977/2000
Capacity (Mt)	100	177	250	560	7.8
Capacity utilisation ratio (%)	76	77	79	83	—
Steel Production (Mt) in rounded figure	76	140	200	470	8.0

Table 2   THE INCREASE OF THE STEEL PRODUCTION OF THE GROUPS OF COUNTRIES
AS CLASSIFIED BY THE 1977 PRODUCTION LEVELS

Group of countries		Steel production (share in total Dg.C.)			Ratio	Ratio
		1967	1977	1978	1977/1967	1978/1967
Group A (10-28 Mt in 1977) China, Brazil, India	Mt	24.7	48.9	55	1.98	2.2
	Share	(67.5%)	(64%)	(58%)		
Group B (1-6 Mt in 1977) Mexico, Korea R Yugoslavia, Korea DPR Turkey, Argentina	Mt	8.9	20.2	23.3	2.7	2.6
	Share	(24.5%)	(27%)	(24.5%)		
Group C (less 1 Mt in 1977) Other developing countries	Mt	3.0	5.8	9	2.27	3.0
	Share	(8%)	(9%)	(9.4%)		

Estimated capacity figures for the years 1985, 1990 and 2000 were obtained as follows:

For 1985: Existing capacity plus capacity increase envisaged from the announced project capacity analysis as outlined above.

For 1990: Rough estimates based on 1985 capacity, announced project capacity, and potential of countries for the steel industry development.

For 2000: Very rough guess mainly based on potential of countries for the steel industry development, partly based on announced macro steel development programmes.

Table 1a shows the summary of estimated capacity by region in 1985, 1990 and 2000 which does not take into account the project realisation ratio.

Table 1a and leaving 2000 capacity as it is, one can obtain a probable steelmaking capacity of developing countries in future years as shown in Table 1b. Assuming the capacity utilisation ratio of 77, 79 and 83 per cent for the years 1985, 1990 and 2000, the steel production was calculated and shown in the Table 1b.

Table 2 also indicates that the steel production in the developing countries has expanded substantially. The top three developing countries have increased the production by 24 Mt. Next six countries have expanded their production at a higher rate as a whole than the top three countries. These nine countries produced 91 per cent of steel of developing countries (it was 92 per cent in 1967). However, it is noteworthy that the rest of the countries have developed their steel industry considerably in the last decade, although the production is still less than 7 Mt as a whole.

This fact that the rest of countries are expanding their steel industry is well supported by the figures shown in Table 3. Total number of countries which have steelmaking capacities has increased by twenty in ten years and reached the level of fifty. Particularly, the number of countries producing between 0.1 to 1.0 Mt increased very much. Although data for small steelmaking countries are not certain, it is clear that those countries are rapidly increasing their steel capacity.

This is a quite encouraging trend for the future development of the steel industry in the developing countries, since the development of steel industry is a rather slow process, the existence of even a small plant plays an important role for future development in acquiring knowledge of steelmaking, and these countries will emerge as significant steel producing countries in the long run.

However, it should be noted that the total number of steelmaking countries is still fifty to seventy whilst other developing countries are still remaining as "zero steel producing countries".

Table 3   NUMBER OF COUNTRIES AS CLASSIFIED BY STEELMAKING CAPACITY
1967 AND 1977

Capacity	Number of countries	
	1967	1977
Over 5	2	4
1 to 5	5	8
0.1 to 0.99	8	19
less 0.1	ca. 15	ca. 19
Total	30	50

One of the main driving forces for the development of the steel industry in the developing countries in the recent past has been the increasing domestic consumption in many of these countries. In countries whose economy has expanded dynamically, "steel calls for steel" phenomenon has been observed, namely the more they produced steel, the faster the consumption grew. A few selected examples are shown in Figure 1.

The steel consumption — production balance in developing countries excluding CPE Asia is shown in Figure 2. It is clearly shown that consumption has grown at the rate which surpass the production growth. The net deficit of those developing countries has widened to 35 Mt in 1977 from 15 Mt in 1967. This indicates that steel consuming industries have developed at the same or higher rate as the steel industry. This high rate of growth in steel consumption is understood as an indication of the economic "take-off" of the many developing nations. As well known, the nations at the economic take-off stage and right after take-off require high rates of steel consumption growth.

Consumption of steel shown in Figure 2 has the following features.

   i)   steady growth of consumption from 1967–1972
  ii)   sharp increase in consumption during 1973–1974
 iii)   completely flattened consumption during 1974–1977.

Figure 1

A FEW SELECTED EXAMPLES OF STEEL CONSUMPTION PRODUCTION RELATIONSHIP

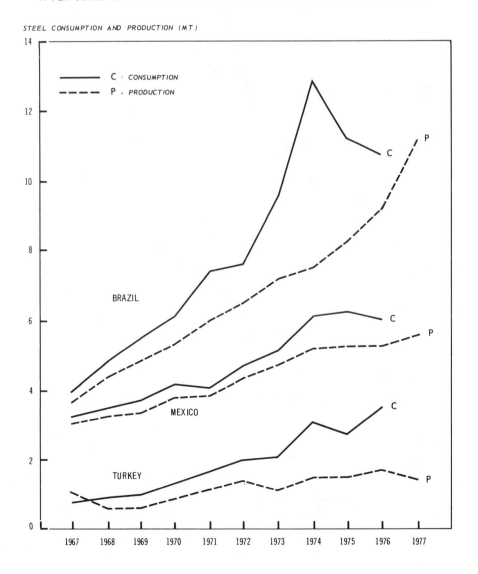

117

Figure 2

STEEL PRODUCTION-CONSUMPTION BALANCE IN DEVELOPING
COUNTRIES EXCLUDING CHINA AND N. KOREA
(Centrally planned economy countries in Asia)

STEEL PRODUCTION, CONSUMPTION
AND NET IMPORT (MILLION TONS)

SELF-SUFFICIENCY (%)

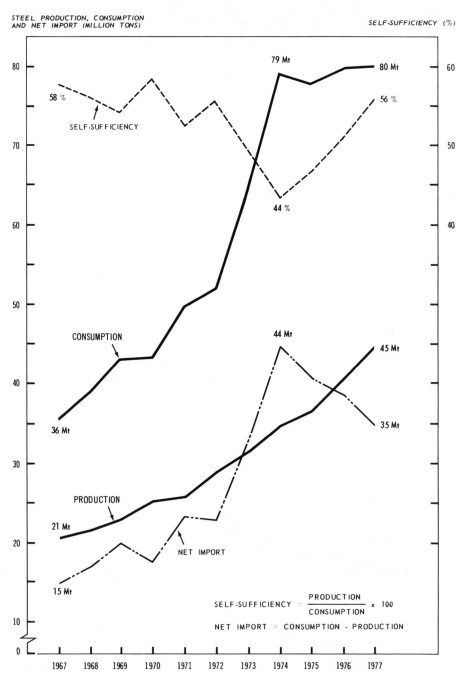

$$\text{SELF-SUFFICIENCY} = \frac{\text{PRODUCTION}}{\text{CONSUMPTION}} \times 100$$

$$\text{NET IMPORT} = \text{CONSUMPTION} - \text{PRODUCTION}$$

Data source : Various issues of steel market, ECE (1977 : partly data, partly estimation).

While there may be many ways of interpreting these features, one way would be as follows:

i) flattening of consumption after 1974, is a combined effect of (*a*) worsened economic situation of most of developing countries other than OPEC countries after sudden jump of oil prices at the end of 1973, (*b*) too much of import of steel during 1973–1974 which exceeded the "steel absorption capacity" of developing countries, and (*c*) inflated steel prices after 1974.

ii) 1973–1974 sharp increase was the result of (*a*) dynamic economic activities and resultant "steel boom" world wide in 1972–1973, (*b*) rushed orders for steel in 1973 in the fear of lack in steel supply of which a larger part was delivered to developing countries in 1974, and (*c*) sharp increase in steel consumption of OPEC countries during 1974.

iii) 1967–1972 growth was a normal pattern of growth in steel consumption which is in line with the steel consumption capacity in the developing countries as a whole.

Since the consumption of steel requires many downstream manufacturing building and construction industries as well as infrastructure such as steel handling and transporting facilities, the concept of the "steel consumption capacity" can be well understood. The normal growth rate of this capacity in developing countries as a whole seems to be of the order of 8 per cent (calculated from 1967–1972 consumption growth).

If it is so, the consumption of steel might grow again from 1979 onwards when the extrapolated consumption from 1967–1972 curve reaches the current consumption growth, provided that the general slump of economy and inflated steel prices do not hinder too much the need for steel of developing countries.

Table 4   EXISTING STEELMAKING CAPACITIES
IN DEVELOPING COUNTRIES (1977)

| Region | Number of countries | | Total steelmaking capacity |
	with steelmaking capacity	with integrated works[1]	
Africa	9	(1)	0.88
Arab	8	(3)	2.37
America	15	(7)	31.03
ME Asia	11	(4)	21.50
CPE Asia	3	(2)	37
Europe	4	(2)	6.90
Total	50	(19)	100[1]

[1] Inclusive figures.

Of 100 Mt capacity, 90 Mt exist in America and Asia. Although the number of countries which have steelmaking capacity in Africa and Arab amounts to 15, capacities of most countries (except Egypt, Algeria and Rhodesia) are very small.

When CPE Asia countries are excluded, steelmaking capacity is distributed by process route as follows:

Total capacity	BF-BOF	DR-EF	Scrap based EF
62.68 Mt (100%)	45.42 Mt (72.5%)	2.99 Mt (4.8%)	14.27 Mt (22.8%)

It is seen from here that BF-BOF (including OH, TC, etc.) integrated steel works count for almost three quarters of the total capacity. Although DR-EF integrated steel works contributes only 5 per cent in 1977, its contribution will grow rapidly in future as discussed later. Scrap based EF route (including OH) occupies significant portion in Africa and Arab where large scale integrated works have not been fully developed. In other regions, this route is less significant but contributes substantially to raise the steelmaking capacity. Particularly it should be noted that thirty one countries are wholly dependent on this route for their steelmaking.

The numbers of integrated works based on BF-BOF (OH,TC) and DR-EF route are 44 and 6 respectively. Of 44 BF-BOF integrated works, only 8 have a capacity of more than 2 Mt and the average capacity is only 1 Mt. This characterises prevailing small scale works in developed countries (for detailed size distribution refer to Table 9). Furthermore it is noted that a few of BF-BOF integrated works are modern type and many OH and TC are still playing an important role.

Comparison of the steelmaking capacity discussed in this section and the steel production in previous sections has been made in Table 5, which shows capacity utilisation ratios. Even if we consider the fact that much new capacity is coming into operation each year and further that it takes a long time to operate the steelworks at the rated capacity, it is stressed that the capacity utilisation ratio shown in Table 6 is low and needs further improvement. At least 85 per cent utilisation ratio should be aimed for, when we consider the various implication of the low utilisation ratio and the tremendous amount of import of steel from developed countries.

1.2 *Future steelmaking capacity*

Referring to announced steel projects in developing countries the total capacity amounts to 203 Mt which is more than twice the existing steel capacity (100 Mt). This figure includes both expansion and greenfield site projects, also conceptual to under construction projects. All the projects were examined in the light of the current status of projects and classified into two categories, namely (*a*) projects probably completed by 1985 and (*b*) projects which might be realised after 1985. Since it is not possible in the classification to make judicious techno-economic assessment of each project, rather generous assessment was made for the projects to be completed by 1985. In reality, it is highly probable that many projects

Table 5   CAPACITY
UTILISATION RATIO
OF EXISTING
STEELMAKING FACILITIES IN
DEVELOPING COUNTRIES
(1977)

Region	Utilisation ratio
Africa[1]	74%
Arab[1]	51%
America	70%
ME Asia	77%
CPE Asia[1]	84%
Europe	69%
Average	76%

[1] Because of uncertainty in exact capacity data, these figures remain as indicative ones.

will be delayed by a few years. It is believed that the project realisation ratio of those classified as probably completed by 1985 might be of the order of 70 per cent as discussed later.

Regional distribution of the capacity of projects is shown in Table 6. The project capacity probably completed by 1985 is 106 Mt out of a total of 203 Mt. Of 106 Mt, America and Asia (both ME Asia and CPE Asia) occupy 84 Mt (79 per cent of total). Also substantive increase in capacity is envisaged for Europe and Arab.

Process routes of the projects probably completed by 1985 are shown in Table 7 which exclude CPE Asia where the data are uncertain. Compared to the existing capacity of DR-EF route (3 Mt), quite dynamic expansion of this process route is envisaged during 1988–1995, although BF-BOF route will still occupy a major part of the capacity increase (62 per cent). It is noteworthy that DR-EF route may play a more important role than BF-BOF route in Africa and Arab where the markets are generally small and natural gas is abundant.

Table 6 PROJECT CAPACITY IDENTIFIED BY REGION AND CLASSIFIED BY TIMING OF PROJECT REALISATION (MT)

Region	Project capacity probably completed by 1985	Project capacity might be realised after 1985	Total project capacity
Africa	2.92	4.55	4.47
Arab	7.94	22.65	30.59
America	28.52	29.18	57.70
ME Asia	25.53	23.50	49.03
CPE Asia	30	7	37
Europe	11.35	29.50	40.85
Total	106	116	223

Table 7 PROJECT CAPACITY BY PROCESS ROUTES FOR THE PROJECT PROBABLY COMPLETED BY 1985 (EXCLUDING CPE ASIA)

Region	Project capacities (Mt)		
	BF-BOF	DR-EF	Scrap based EF
Africa	1.00	1.86	0.06
Arab	3.25	3.59	1.12
America	18.67	7.97	1.88
ME Asia	14.55	6.50	4.48
Europe	9.70	0.14	1.51
Total	47.17 (62%)	20.04 (26%)	9.05 (12%)

When these capacity increases are added to the existing capacities, the shares of each process route will be as follows in 1985:

BF-BOF	DR-EF	Scrap based EF
66.6 per cent	16.6 per cent	16.7 per cent

As for the classification of projects by expansion of existing works or greenfield site, Table 8 shows that 63 per cent of capacity increase envisaged by 1985 will be based on the expansion of existing works. This is very significant and encouraging because expansion projects are generally much easier to carry out and much less in investment costs than greenfield site projects. In this regard, Africa and Arab where new site projects are significant have to overcome many difficulties in order to realise the projects.

Table 8   PROJECT CAPACITY AS
CLASSIFIED BY EXPANSION AND
GREENFIELD PROJECT
(EXCLUDING CPE-ASIA)

Region	Project capacity probably completed by 1985	
	by expansion project	by greenfield project
Africa	0.03	2.89
Arab	2.42	5.52
America	17.70	10.82
ME Asia	16.49	9.04
Europe	11.34	0.01
Total	47.98 (63%)	28.28 (37%)

The size distributions of existing (1977) and future (1985) integrated steelworks were examined and presented in Table 9. The table is self-explanatory; but the following points are worth mentioning:

i)   A distinct trend for larger plants is seen both for BF-LD and DR-EF routes; the number of BF-BOF integrated works of more than 2 Mt capacity will increase to 17 from 8 existing; a number of over 1 Mt capacity DR-EF plants will emerge.

ii) 10 new BF-BOF works and 23 DR-EF works will be constructed by 1985, and the total number of integrated works will increase to 83 compared to the present number of 50.

iii) Of 54 integrated BF-BOF works, only one will have a capacity of more than 5 Mt which is generally considered the minimum economic size for the newly built integrated works.

iv) Average size of plants will increase considerably by 1985, but still remains at 1.7 Mt for BF-BOF route (1.0 Mt in 1977) when number of very small (less 1 Mt) works of which the majority are charcoal based works.

The points above raise the question of appropriate size of works for developing countries. The size should, generally speaking, be decided by the consideration of (a) market potential (b) economy of scale (c) technology level (d) management skill (e) availability and quality of raw materials (f) capability to procure finance, (g) infrastructure level and (h) sometimes regional development of the country. The trend shown above that the sizes of works will gradually increase toward 1985, is a welcoming and encouraging one for the developing world in the sense of not only economy of the industry but also as a proof for gradual level up to technical as well as management.

Since there are many factors to be considered in deciding the optimum plant sizes as mentioned above, a detailed techno-economic study has to be made project by project, with a thorough prospect in mind with regard to long term and ultimate future expansion. Also very careful assessment of management of the project and the plant should be made because the financial implication of the delay in construction and the low capacity utilisation ratio is a very significant factor which pushes up the capital investment resources required as also production cost. Highly ambitious project for a modern large scale steelworks and mechanised automated sophisticated equipment should be, generally speaking, implemented on later phases/stages for countries with little experience in the steel industry.

Table 9 SIZE, DISTRIBUTION OF INTEGRATED STEEL-WORKS IN DEVELOPING COUNTRIES EXCEPT CENTRALLY PLANNED ASIA (1977 AND 1985)

Capacity (Mt)	1977			1985		
	BF-LD	DR-EF	Total	BF-LD	DR-EF	Total
less 0.5	17	5	22	16	13	29
0.5-0.99	9		9	4	10	14
1.0-1.49	5	1	6	9	2	11
1.5-1.99	5		5	8	1	9
2.0-2.49	3		3	5		5
2.5-2.99	4		4	1	1	2
3.0-3.49				3	1	4
3.5-3.99	1		1	2	1	3
4.0-4.49				2		2
4.5-4.99				3		3
5						
6						
7						
8						1
Total number	44	6	50	54	29	83
Total capacity (Mt)	45.42	2.99	48.41	92.59	23.03	115.62
Average capacity	1.03	0.50	0.97	1.71	0.79	1.39

1.3 *Steel production of developing countries and their share in the world in the future years*

The future steel production level of developing countries would be decided by the project realisation ratio of projects which are reviewed in previous sections and capacity utilisation ratio. Although announced projects were already classified into two categories in previous section, it is still necessary to consider project realisation ratio even for the projects which are classified as probably completed by 1985, and 1990. The reasons are that rather generous classification was made as already mentioned, for individual project, and that it often happens that even advanced stage projects are delayed a few years by a number of reasons.

Here it is rather critically assumed that the project realisation ratio for the projects classified as probably completed projects by 1985 and 1990 in Table 6, would be of the order of 75 per cent. For the year 2000, it is assumed that the capacity estimated could be realised, because the capacity estimates for the year 2000 were made on the basis of various socio-economic considerations for each country and not merely on announced project basis. Of course, the capacity in 2000 represents approximations and in reality some countries would far surpass the estimated one whilst others would not reach the estimated level.

As for capacity utilisation ratio, it is assumed that the current capacity utilisation ratio of some 76 per cent would improve rather slowly, namely 77, 79 and 83 per cent for the year 1985, 1990 and 2000 respectively. The reasons for this rather pessimistic view are (a) the time required to achieve full capacity utilisation

123

for new plants, (*b*) many countries are going to have a steel industry for the first time and (*c*) very long time required to acquire the requisite management, maintenance and operation skill.

The future steel production, then, will be as shown in Table 10.

Table 10   STEEL PRODUCTION AND CAPACITY IN DEVELOPING COUNTRIES, ADJUSTED BY PROJECT REALISATION RATIO INCLUDING CPE ASIA

Growth rate (%/y)

	1967	1977	1985	1990	2000	1977/67	1985/77	2000/77
Capacity (Mt)	49	100	180	250	560	7.4	7.6	7.8
Capacity utilisation ratio (%)	75	76	77	79	83	—	—	—
Production (Mt)	36.6	75.9	138.6	197.5	464.8	7.6	7.8	8.0
Rounded production (Mt)		76	140	200	470			

Developing countries' steel production would grow further towards the end of the century at almost the same or at a little higher rate than the last decade, and may reach 140 Mt by 1985 and 470 Mt by 2000. The figure of 470 Mt in 2000 seems to be on the lower side when one considers consumption figures discussed in the next section and this should be regarded as the minimum level desired.

These figures are somewhat lower than those discussed at the First UNIDO Consultation Meeting on the Iron and Steel Industry (it was expected at that time that developing countries steel production might increase to 150 Mt by 1985 and 525 Mt by 2000), reflecting a continuing slump of the steel industry and of the economy in general of the developed Western world and the resulting effects on the steel growth of developing countries.

What will be the share of developing countries in the future in the world of steel production? Since it is extremely difficult to make forecast with any certainty under present status of the economy and steel industry of the world, only the orders of magnitude will be discussed based on the simplest assumptions.

First growth rates of consumption were assumed as shown in Table 11. The basis for this assumption is:

i)   The demand for steel products in developing countries will continue to be high because of their absolute needs for steel in order to promote their overall economic development. Developing countries with applicable steel consumption per capita at present are on the stage of "economic take off" and where the steel demand is the highest. Many countries with less consumption per capita would start taking off in a decade or two.

ii)   As seen before, however, the consumption of steel in developing countries as a whole has not increased during the last three years or so. Therefore, the growth rate for coming years cannot keep the same rate of 8 to 9 per cent per year for the last ten years and will decrease slightly. There is a possibility that the decrease will be substantial if the current depressed economic conditions in the western developed countries continue over the future years.

iii)   The consumption of Western developed countries will grow at a very moderate rate even if full recovery is made from the current economic slump. The structural changes in the steel consumption patterns seem to have taken place in those countries. The growth rate will further decrease in the long run.

124

iv) The Eastern developed countries have shown a steady consumption increase for many years. However, the general trend is that the growth rates are decreasing in recent years (74/73: 6.6 per cent, 75/74: 3.2 per cent, 76/75: 3.0 per cent, 77/76: 2.6 per cent).

Table 11 ASSUMED STEEL CONSUMPTION GROWTH RATE PER ANNUM

	Actual	Assumed	
	1976/1967	1985/1976	2000/1985
Developing countries	8.25%	7.5%	7.0%
Western developed	1.73	2.7	2.0
Eastern developed	4.75	3.2	2.5

Consumption of steel in future years is shown in Table 12. The developing countries could consume about a quarter of the world steel by 1985/1990 and more than one third by 2000.

Table 12 CONSUMPTION OF STEEL IN FUTURE YEARS (a)

	1967	1976	1985	1990	2000
Developing countries (Mt)	54.5	111	213	298	587
(Share in the world)	(11%)	(17%)	(23%)	(27%)	(37%)
Developed countries (Mt)	442	562	726	808	1003
Western (Mt)	309	36	459	506	617
Eastern (Mt)	132	201	267	302	386
World total (Mt)	49	673	939	1106	1190
(Rounded figure)			(950)	(1100)	(1550)

(a) Because 1977 consumption figures are still uncertain 1976 figures are used as basic.

Table 13 PRODUCTION OF STEEL IN THE PAST AND FUTURE YEARS (ROUNDED FIGURES)

Growth rate/yr

	1967	1976	1985	1990	2000	77/67	85/77	2000/77
Developing countries	36.6	76	140	200	470	7.6	7.8	8.0
(Share in the world)	(7.4%)	(11.3%)	(15%)	(19%)	(30%)			
Developed countries	460.8	598.4	810	900	1080	2.7		
Western	325.4	394.2	520	600	690	1.9		
Eastern	135.4	204.2	270	300	390	4.2		
World total	497.4	674.4	950	1100	1550	3.1	4.0	3.7

Production figures are calculated and shown in Table 13 on the following assumptions:
  i) Developing countries' steel production is projected in Table 10.
 ii) Eastern developed countries as a whole would be self-sufficient in relation to their steel consumption.
iii) Western developed countries would produce steel to satisfy their internal/domestic markets and for export to developing countries that are deficient in steel production and cannot meet their domestic market requirements.

The share of developing countries in the world steel production could increase to 30 per cent by 2000 from the current figures. This share of 30 per cent should be regarded as the minimum target for the developing countries on the basis of the Lima Declaration, taking into account the significance of the steel production for the overall industrial development and economic growth in the developing countries. Particularly when we look into the steel import required (consumption minus domestic production) in the future years as shown in Table 14, one might conclude that more dynamic expansion of the steel industry in the developing countries is urgently required.

Table 14   THE STEEL IMPORT REQUIRED BY DEVELOPING COUN-
TRIES IN FUTURE YEARS

	1967	1977	1985	1990	2000
Consumption (Mt)	54.5	115	213	298	587
Production (Mt)	36.6	76	140	200	470
Import required (Mt)	17.9	39.1	73	98	117
Self-sufficiency	67%	66%	65%	67%	80%

Self-sufficiency in supply (production divided by consumption) will hardly improve by 1985 and reach only 80 per cent by 2000. The absolute amount of steel to be imported by the developing countries will increase steadily and reach more than 100 Mt in 2000. The question arises, if it is realistically possible for developing countries as a whole, to import these tonnages of steel with their extremely limited foreign currency resources. In this connection, the slowdown in the steel consumption in the recent years (1974–1977) should be regarded as a significant alarm signal. One might draw a very gloomy picture for the development of the steel industry in the developing countries which is quite different from the one shown above. *Factors affecting the expansion of the iron and steel industry in the developing countries* are outlined below:

The steel market and steel industry in the developing countries in the future will undoubtedly play a very significant part in the world.

The effects and consequence of the recent slowdown of global economic growth and in particular of the slump and structural changes taking place in the steel industry of the Western developed countries, could hamper the sound development of the developing countries' steel industry very much, if

a)   economic strength of some developing countries becomes so weak that they can no more sustain their dynamic economic growth including the growth of steel and related industries;

b)   financial assistance and involvement of some developed countries are not fully forthcoming for the growth of the steel industry in the developing countries for various reasons;

c)   technical assistance and supply of advanced technology and technical know-how and capital equipment are restricted due to various reasons;

d)   protectionistic attitudes develop in some developed countries with established over-capacity in the steel industry over and above the domestic needs to the extent that even small exports from the developing countries to the developed countries are seriously restricted.

HOW STEELMAKING ENTERPRISES
CAN BECOME INTERNATIONALLY COMPETITIVE

by
Dott. Franco PECO
Delegato della Presidenza
Associazione delle Industrie
Siderurgiche Italiane
"ASSIDER"

SUMMARY AND CONCLUSIONS

Any debate concerning prospective trends and developments in the steel industry during the 1980s — a decade that promises to be extremely difficult and highly competitive for the industry — calls for consideration of the ways in which steelmaking enterprises themselves are expected to respond.

If there is one fundamental aim that would seem to embrace all the issues of concern, it is that each enterprise should progressively make substantial improvements in its own competitiveness on the international market.

The international competitiveness of an enterprise implies both an increase in revenue and a reduction in costs within the context of the general economic trend, the structures of the sector and the efficient management of the enterprise itself.

The increase in revenue will have to be sought on a world steel market which is a classical example of a marginal market of a residual character on which supply and demand fluctuate widely over the different phases of the business cycle, a world market that is influenced, moreover, by the structural imbalances existing in the steel sector.

The position of net exporting countries has been changing as time goes by and the total number of countries exporting steel is increasing continuously. International competition can often become pathological and may be conducive to new forms of protectionism. During the 1980s, steelmaking enterprises will be faced with the need to obtain maximum total revenue on an increasingly difficult market which will require a continual adjustment of supply to trends in demand in the different consumer regions and markets.

Where the need to reduce costs is concerned, the magnitude of fixed costs in the steel industry means that enterprises must strive to increase the technical efficiency of plant and continue to invest in measures to step up productivity.

One particularly disruptive factor in the present world economic crisis is undoubtedly the existence of large surplus production capacities, and this applies very much to the steel sector. However, to get a clearer idea of the action needed to restructure the sector, it would seem advisable to differentiate between:

— temporary over-capacity owing to the situation on the market, which will be reabsorbed in due course as consumption picks up again and expands;

— obsolete over-capacity which is partly an inheritance from the time when steel consumption was rising continuously and which must be eliminated for technical and, in many cases, geographical reasons;
— real excess capacity, which is essentially determined by the effective potential for future exports on international markets where competition is increasingly difficult.

Industrial relations is another factor which has to be taken into account in assessing the international competitiveness of enterprises, particularly insofar as the different kinds of response from trade unions and employers liable to affect production costs vary diversely as between countries.

All these technical and social factors together influence the international competitiveness of enterprises by determining their level of profitability in terms of the gross producer's margin, which seems to be the most significant criterion for identifying the most efficient enterprises on which any structural adjustments and remedial action should be centred.

Once the gross producer's margin has been ascertained, the other factors to be taken into account are of a financial nature, essentially external to enterprises, and depend more on the general economic policies of individual countries and international monetary and financial trends.

Even the capital structure of an enterprise as determined by the behaviour of shareholders, is bound up with responsibilities which are external to productivity, although it does influence the level of conpetitiveness on the world market. Structural adjustment measures or remedial action clearly have to be taken wherever they are called for, be it in the technical, social or financial spheres.

To sum up, the aim of the studies carried out and the initiatives taken by the OECD and other organisations, such as the GATT, should be to establish a new order on the world steel market, where the level of competitiveness of enterprises can play the decisive role which it should.

In order to achieve this objective, the steel industry itself must undertake to do its part by ensuring that enterprises are technically efficient, that their production capacities are economically justifiable over time, and that they compete correctly in the market.

On the other hand, the steel industry must be able to rely on a formal undertaking by trade unions and employers to create the economic climate needed to ensure an open but orderly market, a normal financial situation for enterprises, an international harmonisation of industrial relations, a priori transparency of investment, and co-operation on a world-wide scale.

In this spirit, the results which can be expected from the work and initiatives of OECD and its Member countries must be regarded more as a point of departure towards the development of international trade in steel to the general good of the world economic system.

FOREWORD

During the first day of the OECD Steel Committee's Symposium on the Steel Industry in the 1980s — and in preparation for the main item on the agenda for the second day: "The policy responses to the problems of the world steel industry" — the debate will be centred on current trends in the steel industry and developments in the 1980s.

Some of the subjects to be discussed in the course of the first day's preparatory work are very broad in scope, ranging from the general economic outlook in the OECD area and in the world as a whole, to the probable growth of steel demand, changes in production capabilities and supply potential, and the future supply/

demand balance and its implications, especially as regards international trade in steel products. Within this range of subjects, consideration must necessarily be given to the ways in which steelmaking interprises will respond in this general context. It seems to us that this response in the decade just starting will have one fundamental objective central to all specific measures, namely the substantial and continuous improvement of each individual enterprise's competitiveness to the point where it can, at the very least, face up to what clearly promises to be both extensive and difficult competition on the international market.

Indeed, international competitiveness must be understood as the capacity of each enterprise to successfully meet the challenge of the other producers on the market. It calls for a combination of two processes: first, the enterprise must increase its revenue and, secondly, reduce its costs within the context of general economic trends and the structures of the sector and on the basis of efficient management.

A. Markets and revenue

The increase in revenue must be sought on the world market, i. e. every market on which steel products are internationally traded, and also with due consideration of the effects of this world market on the various domestic markets. The first step in a revenue policy is, however, to study the world market which, incidentally, is still expanding (see Table 1).

1. A marginal-type world market

The world steel market is a classic example of a "marginal" (in Marshall's sense) market of a residual character in which desires to sell by producers fluctuate widely according to the business cycle whereas, on the other hand, residual demand for imports remains more rigid.

This world market accounts for at least 50 per cent of trade in steel products and consists of exports by countries whose output exceeds domestic consumption, with the corresponding imports by countries whose consumption exceeds domestic output.

The other 50 per cent (approximately) of world exports represents trade among producer countries for reasons to do with tradition, quality, financial links between producers and consumers, or pure market competition.

Table 1 WORLD STEEL CONSUMPTION
(PRODUCTION) AND EXPORTS

Million tonnes

Year	Exports	Consumption	Exports as % of production
1950	20.5	192.0	10.7
1955	34.0	270.5	12.6
1960	52.7	345.5	15.3
1965	78.5	457.0	17.2
1970	117.5	595.4	19.7
1971	125.5	582.6	21.5
1972	133.1	630.3	21.1
1973	147.4	698.4	21.1
1974	169.6	708.9	23.9
1975	147.7	645.6	22.9
1976	162.8	676.7	24.1
1977	165.2	672.3	24.6

In periods when there is a sellers' market, i.e. when business activity is buoyant, export supply tends to contract to give priority to domestic requirements, while demand rises owing to the difficulties which consumers have in meeting their own requirements from inadequate domestic supply.

In the opposite case of a buyers' market, when business is slack, while demand is reduced strictly to the need to meet requirements which are smaller but not covered by domestic output, export supply tends to mop up the residual production capacities of all producers who do not wish to cut their plant utilisation rate to match the reducer level of national consumption.

While it may be assumed that an overall quantitative equilibrium is established between import demand and export supply when business activity is good but no shortage as yet exists, an excessive imbalance develops when business activity is slack, even if the point of crisis has not yet been reached as regards consumption. In these circumstances, world steel supply can amount to more than three times the effective demand on the world market.

Price swings on the world market, which are exceptionally extensive over the path of a whole business cycle, are easily explained with the opposite slopes of the demand and supply curves.

While it is customary to reason in global terms or, in other words, to give a single aggregate for crude steel covering all trade in steel products expressed in terms of the ingot tonnage equivalent, it must be stessed that each product has its own points of structural equilibrium which are determined by specific production capacities and technical capacities for consumption in the manufacturing industries, notwithstanding the fact that each product obeys the general laws of supply and demand.

As shown in Table 2 below, significant changes occurred in the position of net exporting countries during the 1970s. In the 1980s it seems that there will still be substantial imbalances between supply and demand on the world market, despite the fact that it is hoped to reduce them, even if only after a few years.

Table 2 NET STEEL IMPORTS AND EXPORTS BY MAIN GEOGRAPHIC AREA[1]
Million tonnes ingot equivalent

	1960	1965	1970	1974	1975	1978
EEC countries	17.7	22.5	14.6	34.7	27.6	31.4
Oher European countries	− 3.9	− 7.4	− 7.5	− 8.5	− 6.0	− 0.3
East European countries	− 1.2	2.1	4.2	− 3.9	− 2.5	− 3.3
China and North Korea	− 1.2	− 0.1	− 6.9	− 11.0	− 9.4	− 15.7
Middle-East	− 2.0	− 2.8	− 4.3	− 10.6	− 12.9	− 13.8
Other Asian countries	− 5.3	− 6.7	− 9.8	− 15.8	− 14.5	− 27.5
United States	0.1	− 9.2	− 8.0	− 11.1	− 11.0	− 22.5
Japan	2.6	12.7	23.4	41.4	37.6	40.6
Other industrialised countries	− 1.4	− 2.4	− 0.1	− 2.0	0.4	3.4
Latin America	− 3.1	− 3.3	− 4.6	− 13.2	− 10.1	− 6.3
Africa	− 2.2	− 3.6	− 3.9	− 6.3	− 6.1	− 4.8
Other	− 5.0	1.9	0	0	0.6	− 0.6

[1] Production/consumption deficit (−) or surplus (+).

2.   Pathological competition and new forms of protectionism

Where the imbalances between market supply and demand are most evident, competition may sometimes take on pathological aspects. First, to stress a point that is common knowledge, market prices are not formed at the time products are delivered, nor do they correspond to the quantities subsequently shown in the international trade statistics. When imbalances occur, whoever is stronger on the market — the seller or the buyer — will try to alter prices to his own advantage.

Prices are not formed at the time that trade contracts are concluded, since at this point agreement has been reached between seller and buyer; it is the pressure brought to bear by total supply and total demand on the market which determines the level of prices, on which the contracts may subsequently improve. This is an important point because it explains a subsequent imbalance which turns against competing sellers. If there is a single buyer who maintains his demand based on real requirements — increased at most by future requirements because he wishes to rebuild his own stocks — the seller tends to increase the supply, either by trying directly to place the capacity remaining after production for national consumption (and sometimes adding potential future output if he wishes to fill his order books to the maximum), or by using a large and complex international marketing network, owned by the seller or by third-parties, whereby supply is increased very substantially in the hope of negotiating contracts for sales more commensurate with the seller's ambitions.

When the supply curves determined by the short-term business situation are at the upper or lower extremes, competition may become pathological insofar as high supplements are charged when business is buoyant and very substantial discounts are granted in periods of crisis.

This phenomenon encourages the continuing increase in the number of countries exporting steel on the world market, an increase due to the fact that a number of governments regard a national steel industry as fundamentally strategic in character, i. e. as a mark of industrialisation, even if there is no domestic consumer market. It should also be pointed out that the steps taken by enterprises to achieve economies of scale mean an increase in the size of steelworks, with the result that export activity often becomes a necessity. In 1914, on the eve of World War I, there were only about 12 steel exporting countries and this number remained

Table 3   STEEL EXPORTING COUNTRIES

1930	1950	1965	1965/1975
Belgium	idem +	idem +	idem +
Luxembourg	Australia	Denmark	India
Germany	Italy	Finland	Switzerland
Netherlands	Norway	Greece	German Demo-
Austria	USSR	Hungary	cratic Republic
Canada		Portugal	Chile
Czechoslovakia		Romania	Mexico
Japan		South Africa	Bulgaria
Poland		Spain	Brazil
Sweden		Yugoslavia	Argentina
United			South Korea
Kingdom			Ireland
United States			Egypt
			Hong Kong
			Venezuela
			Turkey
			Taiwan
			Colombia
			Singapore
			Malaysia
			Philippines
			New Zealand
			Thailand

131

almost unchanged over the long inter-war period. Since 1945 the world-wide process of industrialisation has led to the rapid development of new steel industries which are, however, obliged to export all or some of the steel produced. In 1950 the exporting countries already totalled 16, were 25 by 1965 and are now approaching the 50 mark (see Table 3).

The relentless trend towards a pathological type of competition is causing what are often excessive reactions from importing countries which have their own local production to safeguard. This applies where the world market is concerned, especially during periods of economic difficulty.

While the GATT rules clearly cover cases of unfair competition and dumping, it is not difficult to envisage bans or strict quotas on imports of steel products on the world market. A more objective form of intervention, but likewise open to the risk of creating unduly restrictive forms of protectionism, may be found in the United States Trigger Price Mechanism and, with the risk somewhat reduced, in the European Community's basic import prices. In the United States the market could be really protected by a restrictive application of the Trigger Price Mechanism with price levels sometimes higher than those on the domestic market. In the EEC countries, the external measures under the Davignon Plan must be regarded more as a necessary complement to the internal measures to re-establish equilibrium between demand and supply, so as to prevent the balance from being destroyed by increased imports from Third Countries at prices considerably lower than the EEC guidance prices.

However, it must be admitted that there is a risk of seeing the world steel market split up into areas of free competition and areas of increasing neoprotectionism, in other words, a tendency towards a system which would do nothing to help resolve the serious problems in the present world economic and monetary crisis. The 1980s will provide the real test for steelmaking enterprises on the world market.

3. *Revenue policy*

In such a complex difficult context, steelmaking enterprises must try to obtain the maximum total revenue if they wish to remain sufficiently competitive on the world market in the coming years.

The revenue policy to be established for each steel product traded internationally must take account of the real potential for selling the product on both the domestic market and on the various export markets. The best results are achieved by continuously adjusting supply to the short-term trend in demand, thus making it possible to continuously adjust quantities to the requirements of the different consumer areas or regions.

Sales organisations must be firmly controlled by the producer who runs the enterprise, the aim always being to obtain the maximum revenue for a volume of sales which enables the producer to maintain a rate of output ensuring the optimum ratio between costs and earnings. It may very well be that the 1980s will see steelmaking enterprises strenuously pursuing a revenue policy aimed at ensuring the necessary increase in prices in close linkage with the foreseeable rise in costs, even if the risks of pathological competition and a real price war on the world market remain very serious.

B. *The reduction of costs*

An enterprise may remain internationally competitive as the result of a careful revenue policy, but it builds up a competitive position by continuously practising a policy of reducing production costs with a view to achieving long-run viability.

1. Fixed costs and technical efficiency

The need to achieve a high degree of plant efficiency is even more important in the steel industry owing to the fact that fixed costs are extremely high. One fundamental factor of competitiveness is regular adjustment to technological progress by means of an ongoing policy of productivity-boosting investment. There are examples of steel industries in the world where plant is less than ten years old on average thanks to rapid introduction of the most recent technological know-how: blast furnaces, integrated or electric steelmaking plant, continuous casting, rolling mills, etc. Attention should be drawn to the physical potential for high productivity: economies of scale based on the size of steelworks, reception and delivery infrastructures, savings on energy consumption and adjustments to ecological requirements, all of which are measures that can enable an enterprise to achieve the kind of performance which makes it highly competititve on international markets.

2. Surplus production capacities

In the present world economic crisis, which in no way seems to be abating as the 1980s begin, one of the major disruptive factors is certainly the existence of large surplus production capacities. All the traditional sectors of industry are affected. Moreover, the consequent decline in investment is reflected in consumption of steel products which, as we know, are to a very large extent used in capital investment projects. The steel industry is also burdened by surplus capacity but, if we are to get a clearer idea of the measures required to restructure the sector, it seems advisable to define these surpluses more precisely.

a) Temporary over-capacity in relation to the market

This over-capacity results from a fall in market consumption, more particularly on the domestic markets of the countries concerned.

If, for example, there is a 20 per cent reduction in national steel consumption as a result of the current economic crisis, a steel industry that responds positively to the abovementioned requirements for technical efficiency will find itself with surplus capacity temporarily. Indeed, it is reasonable to expect that consumption will subsequently climb back towards its earlier levels. It seems inconceivable that user industries will not take the earliest possible steps to make the most of existing technologies in operating their plant.

b) Obsolete over-capacity

This over-capacity must in fact be eliminated and a number of countries have already introduced large-scale programmes of structural adjustments and remedial measures for this purpose. This type of over-capacity did not become evident as long as the expanding market enabled enterprises to continue using plant which entailed high — or even higher than average — production costs. However, the continuation of the present serious crisis has brought these facts to light and shown the absurdity of keeping such plant in operation.

The elimination of obsolete capacities often involves the conversion of traditional steelmaking regions — where economic activity has centred on this industry alone — and gives rise to particularly serious and distressing social problems which are nevertheless difficult to avoid in the present circumstances.

133

b) Real excess capacity

There are some steelmaking facilities which, though efficient and presenting no problems of conversion, are located in countries and regions where steel consumption is of minor importance or non-existent, so their survival depends on export activity. In the present period of international crisis, they are in the most delicate position and must inevitably turn towards the world market which, as has already been seen, is the most difficult and the most competitive, often to the point where competition becomes pathological.

Restructuring is probably a less urgent requirement in this case but the need for international competitiveness is greater, so longer-term action is also necessary in order to find solutions to the more delicate type of problem presented by these steelmaking facilities, precisely because of their need to be internationally competitive.

To sum up, surplus production capacities exist in the world but they have been created in different ways and there are different ways of resolving them over time, not the least of which will be adjustment to market trends in the 1980s which, in the last analysis, will determine the response of enterprises.

3. Labour costs and industrial relations

Labour is the factor of production that must be examined in the broadest possible terms owing to the political and social problems involved. In the steel industry it is becoming a smaller component of production costs as time goes by, but it is certainly an increasing source of disequilibrium which weighs heavily on the international competitiveness of enterprises.

a) Wage costs

The first set of problems concerns wage levels which differ — sometimes very markedly — from country to country, especially in relation to total labour costs.

No survey in this connection can be regarded as reliable and comprehensive unless, in addition to a straightforward comparison of steel workers' wages, it covers all the other labour cost components like social insurance and tax payments, productivity bonuses and deals, and so on. There are still too many differences which, in turn, involve too many risks of distorting competition when steelmaking enterprises offer their products for sale on international markets.

b) Industrial relations

In addition to the question of differences in wage costs in the broad sense, there are also differences in the behaviour of trade union organisations which themselves differ widely in character in the many countries with steel industries.

At one end of the scale are countries in which work continues during strikes, while at the other we find countries in which strikes are pushed to the point of maximising the loss of production in relation to loss of wages. This range covers a wide variety of industrial relations in terms of productivity, mobility of labour, absenteeism, contractual working hours, normal hours or overtime, as well as the whole gamut of trade disputes, which in some countries are non-existent and in others incessant.

During the 1980s it will certainly be necessary to face up to this problem which is so complex and tricky that there can be few illusions about the time it will take to reconcile such widely divergent positions at international level.

C. *Profitability and international competitiveness*

Where the individual steelmaking enterprise is concerned, it is becoming more and more difficult to achieve anything outside the firm itself in terms of improved revenue on an essentially competitive market. All that can be done is to cut costs within the enterprise by achieving the most productive possible mix of inputs so that the enterprise attains a level of profitability and a degree of international competitiveness comparable with those of competing enterprises from other countries.

However, the question of profitability must be examined in somewhat greater depth.

1. *The gross producer's margin*

Total revenue less production costs gives us what is known as the gross producer's margin. It indicates the enterprise's degree of technical competitiveness on the basis of the efficiency of its production facilities, the cost of the labour factor in the broadest sense and the effectiveness of the industrial management. Table 4 gives the published data on the overall efficiency, i. e. gross producer's margin, of a number of steel companies, but these figures also include depreciation. The figure for the gross margin alone would be higher, but not uniformly so. When restructuring of the steel industry is discussed, a great deal is said about the advisability of centring production on the most efficient facilities and steelworks, and about cutting out dead wood and setting off towards new objectives with structurally sound enterprises. It seems then that there is no more significant criterion for selecting the most efficient enterprises than the gross producer's margin. On the other hand, all the financial problems that arise at this stage with regard to determining the final profit margin, depend more on external factors to do with a country's general policy and with prevailing monetary and financial conditions than on the steelworks itself, i.e. the soundness of its decisions as regards technology and location, the skills of its workers and managerial efficiency.

Table 4 OVERALL EFFICIENCY INDICES
Average 1972-1978

Enterprise	Index	Enterprise	Index
Cockerill	98.6	Estel	106.7
Hainaut-Sambre	97.3	Italsider	109.4
T.M.M.	95.9	B.S.C.	100.7
Arbed	102.8	Voest Alpine	103.4
Usinor	103.2	Ensidesa	113.8
Sacilor	98.4	Bethlem Steel	108.0
Khockner	106.5	U.S. Steel	109.6
Krupp	110.1	Kawasaki	104.7
Peine-Salzgitter	114.4	Nippon Steel	106.6
Thyssen	116.1	Sumitomo Metals	114.2

Source: Eurofinance.

1. Index = $\dfrac{\text{Total product}}{\text{Sales costs} + \text{Labour costs} + \text{Depreciation}}$

Another sure way of improving international competitiveness is to try to increase the value added within the enterprise both by improving product quality and developing vertical integration. While realising the importance of these possibilities, we prefer to say no more about them here since any further development of ideas of this kind might take us too far afield.

2. *Depreciation and financing costs*

The present logic of economic management requires that plant be written off in regular annual instalments. This is done in accordance with the laws in force in each country, which are not harmonised at international level, nor even within the European Community. However, where an enterprise experiences difficulties owing to financial policies or inflationary factors which may compromise the gross producer's margin, the necessary steps should be taken to adjust the financial structures.

a) *Capitalisation and self-financing*

A correct financial policy for an enterprise should prevail on shareholders to provide risk capital equal to at least one-third of the capital invested. Higher ratios are conducive to particularly healthy situations which facilitate the self-financing of new investment, the replacement of equipment, or projects for expansion, whereas lower ratios expose the enterprise to undue financial risks, especially during periods of monetary disorder, high interest rates and high rates of inflation. A steelmaking enterprise cannot achieve an appropriate level of international competitiveness unless there is a high enough ratio between its risk capital and invested capital.

b) *Indebtedness*

In addition to using its own available resources, an enterprise finances its investments by short or long-term borrowing on the capital market. In the present state of crisis in the system, the enterprises in difficulty are mainly those burdened by financing costs in the form of debt repayments and servicing (it is claimed that these outgoings amount to something over 5 per cent of the enterprise's turnover).

In this connection, one cannot help calling attention to the fact that, until the energy crisis broke, the steel industry was carried on a wave of continuously expanding consumption which called for considerable efforts to increase production capacities. Those enterprises that did more than others to keep pace with the strong uptrend in order to develop their own domestic consumer markets, now find themselves with a particularly high level of indebtedness, which is often linked with an inadequate ratio between own risk capital and invested capital. The cost of money, which is rising in this period of international monetary disorder and economic instability, and the all-round rise in inflation — though rates differ from country to country — have raised financing costs to intolerable levels, making it urgently necessary to adjust the financial structures of enterprises. There have already been some particularly big and spectacular instances of financial restructuring with the help of national governments, and all steelmaking enterprises are struggling to establish new patterns of equilibrium.

If the gross producer's margin is still the most significant criterion for determining the degree of efficiency of an enterprise and its chances of holding its own on a competitive international market, it seems worth adding that the more production facilities are up to date and technologically sophisticated, the greater must be the gross margin in order to allow for heavier financing costs.

3. *Profits*

Where the gross margin is big enough and there are no serious financial problems, an enterprise can readily achieve satisfactory levels of profit enabling it to be sufficiently competitive on the world market and making it more viable even when, as at the present time, this market is fraught with difficulties.

On the other hand, where profits are inadequate or, worse still, heavy losses are recorded, it is absolutely essential that immediate remedial measures be taken by the public or private bodies controlling such enterprises, either in accordance with the normal legal procedures governing joint stock companies or under special national plans for industrial restructuring and adjustment.

The rapid attainment of a sufficiently high level of international competitiveness is the only means of ensuring the survival of steelmaking enterprises afflicted by this long crisis and with the prospect of at least ten more very difficult years ahead.

D. *Conclusions*

In the preceding sections we have set out the requirements for building a steel industry that will be internationally competitive, and we must now try to present a few conclusions consistent with the objectives of the OECD's studies and research.

1. *The general framework for action*

All that has been said so far seems quite in line with the objectives set for the OECD initiative.

a) In setting up the Steel Committee, the aim was to work towards and, possibly, establish a new order on the world steel market as a means of overcoming the persistent difficulties and disorders encountered on this market.

b) The Symposium on Steel in the 1980s has been organised to take up all the relevant problems as systematically as possible so as to facilitate the efforts to find sound solutions.

c) The panel on the short and medium-term outlook in the 1980s has indicated the probable scenarios that will provide the context within which the world's steel industry will have to re-establish itself over the next years and find lasting solutions to its problems, be they structural or cyclical.

This paper has identified international competitiveness as the necessary and sufficient condition for a steelmaking enterprise to develop a free, competitive and profitable activity for the future.

By way of conclusion, it may be pointed out that, if steelmaking enterprises are to become internationally competitive in the context of all the above-mentioned problems and the efforts to find economic solutions to them, the steel industry itself must pledge to do its part within the limit of its possibilities, and at the same time make exactly clear what sound and objective help it needs in order to achieve its aims.

2. *The steel industry's pledge to do its part*

The steel industry itself can contribute a great deal with far-reaching consequences:

a) *The technical efficiency characteristic of high-productivity enterprises*

If an enterprise progressively introduces technological innovations, eliminates obsolete plant and out-dated sites and achieves high labour productivity and thoroughly efficient management, it can obtain gross margins which are high enough to secure a degree of international competitiveness.

b) Correct participation in a competitive market

One fundamental rule that must always be observed is that no steelmaking enterprise can be run on a sound basis in the longer term unless its average earnings cover its average costs. The GATT makes this point quite clear and should be a source of reference for all.

Although the cyclical difficulties which are one of the main factors in pathological competition on export markets can nevers be avoided, it is necessary for all to work together to reduce the differences as far as possible in the interests of everyone concerned.

c) Economically justified production capacities

This will become an increasingly important requirement in the future, as is already apparent in this difficult decade that is beginning now, a period when the high growth rates in steel consumption to which the industry had become accustomed will perhaps come to an end once for all. The threat of surplus production capacities must cease to be the most important factor of disorder on the markets.

An industry does not produce to sell but sells to produce, so it is consumption that determines the level of production and not the other way round. Investment decisions should therefore always be justified by the fact that new requirements have emerged on consumer markets: first, on the individual domestic markets and then in the case of exports, in the shape of proven outlets on specific foreign markets capable of absorbing further output.

3. What the steel industry is entitled to ask in return

If steelmaking enterprises themselves make a commitment, it must necessarily be accompanied by an equally firm undertaking by all others concerned, i. e. employers, trade unions, public authorities, steel distributors and users, to create the kind of economic climate to which a responsible and co-operative steel industry is entitled.

a) An open but orderly market

Protectionism has laways been inward-looking and a potential cause of economic and political disaster. Unilateral arrangements by each country to protect its own domestic market are not the way to solve the problems. What is needed, rather, is to try to build a sectoral type of international trade system, our concern here being the steel sector, which can ensure that the rules of competition function correctly. The OECD and the GATT would seem to be the organisations best qualified to work towards establishing these fundamental guarantees.

b) A normal financial situation in enterprises

This might be thought to be a specific task for the enterprise itself if financial equilibrium did not always depend more on external factors bound up with the decisions of shareholders and, to an even greater extent, the vicissitudes of the money markets and government financial policies.

It is therefore necessary to ensure that enterprises have the right kind of capital structures and that there is no indiscriminate government support, or support which cannot be justified in terms of helping to solve social and industrial conversion problems and the technological and economic problems involved in adapting the technical, social and financial structures of enterprises.

c) International harmonization of industrial relations

It will be increasingly difficult to allow the competitive equilibrium achieved by dint of constant striving towards international competitiveness to be upset by the existing disequilibrium in industrial relations and in the responses of workers' organisations, which differ so widely from one region of the world to another. It will be very hard to make headway here, but it is important to realise the seriousness of this disequilibrium, which takes the form of differences in the legislation and practices which determine labour costs, their wage and social insurance components, rules governing strikes, physical productivity, mobility and absenteeism.

d) A priori transparency of investment

To be quite frank, what is needed is an effective way of co-ordinating new investment throughout the world, but such a prospect is little less than utopian. Steps should at least be taken in this direction, however, so that investment decisions throughout the world can be made with full knowledge of all the market factors involved, everyone knowing from the start what outlets may be available for additional output and the potential risks in terms of the cost/revenue balance.

e) Co-operation on a world-wide scale

Much has been said about the need for co-operation between employers and trade unions, but it must be borne in mind that the world steel market is open to all: all enterprises participate and all are liable to suffer the consequences, even on their own domestic markets. The international initiatives with respect to co-operation, co-ordination, the application of codes of conduct and the necessary surveillance measures should be extended to cover all steel-producing and/or consuming countries, be they industrialised or in process of industrialising, market or centrally-planned economies. Governments, like enterprises, must consider themselves directly concerned, especially since the attempts to establish some sort of order that is not based on "dirigisme" but in a set of rules to be observed (a kind of highway code for steel) can only take shape by means of government measures. In this connection, it must not be forgotten that the bulk of world steel production, i.e. more than 50 per cent, is now directly or indirectly controlled by public authorities.

4. Purpose

The work, decisions and findings of the panel and the Symposium, as indeed the activity of the OECD Steel Committee and the goodwill of Member countries, should make it possible not so much to reach a final goal as to establish a point of departure. In other words, it seems to me that what we have to do is simply define a few very explicit guidelines for ensuring orderly conditions on a complex of markets which should remain fundamentally free and which should permit profitable commercial exchanges to the general good of the world economic system.

THE ECONOMICS OF
THE CURRENT STEEL CRISIS
IN OECD MEMBER COUNTRIES

by
Dr. Robert W. CRANDALL
Senior fellow
The Brookings Institution

The casual reader of the steel trade press might be excused for being confused by the current coverage of the steel "crisis" in western capitals. On the one hand, he is told that Europe and the US have excess capacity in outmoded and obsolete facilities. The escape from this problem lies in "rationalization" of capacity through plant closures. But, he is told, less developed countries are subsidizing their steel industries, adding to the short-term pressures on the western economies which have the excess steel capacity. This "unfair" competition must be addressed by the more developed countries if their steel industries are to make the adjustment to less capacity and more efficient operation. If the LDC's are not restrained, capacity declines in the west will be excessive and the next "shortage" will be extremely severe.

The reader is confused. Is the problem too much or too little capacity? How can one talk of excess capacity and shortages at the same time? If shortages are just around the corner, how can the US and European industries continue to consider the retirement of capacity and counsel restraints for the LDC's? This paper attempts to place the current crisis in perspective, arguing that the economic forces which discomfited the European and US steel industries have been forming for two decades. These forces will continue to drive steel capacity away fron Europe and the US unless restrictive trade policies are implemented.

THE ROOTS OF THE "CRISIS"

The present crisis did not begin in 1977 or even in 1975. It started in the late 1950's when ocean transportation costs began to fall and world iron ore prices began a ten year decline. By the 1960's virtually any country in the world with a deepwater port could obtain its basic raw materials at costs competitive with the United States or Western Europe. Moreover, these costs fell for nearly a decade in world markets while US costs continued to rise because of the American industry's reliance upon domestic sources of raw materials. The European experience was between these two extremes.

The fundamental change which occurred in the mid 1950's is best illustrated by the Japanese experience, for Japan seized the opportunity afforded by declining raw materials costs. Its cost of producing steel fell for 11 years — from 1957 through 1968 — because raw materials costs were declining. (See Table 1.) The common belief is that it was ingenuity, hard work, and the installation of ever better equipment which caused the Japanese costs to remain stable for more than a decade. In fact, labor productivity did rise rapidly, offsetting wage increases for part of this period. Labor costs per ton stabilized in the mid 1960's, however, while raw materials costs continued to decline for the rest of the decade. As Table 1 demonstrates, the raw materials cost declines were more than double the decline in labor costs over the period.

During this period, the US and European industries continued an aggressive capital spending program, investing $ 21.7 billon and $ 18.5 billion, respectively, over the 11-year period, 1961–1971. (See Table 2.) The European industry (ECSC-6) added 55 million tons of capacity in the 1961–71 period while the US companies directed most of their investment towards modernization of existing mills. The US industry added only 6 million tons to its 1961 capacity on the 11-year period.

Table 1 JAPANESE AND U.S. COSTS OF FINISHED STEEL: BASIC MATERIALS AND LABORS, 1956-76, (METRIC TONS)

Year	Japanese labor cost per ton	U.S. Labor cost per ton	Japanese material cost per ton	U.S. Basic material cost per ton	Ratio of japanese to U.S. materials cost
1956	26.66	54.67	93.17	56.17	1.66
1957	26.79	60.24	106.42	49.76	2.14
1958	30.12	70.09	68.53	52.10	1.32
1959	25.02	66.67	65.03	47.31	1.37
1960	23.01	71.83	62.07	48.35	1.28
1961	21.94	72.36	69.66	50.14	1.39
1962	24.10	71.36	57.49	47.38	1.21
1963	23.76	69.62	55.27	46.39	1.19
1964	20.97	67.00	54.23	47.97	1.13
1965	22.11	65.06	54.27	47.93	1.13
1966	20.68	65.93	51.18	47.28	1.08
1967	19.93	69.88	49.60	47.82	1.04
1968	20.83	70.35	46.95	49.05	0.96
1969	21.20	75.81	48.73	50.07	0.97
1970	23.22	80.81	54.83	56.42	0.97
1971	27.98	85.03	53.30	60.95	0.87
1972	31.97	89.52	51.59	65.59	0.79
1973	35.32	87.31	65.65	73.90	0.89
1974	42.60	100.91	104.70	114.64	0.91
1975	49.93	132.87	109.33	137.30	0.80
1976	49.64	143.55	112.29	151.10	0.74

Note: Basic materials include only iron ore, scrap, coal, oil, natural gas and electricity.
Source: U.S. Federal Trade Commission, Staff Report on the United States Steel Industry and Its International Rivals: Trends and Factors Determining International Competitiveness, November 1977, Table 3-1.

Table 2 INVESTMENT IN FIXED ASSETS IN THREE AREAS, 1961-71 (BILLIONS OF U.S. DOLLARS)

Millions metric tons

Area	Reported investment 1961-71	Coverage ratio	Estimated investment 1961-71	Capacity in 1961	Capacity in 1971
U.S.	19.5	0.90	21.7	136	142
ECSC (6)	16.5	0.89	18.5	80	135
Japan	12.8	0.81	15.8	30	110

Sources: IISI, Financing Steel Investment, 1961-71, 1974. ECSC, Investment in the Community Coalmining and Iron and Steel Industries, 1977. AISI, Annual Statistical Reports. U.S. Council on Wage and Price Stability, Report on Prices and Costs in the United States Steel Industry.

Surprisingly, the Japanese industry was able to expand its capacity by 80 million tons at a total outlay of less than the ECSC or the US capital expenditures during this 1961–71 period. Even if no expenditures were required for maintenance or modernization of older facilities, the Japanese cost per metric ton of raw steel capacity was less than $ 200 during the period. The US industry (principally Bethlehem) required at least $ 350 per new ton of capacity during the same period. (See Table 3.) The ECSC experience was undoubtedly between these two extremes, but it is clear that the low-wage country, Japan, enjoyed the lowest costs of building steel capacity. This advantage has persisted to the present although locational constraints and increasing labor rates have narrowed the advantage considerably.

Why did Japan succeed so spectacularly in the 1960's? The explanation lies less in the form of industrial organisazation in Japan or the close relationship between government, the financial sector, and the steel producers than in the clear comparative advantage which developed for low-wage countries. Raw materials supplies were no longer a constraint upon LDC development of steel. Technology was available from Europe and the US, and wage costs were dramatically lower than in Europe or the US. Wage rates which were as little as 15 or 20 per cent of US wages provided the Japanese with an enormous advantage given that US firms were paying more than 60 per cent of value added to labor. Steel production grew at a rate of 14 per cent per year through the 1960's in Japan. In the 1970's, the Japanese began to slow their expansion, but only because their export markets began to recede with declining world demand growth and political cries for protection in the US and Europe.

Table 3   ESTIMATED INVESTMENT PER TON OF CAPACITY, 1961-71

$/Metric ton

Area	Investment (billions of U.S. $)	Capacity 1961	New capacity 1961-71	Estimated investment cost per ton for:	
				New capacity	Maintaining old capacity
U.S.	21.7	136	6	350	13
ECSC (6)	18.5	80	55	210	8
Japan	15.8	30	80	175	5

Sources: See Table 2.

How did the EEC and US producers respond to these fundamental changes in world steel economics? The US industry, very much responsive to private investors' willingness to subscribe capital, was forced to pull back. Expansion was limited to only one major new facility and no more than 6 million new raw tons. The European industry, on the other hand, expanded at a rather substantial pace even into the 1970's. Its annual rate of growth was approximately 5 per cent, and it has almost maintained this rate through 1978 as Table 4 demonstrates. At the same time growth in production in the US and the EEC was modest as their home markets slowed appreciably. In the 1970's, steel consumption has grown at less than 2 per cent per year in the US and has actually declined in the EEC. Exports from the US are an inconsequantial share of production, but the EEC continues to export from its overbuilt industry.

The reluctance of the US producers to close inefficient facilities and the continued expansion of the European industry (until recently) have undoubtedly impeded investment in a number of less developed countries and in Japan. Had US and European capacity grown more slowly since 1960, steel investment in Latin

America and Asia would have preceded at a more rapid pace in recent years. Depsite the obvious overhand of excess capacity, the growth rates in production in these latter two areas have remained in the 7 to 11 per cent range (Table 4). As the US and Europeans rationalize their capacity, rational investors will be sure to increase the rate of growth of steel capacity in the LDC's and perhaps even Japan.

Table 4   TRENDS IN CAPACITY AND
PRODUCTION BY REGION, 1960-1978
(MILLIONS OF METRIC TONS)

Year	Capacity		
	U.S.	ECSC (6)	Japan
1960	135	76	25
1970	141	127	104
1978	143	174(e)	144(e)
Annual growth rate 1960-78	0.3%	4.6%	9.7%

Year	Production						
	World	U.S.	ECSC (6)	EEC (9)	Japan	Latin America	Other Asia (excluding North Korea and China)
1960	345.5	90.1	72.8	97.9	22.1	4.9	3.4
1970	595.4	119.3	99.1	137.6	93.3	13.2	7.9
1978	712.8	124.0	111.3	132.6	102.1	24.3	19.5
Annual growth rate 1960-78	4.0%	1.8%	2.4%	1.7%	8.5%	8.9%	9.7%
Annual growth rate 1970-78	2.2%	0.5%	1.5%	−0.5%	1.1%	7.6%	11.3%

Source: OECD; U.S. Council on Wage and Price Stability; E.C.S.C.
e = estimate.

PERSPECTIVE ON THE CURRENT CRISIS

There can be little doubt that part of the problem faced by steel producers in the developed western world derives from the slow growth in steel consumption since 1970. In the 1970's world steel consumption has increased at less than one-half of its 1960's rate. Just as important, however, is the fact that the steel producing assets are in the wrong locations. European and US steel producers would be facing difficulties even if world capacity were much more in tune with current steel demand. They simply would find it increasingly difficult to compete with the emerging steel producers and Japan.

In both Europe and the US, inefficient, old plants have not been closed as more modern capacity has come into production. The United States is in a much better position to face the 1980's than the EEC because its efficient mills are located at inland locations, close to raw materials and markets, but far from the competition.

The more efficient (or potentially more efficient) EEC plants are located at coastal sites — Dunkirk, Ijmuiden, Fos, Taranto, Bremen, or Redcar. These plants will have to compete with imports from Japan and the developing nations in the next few decades, a difficult task given the difference in wage rates between the EEC and the developing world. These coastal plants were built to take advantage of the burgeoning world trade in raw materials and declining shipping costs, but such locations only provide them parity with LDC facilities in purchasing basic raw materials. Unless the capital charges for new LDC of Japanese plants are high, the EEC plants will find it difficult to compete with the emerging steel producers and with Japan.

Because of the politics of trade negotiations, much attention has been given to the differences in the average cost of steel production in various regions of the world. It is often asserted that the Japanese industry has the lowest unit costs, that the US industry has at least a 10 to 15 per cent disadvantage, and that the EEC industry has even higher costs than the US. These comparisons are useful only in the political arena; they have no economic content. Average costs reflect a diverse mix of plants and operating rates. The important questions are: (*i*) What is the incremental cost of producing steel from existing plants? and (*ii*) How much does it cost to produce steel from an optimally-designed new plant? In the long run, the answer to the last question will provide the basis for the determination of world prices and the location of production unless governments intervene in a wholesale fashion. It might be useful, therefore, to compare the production costs of an efficient existing mill in the US or the EEC with the costs of new facilities in various parts of the world.

Current costs

For the most efficient flat-rolled, carbon steel plants in the US, Germany, or the Benelux countries, I assume that 8.8 manhours of labor are required per metric ton of finished products. This might be termed "best practice", reflecting the attainable productivity in the newer, integrated carbon steel mills at 90 per cent capacity. In addition, I assume that the EEC and US face the same world materials

Table 5 COST OF PRODUCTION
IN EFFICIENT EXISTING MILLS
AT 90 PERCENT CAPACITY
UTILIZATION
(1978 U.S. $ PER METRIC TON)

	U.S.	Japan	Germany, Benelux
Labor	114	74	108
Raw materials and miscellaneous	220	200	230
Capital maintenance	30	30	30
Total	364	304	368
Manhours per ton	8.8	8.25	9.0
Total employment cost per manhour	$14.00	$9.00	$12.00

Source: Author's estimate from forthcoming study.

144

prices, but the EEC plants suffer a slight locational disadvantage of $ 10 per metric ton. In 1978 US$, these costs are approximately $ 220 per metric ton of finished flat-rolled products. Finally, I assume that these mills can be kept operating at relatively constant efficiency for an annual investment of $ 30 per metric ton. Thus, in 1978 dollars the cost of a mix of flat-rolled carbon steel products ranges from $ 360 to $ 370 per ton in existing efficient US and European mills. (See Table 5.) These estimates are based upon the assumption of a high operating rate — which is no more than the assumption of economic rationality. The efficient mills should be kept at full operation, even in Europe, when demand falls. Efficient and inefficient mills may be cut back equally during the crisis for political reasons, but this is not a sensible economic policy.

Existing capacity in Japan can produce steel at a substantially lower cost than German, Benelux, or US plants because of greater labor effiency, better yields, more continuous casting, and lower wage rates. The cost of Japanese "best practice" is approximately $ 304 per finished ton, or at least $ 50 per ton better than best US or EEC practice.

Note that my measure of costs from existing plants excludes any measure of depreciation, interest, or profit. This is deliberate because I am focusing upon the competitive positions of steel producers and their ability to supply a world market. Embedded capital costs are simply irrevelant in such an analysis because they do not reflect payments for resources required to maintain production. In the case of European and US steel investments, they reflect charges against original costs of assets which are considerably overvalued in the accounts of the producers. Depreciation, interest, and profit on overvalued assets are not economic "costs" of production but rather a nostalgic reflection of what might have been. Markets, and eventually the bankruptcy courts, ignore this nostalgia as do governments which nationalize steel assets.

New plants

New integrated steel mills producing flat-rolled products can be built in the US and EEC countries at relatively similar costs of about $ 1050 per net finished ton. In Eastern Asia, plants can be built for perhaps 10 to 15 per cent less; in South America, the cost is about 10 per cent higher. Raw materials costs should be approximately equal for coastal facilities around the world. Labor costs, however,

Table 6   PRODUCTION COSTS IN NEW STEEL
MILLS (1978 U.S. $ PER METRIC TON)

	U.S. and ECSC	Japan	Other Asia	Latin America
Labor	108	69	22	44
Raw materials	190	190	190	190
Capital costs	178	152	168	209
Total	476	411	380	443
Assumptions:				
Manhours/ton	7.7	7.7	11	11
Wage rate ($/hr)	14	9	2	4
Cost of facilities/ finished ton ($)	937.50	800	800	1,000
Capital charge	.172	.172	.190	.190

Source: Author's estimates from forthcoming study.

are much lower in less developed countries. I assume that Latin American employment costs are at most $ 4 per hour (1978 US $), eastern (non-Japan) Asian labor costs are $ 2 per hour, and Japanese labor costs are $ 9 per hour, and US and ECSC costs are $ 12 to $ 14 per hour.

A new integrated plant should be able to generate an improvement in labor productivity to about 7.7 manhours per ton for a full mix of flat-rolled products. Productivity in the LDC's might be somewhat lower, even with the same equipment. Interestingly, there is little apparent substitution of labor for capital available in integrated steel-plant construction. The only major substitution is direct reduction for the blast furnace-BOF combination. The risk of building in Latin America or eastern Asia might be marginally greater than in the US or Europe. I assume that this raises the annual "capital charge" by .018 from .172 in Europe or the US. The final result, shown in Table 6, is that Japanese and other Asian plants continue to enjoy an advantage of 15 to 20 per cent over new plants in the US or Europe. Even Latin America plants are less expensive to operate with their assumed lower labor productivity and higher capital costs. If Brazil, Mexico, or Venezuela prove able to build plants at costs comparable to the US and if their productivity approaches US or Japanese levels we can expect the gap to widen from the $ 33 advantage to perhaps $ 60 or $ 70 per finished ton.

If the U.K. can solve its labor problems, operate new basic steel facilities efficiently, and round out its finishing investments, it is particularly well positioned for growth given its low wage rate. Obviously, it has not been able to achieve such results, and it is too early to predict the outcome of the current attempt to revitalize British Steel.

The marginal costs of production from the more efficient current plants and the average cost of producing in new plants will drive world prices in future years unless governments impede the operation of market forces. The above analysis suggests that carbon-steel prices (using US product weights) can fall to approximately $ 365 in 1978 prices before efficient US and EEC producers will begin to allow their facilities to deteriorate. More importantly, western prices are unlikely to rise much above $ 450 per metric ton (including importation charges from Japan or other Asian locations) in the long run if Japanese and other Asian producers are able to continue to expand capacity. At $ 475 per metric ton, even a pessimistic reading of the economics of steel production will allow the Latin American producers to become net exporters in time. But expansion of European or US capacity seems unlikely in the foreseeable future.

If $ 450 per metric ton (in 1978 US dollars) appears to be the long-run equilibrium price in western markets, there would appear to be limited prospects for price increases in the long term in US markets since 1978 flat-rolled prices averaged about $ 430 in 1978. The potential increases in Europe would be somewhat greater, given the discounting which has occurred, but insufficient to allow the industry its growth rate of the part two decades.

RECOVERY FROM THE CRISIS

Up to this point, I have ignored the question of world "supply-demand balance" because I believe that the current problems faced by the US and European steel industries are not simply the result of the slow recovery from the 1975 recession. The US industry has enjoyed near-capacity operation since January 1978 and a substantial modicum of trade protection, but it is not able to earn a rate of return which is more than one half of that available to the average US industrial corporation. The European industry would do no better than its American counterpart even if capacity utilization were 90 per cent.

This is not to say that the current excess capacity in the world industry is without its effects. If demand were 10 per cent higher, the world industry would be close to capacity operation, and world prices would be 6 to 8 per cent higher in normal circumstances. However, the combination of the EEC-Japanese agreement, the US import policy, and the Davignon Plan make this period rather atypical. These cartel-like arrangements have raised world prices (by 8 per cent to the US) so that a resurgence of demand might have only minimal impacts upon prices.

I conclude, therefore, that the current crisis is a product of an imbalance of supply and demand and a reflection of a fundamental shift in the economics of world steel production favoring LDC's. The US industry has slowly come to recognize, if not admit, this shift. There has been no capacity growth in the integrated sector since 1969. The European industry, on the other hand, was far less deterred by unfavorable market circumstances. Its expansion and refusal to close old plants has placed it in difficulties from which it could not be easily extricated even if world capacity were fully utilized. With time, the competition from LDC's, Japan, Eastern Europe and Western-European electric furnaces would so reduce profit margins at many European production points as to assure a perpetual crisis. The economics of existing mills and new mills, detailed above, compels this result.

Retirement of some capacity and cartel-like agreements may rescue some EEC and American mills for a few years, but within a few decades the pressures of competition will make these rescue efforts increasingly costly. These pressures, however, will develop slowly. No one in the less developed world appears ready to repeat the Japanese miracle of 1960–1974 for two reasons: *(i)* there are no large export markets for steel which are open to the extent the US market was open in 1960 and *(ii)* it is unlikely that any of the newer steel producers in the less developed world can overcome the resource and management constraints to rapid growth. Moreover, the Japanese were already net exporters when they had but 20 million tons of capacity. Latin America and Asia are still net importers of steel although their degree of self-reliance is increasing rapidly as Table 7 demonstrates. Despite these constraints and a decidedly more sluggish world economy in the 1970's, these two areas have enjoyed steel production growth of 9 per cent per year since 1970. (See Table 4.) It would not be surprising to see them become self sufficient in the next 15 years and to be net exporters to the developed world thereafter. Obviously, a hostile trade policy in Europe or the US could slow this progress, but only at a cost of 10 to 20 per cent in the cost of steel to steel consumers in these developed areas. We would then expect the growth of LDC's to be based upon internal demand as they invested in steel fabrication and further eroded the US and European durable goods' producers markets.

Table 7 NET IMPORTS AS A PERCENTAGE OF STEEL CON-
SUMPTION, 1970-78

(– = net exporter; + = net importer.)

Year	U.S.	EEC	Japan	Latin America	Other Asia (Excluding China + North Korea)
1970	6	– 11%	– 32%	28%	54%
1974	8	– 29	– 52	42	54
1978	14	– 29	– 54	26	37

Source: OECD

No one can predict the precise shift in steel-making capacity from the developed to the less-developed world. The CIA has estimated that the LDC share of non-Communist world production will rise to 15 per cent by 1985, a substantial increase over its current 10 per cent share. This is not a very hazardous prediction given the lead times in building steel plants. Looking farther into the future is more hazardous because it depends importantly upon the willingness of the developed world to subsidize its steel industry. LDC capacity might expand to a level at which it is capable of supplying most of its own demand for basic steel products by the early 1990's, but further expansion would require exports to Europe or North America. If these markets are closed by governments which are attempting to protect their aging steel producers, LDC growth will be restrained, but only after at least 10 or 15 years of substantial expansion.

There is little doubt that the EEC will not be able to maintain its net exporting position at its 1970's level. It has been shipping more than one-third of its exports to the developing world and almost one-third to North America. The growth of LDC steel capacity and a hostile trade policy in the United States will not permit the EEC to maintain these exports. As a result, it would not be surprising to see as much as a 10 per cent reduction in EEC capacity over the next decade. A resurgence in world demand might slow this contraction, but the induced effect upon LDC capacity will only further imperil the aging, marginal plants in Europe.

In the United States, a more modest reduction in capacity is likely. An aggressive mini-mill sector, fed by abundant scrap, will continue to grow while the integrated sector contracts slowly. Plants in Western Pennsylvania, California, Utah, and the Southeast will either close or be reduced in scale. Some new capacity (but not greenfield plants) will be built on the Great Lakes. The overall prospect is for a very modest reduction in total steel capacity as US steel consumption grows at a very slow rate.

Japan continues to be a good location for steel assets. If it can overcome trade barriers in the US and Europe, it could continue modest growth in the 1980's. A major uncertainty in the Japanese future is the role of China. If China increases its steel capacity in the 1980's and 1990's at a rapid rate, Japan's ability to increase its exports in its major market — Southeast Asia — will be impaired in the future. Nevertheless, Japan will be able to compete with most LDC's in the export market and is therefore in a much better position than the steel producers in the US or the EEC.

In conclusion, I must stress that the current crisis can be ameliorated through a strong recovery of world demand, capacity reductions in Europe and the US or both, but demand growth alone will not solve the long-run problem for the developed western countries' steel sectors. The choice for the United States and Europe is a reduced role for their steel industries, trade protection, or other subsidies which will allow them to fend off the competition from the less-developed countries and Japan.

II

THE POLICY RESPONSES TO THE PROBLEMS OF THE WORLD STEEL INDUSTRY:

A. POLICIES FOR ADJUSTMENT MODERNISATION AND ADAPTATION OF THE STEEL INDUSTRY IN THE LIGHT OF EXPECTED WORLD DEVELOPMENTS

B. THE PROBLEMS OF THE LABOUR FORCE AND READAPTATION AND RE-EMPLOYMENT POLICIES

CHAIRMAN MODERATOR: Viscount E. Davignon, Member, Commission of the European Communities responsible for Internal Market and Industrial Affairs.

I should like to open the discussions on the second day of our Symposium by saying that today it will be difficult to reach the same standard as achieved in yesterday's discussions, because there has been a change of chairman and today's chairman is of definitely inferior quality. But this is the hazard of this type of meeting, and I don't think that Mr Hodges should fail to realise that the best part of the meeting was yesterday, not today. But, I am only referring to the chairman and not to the other speakers, of course.

We have a simple matter to deal with. We shall be discussing our reactions to the crisis. To prepare for our discussion, we have a list of most eminent speakers who will present the various topics...

I would like to begin the discussion by voicing a few thoughts on our attitude towards the situation from the point of view of Brussels.

I think this is a perfectly natural way of proceeding since, out of all the steel producing regions, it is the European Economic Community which has been affected most severely by the crisis.

I think it is important to underline this by means of a few figures.

First, between 1974 and 1979, sixteen per cent of steel jobs left the industry. Sixteen per cent is a high figure, but it is a figure and therefore is abstract. When it is considered, however, that this percentage corresponds to between 120,000 and 125,000 persons, the whole picture looks very different. If we add to this fact the reminder that since 1975 approximately half the drop in world production has been borne by the Community whereas its share of world production is only 22 per cent, then we have another figure which indicates the impact of the crisis at the Community level.

Of course, we must compare these basic figures with another basic element, namely, the steel industry is for the European Community a strategic industry, with the following twofold consequences:

a)   we cannot abandon our steel industry unless we are willing to increase our level of dependence on the outside world, a dependence which is already considerable with regard to raw materials, commodities and energy;

b)   if we have made the choice of keeping our steel industry, and if the industry is not to be a burden on our countries' economies, our steel industry should be adaptive and competitive and should be able to exist on the basis of its own strength.

This is the explanation that the European Community has accepted in deciding to pursue a policy aimed at restructuring its steel industry. The objective is to make the steel industry sufficiently competitive at reduced production capacities to allow it to continue to exist in a world where trade remains open and free, a situation which maintains the Community as a net exporter.

In the report which Mr Florkoski will be introducing very soon, he referred — and rightly so — to one question which is the most important of all, namely, should restructuring take place in the framework of a long-term policy of restructuring of all industries, or should we plan for an approach which is specific to the steel sector?

This question of principle has not arisen for the European Community because the Treaty of Paris established the European Coal and Steel Community. The Community has a special legal and political status which means that the European institutions have to act with regard to their steel industries. The choice was made by those who prepared the Treaty in 1954. Instead of using the legal instruments which are provided at some length by the Treaty, the European Community and the Commission considered that a basic restructuring policy could be effectively implemented only if there was real consensus among industry, trade unions and users within the Community as well as agreement with trading partners outside the Community. In other words — and I shall stress this later on — any unilateral action from whatever sources and for whatever reason is absolutely contrary to the basic objective which is to achieve harmonious restructuring of steel industries in industrialised countries so that the world steel industry can find its rightful place.

To implement this restructuring policy, I should like to say that a number of recommendations to which Mr Florkoski refers in his document, particularly on page 157, are very similar to our actual practices.

On the first point, our approach is tripartite with regard to long-term steel policies. The Treaty of Paris provided for an advisory committee of the ECSC, and I am happy to see the Chairman of that committee, Mr Judith, among us today. The Consultative Committee includes steel producers, steel users and trade unions.

One of the tasks of this Committee is to give its opinion on the General Objectives which the Commission should establish for the steel sector. In this way we have defined the General Objectives for the Community steel industry up to 1985, General Objectives which were subsequently presented to the Council of Ministers of the Community. These objectives indicate the context in which companies and governments — insofar as they are involved in companies — should define their strategic options taking due account of the capacity and productivity levels which the Community should achieve as a whole if the Community steel industry is to remain competitive, and I mention competitiveness as one of our objectives.

Second point: Mr Florkoski recommends that firms should be given the necessary freedom to make the short-term adjustments required by the market situation.

The Community takes the firm view that enterprises are primarily responsible for designing and implementing restructuring programmes. However, the Commission's anti-crisis plan provides for solidarity among Community enterprises so that we can be certain that this enormous restructuring effort goes forward under conditions which are acceptable to the Community as a whole, acceptable to workers and acceptable to certain regions that are particularly affected by the steel crisis.

The one purpose of these provisions relating to the market — and I want this to be unambiguous — is to allow the steel industry to be restructured. Their purpose is not to permit firms or governments to elude their obligation to make the necessary adjustments, necessary because we must have a steel industry which is competitive in the long-term. No one in Europe believes that maintenance of the *status quo* will solve the problem. Everyone is aware that the golden rule for industry is that it should be able to adapt. Any measure which puts a brake on adaptation, whether this measure comes from the inside or the outside, can lead only to confrontation among the major steel producers, the result being to delay adjustment and to make adjustment even more difficult for everyone concerned.

The third point on which we agree with Mr Florkoski is that no steel market, no matter how vast, can define its own policy without taking due account of other markets. It is for this reason that at the beginning of this crisis, and in concert with our partners mainly in the United States and Japan, we tried to ensure that within the OECD there could be established a forum for discussing steel problems. In our view it seemed essential to have a meeting place at our disposal assisted by the excellent work done by the Secretary General of the OECD. I wish now to thank Mr van Lennep and Mr Wootton for all the help they have given us. This enables us to focus on the real problems.

It is amazing that in the industrialised world to which we belong we have had to wait until 1977 and 1978 before being sure of finding a suitable forum where the pressing problems of the steel industry could be examined in depth and in a spirit of solidarity in accord with the interdependence of our economies.

But, better late than never, the Steel Committee was established. Interesting progress has been made at the level of the actual analysis of restructuring efforts made by the steel industry so that we can verify in fact whether this industrial adjustment which allows greater harmony on world steel markets is being achieved and whether the various temporary measures taken help us to meet our objectives.

Here at the OECD every country has opportunity to explain its position to the others, to convince them of the existence of transparency and of the balance between the instruments used and objectives attained.

I think that this is a good definition of international cooperation in the economic field.

*
* *

I would now like to make a brief comparison of the 1977 and 1980 situations from the Community's point of view.

In 1977 the restructuring programme was just an idea. It was based on the General Objectives and supported by major programmes in the main steel producing countries of the Community. Restructuring the steel sector in France, Belgium, Luxembourg, some regions of the Federal Republic of Germany, and the United Kingdom is no longer a matter of theoretical conjecture. Restructuring consists of precise and concrete steps which have led to a change in instruments, a change in the quantitative targets which the various enterprises had established, and in a series of shut-downs, the unfortunately inevitable consequence of this adaptation policy.

Today I should like to state very firmly that no one can tell Europe it is not restructuring its industries. Today no one can tell Europe it has an outdated industry or that nothing is being done to modernise it. The figures are there. None of these can be disputed, whether they concern figures on capacities, job losses, or investment to achieve rationalisation.

Alongside Japan's ongoing efforts to adjust its industry (this has meant reducing its capacity by one third compared to initial proposals) and alongside major efforts made in Sweden, for example, it is the European Community where efforts to adapt are the greatest. This is confirmed in a table set out in one of the reports, showing that capacity in the Community is being reduced by 2 to 3 per cent whereas, for example, in a major North American market — it could well be that of the United States — a 4 per cent increase in production capacity is being planned for 1985. I would like this to be taken into account when we consider the efforts made by various countries and areas.

In order to make the restructuring policy a success, the European Community has drawn up supporting social policies for retraining workers, for early retirement, for European Community aid so as to spread out the effects of job losses on workers' incomes over a long period, and for a large-scale programme to create alternative employment and to train workers for new employment. The problem is a serious one. It involves some of the vital responsibilities of our countries in terms of social measures which are indispensable.

It is often said that the social policy of the authorities with regard to the steel sector has been disorganised and contradictory, and has led to a distortion of competition and has altered the nature of the market. Since the end of last year, the Community has at its disposal an instrument aimed at monitoring public assistance measures to insure that these are temporary and will not result in maintaining the status quo but in adjusting the industry as required by world competition. This is an element of clarity and transparency which we contribute in the hope that we can overcome the difficulties that confront us.

This presupposes that each one of us will resist the temptation to become protectionist. I should like to stress with the utmost vigour that any action in the steel industry which would lead to establishment of artificial protectionism in a market, as compared with conditions of normal competition of exports, at a time when efforts to restructure the steel industry are as major as I have mentioned, runs several risks:

1. This may put in question the restructuring efforts, since the basic conditions will be adversely affected; and
2. What would be considered as true for one market may necessarily be regarded as true for all markets.

From my standpoint, this is a word of warning on which I have duly reflected: if we have a situation of protectionism or a trade war in the steel sector, this will soon be followed by a similar situation in the car industry and after this in shipbuilding and, finally, in all áreas of advanced technology. This would mean that in less than one year after conclusion of the Tokyo Round negotiations, which gave us new instruments for the organisation of trade, we will have affected 60 per cent of the trade covered by the negotiations. Each country would justify its own position saying that it had not been the first to act in this way.

The European Community will not practice this policy. The European Community has so far refused to practice a policy of this sort by concluding arrangements with third countries in order to ensure that commercial transactions take place under the optimum price conditions. We have tried to replace price conditions arising out of the anarchy caused by the steel crisis with more reasonable and satisfactory conditions. Naturally, however, the European Community expects that its other partners, both public and private, will behave in the same way towards the Community as the Community behaves towards them.

Mr. Secretary-General, ladies and gentlemen, the action undertaken by the European Community in 1977 and to be continued for several more years is the most ambitious step ever undertaken by the European Community in an attempt to adjust the conditions of social equity, clarity, transparency and competitiveness of a basic industry. The effort is a painful one. The Community does not expect any congratulations or gratitude from its partners. But, the Community does expect its partners to acknowledge the objective effort that is being made and to work with the Community to enable the world steel industry as a whole to recover normal growth and profitability. This is vital for the economies of our various countries. This is vital for the people of our various countries.

Today's discussions will centre on the extent to which the various measures taken are adequate and valid and enable us to meet our objective. As far as we are concerned, we shall continue to pursue this policy. My invitation as Member of the Commission of the European Communities and as Chairman of this session is as follows: may we all work together to attain this goal which is essential and fundamental, and which ensures that we continue to live in a commercial context obeying rules which are acceptable to all and not governed by unilateral rules dictated by the respective force of those who are involved.

The Community is not without importance or power, but it has decided that the rule of law is more important than the use of its unquestionable power.

Thank you.

CHAIRMAN:

I should now like to ask Mr. Florkoski, who has just presented an extremely interesting paper, to outline its main points so that we can start the discussion.

POLICY RESPONSES FOR
THE WORLD STEEL INDUSTRY IN THE 1980s

by
Edward S. Florkoski, Jr.
OECD Consultant

SUMMARY

The 1980s presage a period of immense uncertainty and economic tension for steel. Government, industry, labor and consumer interests are urged to adopt policies which will facilitate steel industry adjustment. Innovative — even bold — solutions must be developed for five key issue areas.

First, *restructuring* in the dual sense of capacity alignment and economic adjustment needs to be facilitated and fostered. Governments must allow industry more latitude in several areas, including price adjustment. Policymakers must reflect on how to preserve the legitimate interests of Government while allowing industry managements the freedom to make decisions affecting their essential interests.

Second, if the steel industry is to become dynamic, it follows that *labor* will have to become dynamic.

Third, *developing countries* as a group pose little threat to the steel industries of developed nations. Amongst other issues, industrialized countries must seriously consider removal of tariff and non-tariff obstacles to steel exports from developing countries.

Fourth, it is unrealistic to expect the international trading system to ever become devoid of some form of restraints on steel trade. Therefore, it is more reasonable to permit countries to adopt trade policies which have the effect of ameliorating *market disruption* and, in the period ahead, have the added effect of facilitating the adjustments and transitions to more modern, competitive industries.

Fifth, the 1980s are likely to witness efforts to more clearly define the role of *state involvement* in steel industry affairs, with a bias towards allowing managers of state enterprises more autonomy. A key policy issue will focus on the fairness of competition between government-funded enterprises and those dependent on the private capital market.

Finally, the policies and solutions of steel will have to become global. Several international institutional improvements must be made, including bringing developing countries into the policymaking arena and examining ways of handling trade disruption.

Introduction

The purpose of this paper is to serve as a framework for the discussion of policies governing the steel industry and its labour force in the period ahead. As a framework document, there is no attempt to exhaustively treat the subject matter, nor to comment on all of the issues and sub-issues that legitimately influence policy.

Which issues will be critical in the 1980s is a matter of subjective judgement. If this review serves no further purpose than to stimulate constructive thought and debate on that subject alone, it will have served its purpose. But, hopefully, it will do more.

As the world steel industry moves into the 1980s, it will be challenged to respond more effectively to shifts in economic forces and circumstances. Much of what lies ahead is not encouraging from the standpoint of the complexity and uncertainty of issues that beset economic forecasters.

There are basically two options. First, governments as well as industry and labour can continue to engage in topical reactions to the problems of steel, responding on a day-to-day basis. Or, they can adopt a more enlightened approach by identifying key issues and thereby setting policies in motion to help overcome or adjust to what otherwise will become insurmountable problems with a real potential for fueling international conflict.

The intent of this review is to urge national governments, their industry, labour and consumer interests to encourage adoption of policies which facilitate the adjustment process.

Unfortunately, policymakers as well as many academicians labour under the characterisation of the steel industry as a mature or declining industry. This mistaken attitude has tended to reinforce protectionist policies and to thwart the readjustment process.

It is incumbent upon policymakers in the public and private sectors to understand that the steel industry is one of the most important sectors of any industrial economy and is therefore still dynamic. Governmental policies should be adopted which foster, not stifle, the dynamic forces within the industry. That is the challenge of the 1980s.

Economic trends entering the 1980s: An overview

It would be unrealistic and even deceptive to characterise the next decade as *La Belle Epoque,* a term used by Herman Kahn a leading futurist to describe the period recently ended. Instead, Kahn identifies the period 1974 through 2000, and perhaps beyond, as *L'epoque de Malaise.* He defines malaise as " 'not feeling well' i.e., one is not quite sick but not really well either. (However, one who has malaise may be much more prone to serious sickness.)"

This definition is an apt description for the outlook faced by the world steel industry in general, although the scenario differs from country to country. The industry faces the prospect of considerable instability and uncertainty.

Inflation is a plague that imposes a wide range of costs and distortions on all economic activity. Steel is particularly sensitive to inflation in that it is "squeezed" between rising costs and government pressures to contain price increases.

The short-term preoccupation with keeping prices down has contributed in large part to the long-term consequence: the world steel industry is cash short. It cannot pay its mounting bills for energy and other operating costs, let alone modernisation or expansion. The capital shortage affects the steel industry of both the industrialised as well as developing countries. Lack of capital in turn creates pressures for government financing and with it the conditions of public policy that attach as a price for such funding.

Inflation also neccesitates that trade unions protect their membership by cost-of-living linked increases to offset the uncertainties of the future.

Energy supplies are uncertain. Costs are increasing. Consequently, pricing strategies remain unsettled. The United States industry, for example, has energy costs equalling about 20 per cent of production costs. Worldwide, the steel industry accounts for approximately 7 per cent of world energy consumption.

Raw material availability is becoming more subject to disruption than in the past. One analysis supports a conclusion that "the world is in little danger of approaching an absolute physical limit on resource development in the foreseeable future. Increased commodity supplies will continue to be available at some price. What that price will be is the subject of uncertainty and debate."

Adding to the industry woes is the prospect for adoption of deflationary monetary and fiscal policies that will slow economic growth in general and growth of steel consumption in particular. By most assessments, the period of malaise will be a slow-growth period.

Slow-growth will in turn accentuate the problems of excess steel capacity. Developed countries will vie more aggressively for their own and each other's home markets. Increased competition is likely to arise in export markets as the developing countries actively pursue policies to satisfy their own steel requirements, while moving their excess into the world export market.

The developments presage a period of immense uncertainty in the affairs of steel. Coupled with increasing concern for national security, countries will tend towards retrenchment, preservation of existing operations, underwriting of losses from operations, and — regretfully — protectionism.

It is against this background that the international steel community must come to grips with the decade ahead. If left unchecked, policy will drift into becoming more insular. Fear of the unknown, mistrust, and the strong interest to protect market share will deter a positive adjustment process from occurring.

What then is needed? Basically, four sets of policies are at stake.

— First, government — with the co-operation of industry and labour — must establish long term policies for a framework within which the steel industry can plan its future development.
— Second, governments must set in motion domestic policies which will permit their domestic steel industries to freeely make the adjustments in the shorter term that are called for by the marketplace.
— Third, governments, as a matter of policy, must define their attitudes towards trade measures that are designed to deal with market disruption, and the way they are linked to appropriate adjustment programmes; and
— Fourth, governements must continue to undertake co-operative efforts on an international basis to insure that the interaction of national policies does not seriously distort trade flows and thereby impair progress towards international trade liberalisation.

Unless all four sets of policies operate in tandem, steel industry problems are bound to regularly reoccur and to threaten trade confrontations on a large scale basis.

Key policy issues

Looking ahead into the 1980s, what issues are likely to cause the most controversy and call for anticipatory policy action?

Needless to say, one can identify a panoply of issues and sub-issues. Basically, the issues of the 1980s are not vastly different from those which the industry began facing in the mid-1970s. They fall into five categories, which overlap onto each other:

156

- Restructuring and modernisation of industry including financing
- Employment
- Emergence of steel production in developing countries
- Trade and market disruption
- Government involvement

The passage of time makes more apparent that many of the responses of the 1960s and 1970s to these five issue areas will clearly not pass the severity of the test that lies ahead for the next decade. Government, industry and labour must co-operate to find better approaches, more innovative — even bold — solutions to the issues. Without such creativity, the problems of steel will continue to thwart domestic industrial policy and will of necessity inhibit progress towards trade liberalisation.

Restructuring and modernisation of industry, including financing

Given the excess of steelmaking capacity that currently exists in the industrialised countries, the trend towards bringing capacity more in line with steel demand will continue well into the 1980s. This is not an easy process. It involves political, economic and social sensitivities. It involves an ingrained resistance on the part of producers to admit that they misjudged the market and have failed to adjust production accordingly.

"Restructuring" in a broad sense embraces a wide range of steel industry activities throughout the world. The adjustment process differs from country-to-country and in degree. It is judged to be more successful in some countries than in others.

In Europe restructuring is being approached as modernisation and the closing of facilities to bring capacity in line with lower demand for certain products. This is generally referred to as rationalisation of production.

In the United States, the objective is somewhat different. While there are plant closings, the principal aim is to modernise and expand capacity in the years ahead.

Japan is engaged in a modernisation programme that is not aimed at expanding productive capacity, but at improving the quality of products and production efficiency.

Restructuring should be viewed as a response not only to excess capacity but essentially as a response to changing economic conditions. As an example, a number of factors will affect the energy consumption of the industry in the 1980s and beyond, including:

- Total steel output;
- Process used to produce the output, e.g. the trend to non-integrated electric furnace production;
- Availability of higher grade low volatile coking coal;
- Adequacy of the supply of coke, given the impact of environmental control costs on coke production operations;
- Increased use of direct reduction and continuous casting; and
- Use of atomic power in steelmaking.

This range of variables demands in turn maximum flexibility on the part of steel industry managements to adopt new technologies or otherwise adjust their manufacturing processes to cope with the energy situation.

Similarly, industry should be free to offset the scarcity of or increased costs of certain raw materials by use of substitute materials of modified production precesses.

Pollution abatement is a generally desirable social objective but, in the case of steel, with significant economic consequences. Clearly, steel industry management must be afforded adequate opportunity to "restructure" or re-gear to accommodate this legitimate social goal — assuming that the standards are reasonable.

Restructuring in the dual sense of capacity alignment and economic adjustment needs to be further facilitated and fostered. Only if industry has the freeedom to adjust to changing economic realities will it feel less threatened by what might otherwise be construed as bleak prospects or dire uncertainties confronting the 1980s.

Areas where restructuring latitude is required and where governmental pressures have been most pronounced include:

— Price adjustments;
— Health/safety/environment standards;
— Plant closings;
— Employment requirements and wage settlements.

Policymakers should not be surprised if, as a result of persistent governmental restraints on price adjustments, an industry finds itself unable to afford the updated technology needed to compete effectively either at home or in the overseas market. The eventual consequence is an outdated industry. The further consequence is to either *(a)* protect the outdated industry from harmful competition, or *(b)* give no protection at all to the industry on the basis that it has seen its prime and is therefore unworthy of further public resources — a *laisser faire* attitude.

Restructuring and modernisation of steel industries is a costly undertaking, more costly than many industries and even some governments can afford. Financing is therefore one of the major problems confronting the steel industries in industrialised and in developing countries.

The American steel industry estimates its total investment needs for modernisation and expansion at $7 billion per year.

Japanese carbon steel producers for the fiscal year ending march 1980 will spend close to $2.5 billion on plant and equipment — on a construction basis. Most of these investments will involve rationalisation projects.

Arbed, the Luxembourg-based steel producer that is Europe's fourth largest, is in the throes of a radical restructuring programme that by 1984 will have cost more than $1.3 billion and cut the work force of 60,000 engaged in iron and steel operations by more than one person in five.

Costly steel restructuring programmes are underway in the United Kingdom and France that are directed to reducing surplus capacity and increasing modernisation. Smaller steel producing countries are also actively involved. Norsk Jennverk, Norway's state iron and steel concern, was seeking $215 million for planned investment for the period 1980-1984 with most of that targeted for modernisation.

Looking at the financial needs of developing countries, the projected costs are astronomical. The projections vary greatly, however, depending on the economic assumptions that are used. For example, Father William T. Hogan estimated in 1977 that it would cost developing countries somewhere between $300 and $350 billion to increase steel capacity by 310 million tons by the year 2000, the modest scenario. By today's standards, these figures are vastly understated and take no account of the current rate of inflation.

The heavy financial burdens being imposed on steel world-wide to readjust, rationalise and modernise are in turn causing a type of restructuring to take place in the relationships between government, industry, labour and consumers. In a sense, all four parties are bearing a portion of the financial load.

Government financial assistance to steel is increasing substantially. Forms of assistance vary widely from country to country. They also vary widely in terms of specific aids — including aid to workers, conversion of debt to equity capital, arrangements to ease debt servicing, loan guarantees, regional subsidies. Consequently, government involvement in steel is today greater than it has been in the past in most countries, although the nature of the involvement is changing.

The process and cost of restructuring that will continue into the 1980s raise a number of policy issues.

Perhaps the "umbrella" issue is whether steel industry restructuring should be carried out in the context of a long-term general industry adaptive policy or whether a steel-specific scheme is more appropriate.

One cannot fail to note that so many of the current rescue schemes are described as "crisis plans", "emergency measures" and the like. This suggests that both the public and private sectors share blame for lack of either vision or initiative. Clearly, future policies should aim at alleviating, if not avoiding, the building of crisis conditions.

Equally important from a policy viewpoint is the extent to which governments should strive to shift the financial burdens associated with restructuring to industry generally by easing pressures on steel price increases, especially during periods of strong demand, and thereby enabling generation of more profitability.

Empirical evidence suggests that the short run benefits of steel price restraints are being outstripped in the longer-run through the higher costs associated with crisis oriented rescue plans, higher product costs resulting from production inefficiencies, and the hidden costs of trade protection that are ultimately incurred.

Finally, with the public stakes that are involved in steel-industry restructuring and modernisation, policymakers must reflect on how to preserve legitimate government interests while at the same time allowing industry managements the freedom to make decisions affecting the size of the industry, the types of products to be produced, the processes and technologies to be employed in the production of those products, or the markets to be served by the products.

Employment

Two forces are going to operate more strongly in the forthcoming decade to affect workers in the steel industry. First, pressures for rationalisation and modernisation will in turn result in pressures to reduce the labour force in developed country industries. Second, to the extent that market disruption caused by severe international price competition is not contained during periods of recession, prospects for large-scale unemployment become real.

The American Iron and Steel Institute recently projected that upwards of 90,000 jobs will be lost in the United States if American steel production capacity is reduced by 20 per cent in the years ahead, a likely prospect if current trends continue in the United States.

The well-publicised restructuring efforts in Europe have led to substantial reductions of the steel work force, and more reductions are in store for the future if current plans are carried out.

Even in Japan, where life-time employment still predominates, the blast furnace steelmakers have not engaged in lay-offs, but have reduced the size of the steel labour force by relocation of steelworkers and by retirements.

Developed nations will have to set in place practical, foresighted policies to insure that workers are not made victims of restructuring, modernisation programmes, disruptions caused by international market behaviour, or policies by which the industry reduces itself in size by simple attrition.

If a nation undertakes an industrial policy approach which aims at encouraging the adaptability of its steel industry to changes in current economic conditions, then it follows that labour policies must be equally adaptable and flexible.

Specifically, governments and trade unions of those steel producing countries which have tended to resist dismissal or relocation of workers in response to slackening steel demand should by now be aware of the public and private price that is paid for inflexible labour policies. For example, despite large scale renewal of the French industrial base, productivity in the French steel industry has not increased correspondingly, since there is a tendency to retain redundant workers, often under pressure of government social policies.

If the steel industry as a whole is to become dynamic, it follows that its components — including labour — will have to become dynamic. Restructuring and modernisation programmes that must somehow function without any effect on employment are selfdefeating at the outset.

Aside from the employment challenges posed for the industrialised countries, the dimensions of the employment challenge facing developing countries are unprecedented. According to the 1979 World Bank development report, between 1975 and the year 2000 the labour force in developing countries is expected to increase by about 550 million — over twice the increment of the previous quarter of a century.

The report continues, "given the already high levels of under-employment and absolute poverty, the scale of the task of expanding productive employment and income opportunities cannot be over dramatised".

The International Metalworkers' Federation (IMF) set forth its views to the OECD in 1978. The IMF statement included the following points:

— Governments will not meet their responsibilities if they engage in national assistance and pricing practices for exports which disturb the market and result in social repercussions in other countries.

— Subsidies for job security and social guarantees must in no way create market disruption.

— Social plans should provide for the full protection of workers in any situation that might arise and in order to facilitate necessary adaptations. These plans comprise such measures as early retirement with a guarantee of income and full pension rights, additional rest days and special work programmes.

It can be argued, however, that government can impede labour mobility, both occupational and geographic, by policies of extended unemployment benefits. Professor Melvyn B. Krauss has observed:

... if unemployment insurance is available on an extended basis, as is the case in many welfare states, unemployment insurance permits the workers to resist the necessary adjustments that are called for by the economy... More and more, workers have come to expect democratically elected governments to secure those entitlements — to work at his or her old trade, at his or her old geographic location — for them — *a trend that has already reduced the abilities of welfare economies to adjust to the needs of a changing economic environment.* (Emphasis added)

This point was recently highlighted in the United States. A study by the General Accounting Office indicated federal benefits under the trade adjustment assistance programme diminished laid-off workers' willingness to work.

The policy issues regarding employment in steel are tough. They are politically intertwined. Clearly, a prime issue for examination is the extent to which government policies have sanctioned labour-market monopoly practices and thereby restricted factor mobility and wage flexibility.

Deeper consideration must be given to whether the emphasis of trade and labour adjustment assistance should be directed to phasing out sectors of the steel industry or to rejuvenating them.

160

Policymakers must also focus on the structural adjustment issues that are being opposed by organised labour, such as conglomerate mergers. The IMF is suggesting that policies should discourage such mergers on the basis that they draw funds from steel and may benefit stockholders, but not necessarily steelworkers.

And finally, the issue of "who decides what?" is bound to increase in intensity in the next few years. Unquestionably, workers are affected by readjustment programmes. The more difficult issue ist the extent to which labour has a role in formulating management decisions without impairing the flexibility needed by management to adjust to changing economic conditions. Pre-notification regarding management decisions is at one end of the options spectrum. Co-determination is at the other.

Emergence of steel production in developing countries

From the standpoint for adaptation, another issue that will become increasingly more significant in the decade ahead will be the continued emergence of the developing countries as steel producers. This issue often evokes a fear response on the part of developed countries.

In fact, the developing countries as a group pose little threat to the steel industries of Europe, the United States, Japan and the rest of the developed world over the next decade. Michael Elliot-Jones of Chase Econometric Associates observes:

> Capacity growth will resume in the Third World within the next five years, but so will the growth of demand. Moreover, capacity growth in the Third World will not be as large or vigorous as sometimes is feared. Note that these findings are contingent upon a relatively bullish outlook for both the economies of the Third World and the industrial nations.

Much of the fear on the part of established steelmakers has been generated by the March 1975 Lima Declaration. Ambitious plans for industrialising the Third World were set forth. Steel was a target industry. By the year 2000 it was hoped that third World steel production would account for 25 to 30 per cent of total world production versus approximately 5 per cent (8.5 per cent including China) in 1973.

The length and depth of the post-1974 steel recession led to a reassessment of the world iron and steel industry by UNIDO in 1978, and re-evaluation of the feasibility of rapidly expanding Third World steelmaking capacity. The UNIDO study concludes that most Third World countries have neither the natural resources (ore, coal or scrap) nor the markets to justify the construction of integrated steel mills which are capable of producing flat rolled products and specialty steels.

Most observers believe that there will be intensified capacity expansion taking place in several Third World countries. However, approximately three-fourths of the increase is expected to be concentrated in only seven countries: Brazil, India, Venezuela, Mexico, South Korea, Taiwan and Argentina.

A number of factors could influence the rate at which this expansion takes place. Slow world growth and unsettled general economic conditions, for instance, could radically alter the economic feasibility of various projects currently being considered. Second, a shortage of capital or reasonable financing terms could lead to abandonment of steel projects in favour of less costly alternative investments which offer better long-term returns. Third, when the full cost of establishing a steel industry in Third World countries is identified (in terms of the need for a supportive infrastructure and a larger, skilled labour force), ambitious undertakings may be scaled back considerably. Some of these scale-downs have already occured.

Alternatively, increases in world scrap prices, further improvements in direct reduction, and development of "formed" coke produced from metallurgical coal could create incentives for more rapid development of both mini and midi-mills as well as integrated mills in the Third World.

On the demand side, two developments are likely to occur.

a) Developed country producers will lose some of their export markets in more advanced developing countries which are expanding steelmaking capacity, but on balance will still be able to expand exports to the Third World where demand for steel is projected to grow by upwards of 8 per cent per annum.

b) Even with the installation of steelmaking capacity in the Third World, a developing country not infrequently finds itself producing and exporting certain products but importing others. For example, resource-poor countries may be able to enhance their output of long products (e.g. re-bar, machine wire and small merchant goods) as well as second quality flats; they are often unable to produce first quality flat rolled products and specialty steel. This factor should not go unnoticed by those who are overly pessimistic regarding Third World impact.

As noted earlier, the emergence of some developing country producers as steel exporters is regarded by some as a potentially disruptive factor in the world steel market. In general, this fear is exaggerated. But, it has resulted in generating mistrust and deep suspicion on the part of developing countries.

Efforts to persuade developing countries to actively participate in international forums on steel are proving difficult. Vituperative complaints by some industrialised countries against the granting of preferential export credits for financing new steel plants in developing countries contribute to the distrust. Similarly, exclusion of steel mill products from preferential tariff schemes of developed countries and failure to make allowance for developing country exports (which do not have the benefit of historic market share in contrast to exports from more established sources) in the application of quantitative import restraints are cause for legitimate criticism by developing countries.

In fairness it should be noted that the issues noted above apply to industries other than steel. The problem is of wholesale dimensions. For example, developing countries are staying away en masse from signing the Tokyo Round accords for "freeer and fairer" world trade. Their failure to sign the accords should be cause for high-level concern in the developed world.

Attention must be focused on a number of issue areas, some of which are highly controversial and have tended to be shunted aside.

Industrialised countries must seriously consider removal of tariff and non-tariff obstacles to developing country steel exports.

Consideration must be given by both developed and developing countries to the responsibilities of fair trade practices and adherence to international codes of trade behaviour. Developing countries, particularly those that have graduated in terms of economic development and which have sophisticated steel industries, cannot escape the responsibilities that attach to interaction in the international community.

Financing steel capacity in the Third World will demand better policymaking than has heretofore been the case. Specifically, consideration must be given as to co-ordinating investment planning and financing (for example, by the World Bank and national export financing institutions) with the legitimate development aspirations of developing countries, but in light of forecasts and projections being developed by OECD, UNIDO and other international institutions. How this can be accomplished without retarding justifiable steel expansion programmes is a monumental challenge. But, it should be attempted.

Trade and market disruption

While the entry of Third World exports can become a source of market and trade disruption from the standpoint of developed countries, competition among developed countries in the 1980s is also likely to cause market disruption, either as a result of price competition or simply increased quantities being directed into the market at a given point in time.

Why should market disruption be a more serious concern of the future than it has been in the past? The answer depends on the scenario that one adopts in terms of future economic growth. Obviously, if growth in demand is relatively high and industries are operating at high operating rates, there is less likelihood that market disruption will be a serious concern.

However, much of the economic outlook in terms of continued low growth, high rates of inflation, scarcity of capital, and pressures on employment contribute to the judgement that we are headed for at least a half-decade of economic tension and uncertainty, much of which will depend on political conditions.

Economic tension and uncertainty almost insures the inevitability of international trade friction. It therefore seems both logical and intelligent to adopt national and international policies wich will control or alleviate the problem of market disruption.

Currently, there exists a patchwork of measures designed to deal with trade-induced disruption. The united States has its Trigger Price Mechanism. The European Communities operate under the "Davignon Plan". Canada policies imports through an informal "bench-mark system". There are numerous other restrictions, bilateral and unilateral, in effect to control steel market disruption by controlling international trade flows.

From a policy standpoint, it is unrealistic to expect the international trading system to ever become devoid of some form of external restraints on steel trade.

Given this premise, it is more reasonable to permit countries to adopt trade policies which have the effect of ameliorating market disruption and, in the period ahead, have the added effect of facilitating the necessary adjustments and transitions to more modern, competitive industries, for example in Europe and the United States.

In terms of policy responses, consideration should be given to the issue of whether there should be an international market disruption or safeguard standard (flexible enough to permit national variations), rather than individual country or regional schemes (not subject to any international standard).

If an international standard is preferred, policymakers will have to decide whether the standard should be generic applying to all industries or whether it should be steel specific.

Another issue that is likely to receive attention in the future is whether market disruption measures should be price and/or quantity oriented; whether they should distinguish between fair and unfair competition; and whether they should be tied to identifiable restructuring and modernisation programmes?

Government involvement

A major issue of the 1980s will be the extent to which governments will be involved in the affairs of their national steel industries. This issue is intimately connected with restructuring, financing and labour adjustment.

Several notable and interesting shifts are underway. Many countries, such as the United Kingdom, wherein the basic steel industry is government owned, seem determined to reduce the public burdens borne in the financing of their steel industries. Thus, the movement is towards a more arms-length relationship than has existed in the past.

In the United States a contrasting tendency appears to be emerging. The long-standing arms-length relationship between the American steel industry and the United States Government is becoming less pronounced. Creation of the Steel Tripartite Committee in 1979 with government, industry and labour represents a major breakthrough in terms of co-operation on pressing steel issues even if the committee results prove to be less than momentous.

Another major departure in United States Government and industry relations is manifested in the recent granting of United States loan guarantees to several steel firms. This is, however, a hotly debated issue within the American steel industry with perhaps the dominant portion of the industry still maintaining that such government intervention in the affairs of steel should be decried.

In France there appears to be belated recognition that the long term and recent woes of the French steel industry are not due solely to the current worldwide recession in steel, but equally to French government policies that have impacted on price, investment and industrial policy over the years. Although the government has structured a financial rescue plan, its stated goal is to eventually return control to the private sector.

Italy, another nation with heavy state involvement in steel and other industries, is in the process of trying to turn many state enterprises back to private ownership or to abandon them. Mounting losses in steel as well as shipbuilding, aluminum and textiles are burdening the Italian national budget to the point where they are cause of major political controversy.

The 1979 World Bank Development Report provides a lucid exposition on the issue of government support for industrialisation form the standpoint of developing countries. The report points out the deficiencies of state enterprises and suggests principles for their improvement. On balance, however, state enterprises in developing countries do have a substantial role in steelmaking and that role is bound to increase as the financial commitments for new steel-making capacity increases.

Herman Kahn observes: "One of the key characteristics of nationally created industries in South Korea and Taiwan that sharply differentiate them from those of the developed world is that they have been created out of nothing, as opposed to having been consolidated out of old and declining business. And they represent a trend *towards* private enterprise rather than *from* private towards public enterprise."

It seems evident that the 1980s are likely to witness efforts to more clearly define the role of the state in steel industry affairs with a strong bias towards according managers of government enterprises greater autonomy and freedom from political interference.

The debate is likely to focus primarily on the types of governmental policies that will in the first instance lead to more competitive steel-making enterprises — irrespective of the form of ownership.

For steel firms operating in market oriented economies, the policy issues will centre on the fairness of competition between enterprises dependent on the private capital market and those that rely primarily on government funding.

International arrangements and institutions

No examination of steel issues is complete without some focus on international arrangements and institutions. This area is in need of a much higher degree of interest and commitment than governments, industry and labour have been willing to devote up to this point in time.

As the steel industry becomes global in its technology, raw materials supply, and markets, the policies and solutions of steel will have to become global. This in turn implies a strengthening of international institutions and rule-making framework in the decade ahead and beyond. However, it is unfortunate that, for diverse reasons, both government and industry have resisted development of an international approach to basically common issues on steel.

Decision-making can be aided internationally by proceeding two ways, generically and sectorally.

Generically speaking, the international community has developed international rules of trade behaviour and macro-economic examinations of economic policy that can greatly benefit the steel industry. Reference is made here, for example, to the GATT codes governing dumping, subsidisation, government procurement and other topics that affect the trade interests of the world steel industry.

International codes of behaviour should be accorded an opportunity to work. If experience vindicates that adherence to the generic codes fails to provide adequate response to the problems of steel, at that time consideration of a steel sector approach may be justified.

At the same time that governments continue to strive for understanding and co-operation at the macro level, sector activities should be encouraged to operate complementary to the broader efforts.

Governments are generally reluctant to adopt a sectoral approach to international economic issues in the industrial area. Textiles was an exception. In the case of steel there is the fear that doing something for steel may result in having to do something for other industries.

There is another fear, namely, that steel issues are inherently directed towards trade restriction. This notion arose during the Tokyo Round multilateral trade negotiations. There was a fear that any form of steel sector examination would result in trade restricting rather than trade expanding arrangements. Ironically, at the same time that government negotiation officials in Geneva were decrying a sectoral approach as trade restrictive, national governments were emplacing trade restraints on steel imports.

Establishment of the OECD Steel Committee in 1978 represented a major departure in policy and a potentially significant breakthrough. The significance of the committee is four-fold:

For the first time —

1. It requires participating governments to commit themselves to certain principles affecting steel;
2. It operates at the policy level;
3. It provides opportunity to evaluate steel forecasts and developments over the medium and longer term, thereby providing a better framework by which to deal with short-term issues; and
4. It makes transparent steel policies and actions that have heretofore been closeted by national governments.

Unterstandably, there is some impatience with the committee and perhaps skepticism that it can effectively influence national policies. But critics may be well advised to ponder the alternatives at the governmental level — either no international policy forum or a highly structured inter-governmental arrangement governing management-type decisions.

In a recent report issued by the Trilateral Commission, its authors (experts from North America, Western Europe and Japan) cautioned that:

> Governments should take those — OECD Steel Committee — commitments seriously and be prepared to adjust their policies in the light of the review. It is hoped that this will suffice to overcome the crisis. If not, it

may become necessary to move towards some kind of multilateral agreement concerning matters such as levels of trade, prices, capacities and degrees of self-sufficiency.

Where does the international community go from here in terms of future institutional arrangements?

Clearly, the developing countries must be brought into the policymaking arena insofar as steel is concerned. The welfare of the developed world and its industries depends on it. The welfare of the Third World also depends on such co-operation. Mutual trust must be established.

Policymakers should seriously consider the issue of market disruption and safeguards. Since the efforts to realise an international generic safeguards code have not yet borne fruit and since the problem of trade disruption threatems to escalate further through use of national schemes, it may be appropriate to examine the prospect of establishing a steel specific code. The OECD Steel Committee appears to be a logical forum for such an examination and recommendations.

Finally, the institutional process should be dynamic. In the case of the OECD Steel Committee, participating governments should annually review the purpose and achievements of the committee with a view to its being responsive not only to current but also to long-term needs.

In this regard, the "initial commitments" of the Steel Committee need to be examined from the standpoint of whether they should be modified, deleted, or expanded. The thrust of the committee should be re-examined to determine whether a more activist role is justified, a role that goes beyond mere examination, analysis and exchange of information.

Responsibility for international activity should not be viewed as exclusive to the government. Industry, labour and consumer interests are affected by the decision or non-deicisions that emanate from OECD, ECE, GATT, UNIDO and other international oganisations. It is easy to criticize these operations; it is more difficult to constructively influence their activities.

Certainly, a challenge of the 1980s should be to develop more intensive debate at the international level on the issues of steel with broader representation of interest groups and countries and to perfect the institutional mechanisms for doing so.

SOURCES

Jagdish N. Bhagwati (ED.),
Economics and World Order: From the 1970s to the 1990s (New York: The Free Press, 1972).

J. David Carr
"A projection of Steel Technology in the World for the 1980s" *Iron and Steel Engineer* February 1979, page 51.

William Diebold Jr.,
Industrial Policy as an international issue 1980s Project/Council on Foreign Relations (New York: McGraw-Hill, 1980).

Michael F. Elliott-Jones,
"Iron and Steel in the 1980s: The crucial decade" A speech presented at a seminar sponsored by the George Washington University and the American Iron and Steel Institute, Washington, D.C., April 1979.

William T. Hogan,
"Future Steel Plans in the Third World" *Iron and Steel Engineer* November 1977, page 25.

Herman Kahn,
World Economic Development: 1979 and Beyond (Boulder, Colorado: Westview Press, 1979.)

Melvyn B. Krauss,
The new protectionism: the welfare state and international trade (New York: New York University Press, 1980).

National Foreign Assessment Center
The Burgeoning LDC Steel Industry: more problems for major steel producers A research paper (Washington, D.C.: U.S. Central Intelligence Agency, July 1979).

Organization for Economic Cooperation and Development,
Facing the Future An Interfutures Project (Paris: OECD, 1979).

John Pinder, Takashi Hosomi, and William Diebold, Jr.,
Industrial Policy and the International Economy, A report of the trilateral task force on industrial policy, The Triangle Papers: no. 19 (New York: The Trilateral Commission, 1979).

"Steel Trek"
33 Metal Producing A series of articles on Steelmaking in the second millennium, February 1980, Vol. 18, no. 2, page 37.

The American Iron and Steel Institute
Steel at the crossroads: The American Steel Industry in the 1980s (Washington, D.C.: American Iron and Steel Institute, 1980).

The World Bank
World Development Report, 1979 (Washington, D.C.: The World Bank, 1979)

CHAIRMAN:

We have just heard the presentation of the first document which was related to steel in the 1980s. I would now like to call on Mr Bill Sirs who will present a document relating to employment, and reconversion an employment problems. Bill Sirs is of course Chairman of the steel union in the United Kingdom. He is also a member of the Consultative Committee of the European Coal and Steel Treaty...

Mr. William Sirs:

In introducing my paper, I think I would make the point very clearly that the steel industry does have to face up to its responsibilities. As we see it, the present problems do arise from the cyclical recession which was converted for deepening economic reasons into a profound slump. This was an event with far reaching consequences.

In the developed countries there has been a haemorrhage of jobs. In the EEC for instance, one hundred thousand jobs have disappeared in the years 1974 to 1979. Four years on from 1978 looks to be just as difficult, if not worse.

An industrialist can, if he so desires, pull out of a particular site and move on. The steelworker has no such option available to him; his capital is his labour. In a dreamworld free market, he would offer it to the highest bidder. But in the grim real world, mobility of labour means unemployment.

We have in Britain a town called Corby. It was built for steel; it imported its employees from depressed areas in the 30s. Now, within a period of the months, almost all steel making has ended in Corby, and production has moved on. The only legacy is a tube works with an uncertain future.

Mr.Chairman, Corby is a microcosm of the social problems of a great world industry. And, it doesn't really make a great deal of difference whether it is Corby or Youngstown of Lorraine because the problems are much the same. There is an internationalism of hardship which does not recognize frontiers.

The social problems of steel are being concentrated in a very rapid time scale. In my own country 20 000 jobs vanished from the industry in the calendar year of 1978, and contractions of this rate are a backlash of unrest. It is a minor miracle that the disruptive resistance of the French steel workers has not so far been an international phenomenon. One valuable outcome of this symposium today would be a recognition of the self-denying restraint of the trade unionists, the trade unions and the IMF affiliates in the face of what amounts, in some places to a challenge to their very existence.

We know that there are many problems, and today there has been mentioned question of subsidies. This is mentioned by, I think, most people who make contributions. They flow, disguised or transparent, from every government to its steel industry. They are not monopolised by the public sector, but also thrive in the private sectors too. What varies from country to country is the mode and not the fact of state aid. It is a fact of international steel life and no screen of the Friedmanite dogma can hide it.

With this type of assistance, of course, countries do hope that they can retain an adequate steel making capacity. But, what about the broad social implications which are at stake? Is there a role for government here, or are we to argue that capital will be assisted through a time of trial but not labour? We recognise also that no country can insulate itself from world economic trends, outside the centrally planned economies. Is there any significant steel industry that can be indifferent to the fate of its exports or, alternatively, meet all the needs of its home market without needing to buy in?

The developing steel discussions within the OECD seem to be a recognition that no one can opt out. No doubt every country begins from the point where it seeks to prevent any harm to the national interest and perhaps a more ambitious approach has been stimulated by the international discussions since 1978. Moreover we do have international mechanisms to coordinate our views in the International Metalworkers Federation. Through the TUAC we come together from time to time with the OECD to discuss at least the social and employment implications of steel trends.

We also have been participating in the analytical steel working party; this is advisory to the Steel Committee itself. We say that the Steel Committee should be representative of government, employers and trade unions. We have one such body in Britain and this is the Iron and Steel Sector Working Party of NEDO. It has carried out some impressive monitoring work and undertaken some interesting work on plant study comparisons. Its contribution has raised the level of the whole debate about steel in Britain, and there is no reason why the working together of the three parties should not be just as fruitful in an international agency.

So we also say, what do we seek from the conference and what do we seek from the OECD? Well, we want a co-operative effort on the part of steelmakers to help us clamber out of this slump. We say that governments should stop trying to export their own difficulties by unfair trade. We recognise that there is no unchallengeable yardstick by which production costs may be measured. However, if companies and countries do not exercise self-discipline, we are faced with the disastrous possibility of market protection.

Whilst I agree that it is very difficult not to have some degree of protection, to go fully into that situation would be, as Viscount Davignon has pointed out, extremely difficult because of retaliation and global cartels. There are already ominous signs. Dumping, the achievement of artificial exports made possible by a false price, creates in the end its opposite, the increasing confinement of producers to their home market.

It is difficult grappling with the employment consequences of restructuring without carrying the additional burden of a countervailing outlet for inefficient competitors. If this were to continue, it could be the survival of the unfittest. I say that, recognising the tremendous problems we have in the UK and the tremendous changes that we have seen. But we do not want to see the steel industry toppling into one corner of the globe, as has happened in shipbuilding. Free, but fair trade may be a cliché nowadays, but it does, we believe, express a principle which the OECD ought to guard.

We don't see a contradiction between seeking minimal interference with trade and asking for state support to mitigate the social effects of restructuring. It is the second target which we aim at and steelworkers who bear no responsibility for bringing this crisis are living today with the consequences.

Some governments will claim that intervention even on a social matter distorts trade and that market forces, through the operation of some hidden hand, will miraculously give birth to new jobs and new industry in place of old. My experience has just not been that. We don't feel that we would want to be confined only to discussion about the effects of policy; we want to play a part in formulating it. We have in every country clear and definite perspectives based on deep expertise and a genuine commitment to steel. And who is more expert in the workings of the mill, the melting shop — the blast furnace — than the operator himself? And what organisation can more democratically represent him than his own union?

And yet, time and again our views are turned aside. We are sometimes listened to politely and sometimes not listened to at all. But we are to often ignored. And that is the position in many countries. But when we turn out to have been right, we have to pay our share for the price of failure. And where government and industry have miscalculated, they must not avert their gaze from the result. They must also learn for the future. They must harness all the knowledge and support available to them, and we are a major part of that support and knowledge.

There are examples in the world of how a rundown, where it is necessary, can proceed by agreement and without disruption. There are also examples which serve as an awful warning. I want to be specific; the document is specific: We have reached a stage where at the moment we cannot accept further unilateral rundown of steel jobs and the living standards of steel workers. If a country can no longer sustain its industry, it must provide other jobs.

In my own country we are even now appealing to Viscount Davignon against the people who run the industry and against our own government, because of the unfairness of the situation which we feel and we face in the United Kingdom. These responsibilities will not vanish. There are tremendous problems.

Mr. Chairman, there are some detailed proposals set out in the last pages of the following Document. I hope that they will be treated by the Symposium and the OECD as whole as seriously as the traditional pursuits of production and profit.

THE PROBLEMS OF THE LABOUR FORCE AND READAPTATION AND RE-EMPLOYMENT POLICIES

IMF-Report submitted by

Williams Sirs
President of the International
Metalworkers' Federation's Iron, Steel
and Non-Ferrous Metal Department

IMF AND ITS OBJECTIVES OF A WORLD STEEL CONFERENCE

The International Metalworkers' Federation — IMF, is the worldwide body uniting unions which organize workers in all branches of metal manufacture and engineering. In the steel industry of OECD countries and all other countries where democratic unions can operate freely, and average of 90% of all workers belong to organizations affiliated to IMF.

Representing the people in the industry on a worldwide basis, the IMF at its Iron and Steel Conference in Pittsburgh of 28–30 June 1976, issued the call for a world steel conference of governments, employers and trade unions to examine the common problems presented by the steel crisis and to seek cooperation for effective action.

This Symposium is a partial, but welcome, response to such an initiative and dovetails well into the enhanced steel activities of OECD.

It provides an opportunity for a continuous effort on the part of all those concerned with putting the world steel industry on a sound basis for the benefit of both developing and industrialized nations.

THE CAUSES OF THE STEEL CRISIS

The steel crisis, now half a decade old, developed from a severe rupture caused by political world tension and extended from a cyclical recession to a severe structural slump.

The inadequacy of policies, world political events and insufficient counter-measures have created a vicious spiral of lost jobs, over-capacity, under-consumption, astronomical deficits and bankruptcy of steelworks.

The major problem is a lack of growing demand for both capital and consumer goods and concerted actions for development in the area of greatest need.

For years now steel plants including most modern installations suffered losses from competition below production costs whilst the effects of subsidizing unfair price practices has aggravated disruption on the world market.

Ruthless competition, random capacity expansion and often failure to modernize and restructure existing production facilities have led to chaotic situations.

The workers in every country are bearing the brunt of the disruption industry.

In the European Community steel employment from 1974–1978 declined in the average by 14.2 per cent with a particularly heavy drawback in important steel regions, 28.7 per cent in Luxembourg, 24.5 per cent in Belgium, 16.7 per cent in France, 15.7 per cent in Great Britain and 14.8 per cent in the Netherlands. The average figure for the Federal Republic of Germany is 13 per cent whilst the Saarland region takes the brunt with 24.5 per cent. In Japan the decline for the same 4 years was almost 9.5 per cent and 11.5 per cent for the USA. This means a loss of steel jobs for all industrialized OECD countries of over 156,000 until 1978. This has again substantially increased in 1979.

There are also employment fluctuations in the rapidly expanding steel industry of developing countries. Despite a continuing strong growth rate of steel production in Brazil a substantial employment cut-back occurred in recent years and job expansion resumed on a rather modest scale, due to production rationalization and structural change, arising from huge programmes for modern investments.

There is a continuing threat to jobs of steelworkers, particularly in certain traditional steel producing centres, through closures, mass dismissals, loss of income due to unemployment and short-time work and growing insecurity in steel communities.

In assessing the effects of a decline in steel employment one has also to take into account the key position of steel production and its interrelationship with other industries. For instance, the further loss of 53,000 jobs in iron and steel-making as planned by the British Steel Corporation will result in something nearer to 100,000 lost jobs, when the falling demand for coal, affecting miners and new capital equipment hitting the engineering industry and other disappearing advantages of local steel supply, perculate throughout the economy.

There is also to be awareness of the fact that the provision of alternative employment in those areas hit by steel closures is much easier planned than done. An example is Ebbw in South Wales where, despite all good intentions, only 1,850 new jobs have been created to replace over 5,000 jobs lost in the steelworks since the early 1970s. Only about 3 new jobs per week, mostly for women have been created. At this rate it will be 1992 before every redundant steelworker in this region will be back in employment.

ARTIFICIAL DISTORTION OF PRODUCTION COSTS — NO ANSWER

Failures in policies coupled with unfair trade practices in a badly managed free market system are at the root of the abrupt menace to whole steel regions. Yet markets, production patterns, technology and competitive advantages constantly change and there are and will continue to be variations in production costs. These variations cannot be handled to the detriment of employment but call for positive initiatives to steer effectively economic development towards the meeting of human needs.

In this context, trade unions oppose the use of subsidies to artificially lessen production costs in order to achieve unfair competitive advantages, but they do see a role for subsidies in assisting companies to meet their social responsibilities in adapting to market trends and new economic opportunities. Whilst government aid should not consist in artificially covering price differentials it has its full justification and meets a basic need in improving the economic and social infra-structure.

There is no universally recognized measure of production costs. The relationship between government and industry in the area of trade assistance leads all the more easily to endless price cuttings, dumping and unnatural prices defeating economic and social objectives. This handicaps timely action to secure steel jobs by internal readjustment and well-defined policies and plans for adaptation safeguarding overall employment.

SOCIAL ORIENTATION — A KEY TO SOUND INDUSTRY AND NEW EXPANSION

Trade unions accept technological and structural change but insist that it provides for overall social and economic genefits extending also to the workers directly affected.

IMF unions have led a hard struggle to find solutions for the protection of their members and the industry both in the present crisis and for the future. A healthy industry provides the best employment protection, but industry cannot be cured by decimation or even complete run-down.

Steel production and consumption are natural geographical allies along with expanding world trade in steel products. The relative competitive advantage of a steel industry which serves both national and international markets on fair terms, must remain a key feature of national planning for full employment.

At the same time, the world steel economy can profit from natural advantages like raw materials which equally are logical alies of productive capacity. However, steel production in developing countries must be consistent with domestic economic growth and determined planning to meet social needs, whilst benefitting from the integration of their steel industry into the world market under full recognition of progressive social standards.

A primary function of the OECD is to evaluate the pressures which develop from trade flows. Providing more assurance and stability to the international and domestic markets requires a great deal of government negotiations. The OECD is a proper forum for providing transparency of any emerging understanding, especially since the role of trade union movement is increasingly recognized in this forum. However, such international understanding regarding trade and other matters can proceed only if there is also fully developed national social policy based upon the principles elaborated in this document.

THE SOCIAL RESPONSIBILITIES OF ALL CONCERNED

Trade unions did not cause the world steel crisis although the effects bear heavily on their members.

Over the last five years trade unions have provided the impetus for social action and submitted clear and realistic policy proposals to tackle the tremendously difficult economic problem of the industry, without pretending to have all the solutions.

Adjustments have taken place and new ones are being faced often in the same region and the trade union have played their part. They forced companies to accept their social responsibilities and also involved local authorities and governments. However, the development of social and economic policy is the province of governments and the employers have obligations within these policies and must fulfill their entrepreneurial role.

Adaptation policies cannot be a separate exercise but are inextricably entwined with basic steel production, allied activities and general economic development.

Trade unions alone cannot tackle all these problems; government and industry, employers and trade unions, must overcome them together.

Neither can restructuring policies be successfully carried out from a narrow national angle. International cooperation is an absolute necessity and must involve governments, steel industry employers and trade unions in a common effort to find socially and economically adequate solutions.

EMPLOYMENT AND SOCIAL GUARANTEES

The trade unions are prepared to cooperate and the IMF is striving constantly to achieve such cooperation but it must be under certain specific guarantees and with a progressive general objective.

Trade unions cannot accept the reduction of steel jobs with no timely alternative employment available.

Neither can trade unions tolerate a lowering of living standards and the loss of accumulated gains and the right to future progress.

Therefore, IMF unions insist that active manpower policies and general social and economic measures be effectively implemented to achieve these central guarantees.

The IMF will play no part in retrenchment if these guarantees are not observed.

THE NEED FOR POSITIVE ACTION IN MAINTAINING AND CREATING EMPLOYMENT

The IMF refuses to be confined to the sphere of social policy alone and demands the launching of a positive action programme for industrial revival.

Only in such context is readaptation more than an empty promise.

Reconversion, envisaged for several industries in difficulties, must not amount o shifting industrial activity from one sector to another, it must be carried out by creating additional jobs.

There are instructive examples which feature IMF policy applied in the world teel industry today and interesting experience of governement, employer and trade mion cooperation in facing economic and social problems in steel.

THE ADVANTAGE OF CODETERMINATION — A STRIKING EXPERIENCE IN THE FEDERAL REPUBLIC OF GERMANY

In the Federal Republic of Germany competitive adjustments through new chnologies — change over to new production processes and rigid plant -organization — have been a continuous practice. In all these operations, hard egotiations between employers and trade unions were a main feature. There is no ubt that the German Co-determination Act was instrumental in securing a sitive outcome of all these negotiations. In fact, this Act provides trade unions ith co-determining rights in investment decisions. This gives them the possibility eventually prevent or postpone modernization decisions which involve reduc- ns in employment. They have an insight into the planning of investments enabling them to assess the full market, production and employment impact of envisaged measures, to put critical questions and orient actual policies. Consequently German trade unions do not oppose the necessary decisions but insist on the requirement that production investments be accompanied by adequate social investments. This is to guarantee employment and income to all those concerned through joint programmes to effectively meet the social objectives laid down in the re-structuring programmes.

These trade union possibilities were used to such an extent that objective discussions with management extended to both short and long-term job security in the steel industry. Timely measures were taken and a major part of restructuring could be carried out in periods of steel expansion. However, the steel crisis was such that employment difficulties could not be averted altogether although the German steel industry fares much better in the present situation. There has been no social upheaval in the Federal Republic of Germany due to shop closures and massive dismissals. The strike in the steel industry during December 1978/January 1979 was a collective bargaining conflict over wage conditions and the reduction of working time pertaining to general issues of employment and social policy. Obviously, the struggle for shorter work time aimed at safeguarding employment.

Threats of cut-backs in employment in the steel industry should be tackled in the following way. First, the steel company concerned establishes a restructuring programme with full trade union involvement provided for by co-determination and with the objective of permanently maintaining the greatest possible number of jobs in steel. This is combined with efforts to expand company activities to closely connected and/or related fields for job saving purposes. Secondly, such a programme will receive public assistance through such measures as sureties and lower interest rates in as so far as the objectives of restructuring concord with the national and international concept of general steel policy. Thirdly, the programme will include accompanying social measures through agreement with the trade unions for the necessary social guarantees.

The granting of state assistance must meet the condition that such an agreement with the trade unions has been achieved. The fourth set of measures extends to the creation of replacement industries through initial public funding and in the framework of general town, country and industry planning.

TRIPARTITE NEGOTIATIONS:
EMPLOYMENT GUARANTEE FOR STEELWORKERS
IN THE SAARLAND

A specific action along these lines is the programme for the structural reform of the steel industry in Saarland, the Federal Republic of Germany, in which participate — employers side — steelworks belonging to the Luxembourg based ARBED concern, on the public side the Federal German Government and regional Saarland authorities an on the trade union side the Metal Industry Union IGMetall for the Federal Republic of Germany.

This programme comprises an agreement, reached after 100 hours of strenuous negotiations, stipulating a number of basic social guarantees, ensuring that:
— there will be no dismissals during restructuring,
— in case of job transfers, wages and salaries including premiums and bonuses, are guaranteed for 5 years and will be paid,
— workers must accept alternative jobs that can be reasonably expected from them, whereby a joint commission will decide in case of dispute,

175

- steelworks undertake serious efforts for the replacement of jobs lost and that they will be assisted in this endeavour by both he trade union and regional authorities,
- Works Council members will be informed regularly, at least monthly, about market and production plans,
- through better information, coordination of production within the whole steel groupe will facilitate transfers between the various works, to jobs that can be reasonably expected of the workers concerned,
- a precise, comprehensive and regular information to Works Council members in all works within the new steel group is guaranteed by the union/management agreement thus furthering the trade union objective rights from the very beginning of such a plan to obtain the greatest possible equalization of the workload throughout all works to maintain the maximum number of jobs.

Union representatives so far have been able to exercise control on adequate capacity utilization by steelworks in Saarland but claim extension of this information across borders to comprise all units of the ARBED concern.

This restructuring programme is in full application serving the agreed objectives.

IN SWEDEN:
EMPLOYMENT GUARANTEE DURING RESTRUCTURING

The World steel crisis has also heavily affected Sweden. The basic structural counter measure consisted in the merger, 2 years ago, of the 3 biggest steel companies. A joint venture with 50 per cent for state shares and 50 per cent private participation was established.

As a safeguard, the Swedish trade unions succeeded in their demand for an employment Security Act applicable for the new steel company, forbidding dismissals of workers for 2 years. This was coupled with a whole series of complementary measures concerned with education, training and all other aspects of an active labour market policy. The Government provides financial assistance for this whole programme of restructuring.

Instead of first dismissing people and then implementing active labour market policies, the Act provides for effective prior action through application of active labour market measures at the level of each company. In this way, workers are protected during the difficult operation of finding a new job. Aspects of the programme are employment agencies on site and the provision of indemnity for mobility allowances. Vested rights in the form of wages and social benefits remain fully protected.

All these social measures are intricately linked with the structural plan. This means that many steel jobs will disappear with timely alernative employment being made available within the national economy. The key measure is the two year relief time from dismissals to permit structural change in an environment of employment security.

A NEW APPROACH IN BELGIUM FOR COMMON PLANNING
AND SOCIAL MEASURES INCLUDING SHORTER WORK TIME

Belgium with the export orientated steel industry worst hit by the crisis faces particularly serious employment problems in this sector. The Committee for

Concerted Steel Policy (Comité de Concertation de la Politique Sidérurgique – CCPS) with government, employer and trade union participation, despite changes in its statutes, failed in attempts to tackle the problems of the crisis, It was not authoritative enough to steer an effective restructuring programme. In suffered from the inconsistent policies of the Belgian steelmakers and their financial groupings which continued to compete against one another with overlapping investments and real wastage of production facilities. Since 1976, restructuring operations were undertaken often outside the Committee for Concerted Steel Policy without preliminary discussion despite increased public aid.

The Belgian metal unions in Autumn 1977 took a firm position for the future orientation of structures and policies of the steel industry and demanded that new perspectives for a harmonious and balanced development of the specific steel producing regions in the country be defined. They requested the Belgian authorities to assume an initiating, orientating and steering role in determining a common strategy for the whole sector. In their opinion this could only be done in the framework of a public status for the steel sector.

In face of mounting difficulties and trade union pressure the government authorities convened a first round table conference to examine the steel industry situation in taking into consideration the results of a commissioned study by the MacKinsey Institute, analysing basic problems in the steel industry as well as trade union proposals.

Arduous negotiations for a restructuring programme developed. In November 1978, they led to a new Conference which brought about a principal agreement on a clear definition of the avenues for a new national steel policy. This resulted in three Royal decrees for major re-arrangements. One concerns the participation of the state in most of the enterprises of the steel sector and another one the creation of a National Planning and Supervisory Committee for Planification and Control (Comité de Planification et de Contrôle). This new Committee consists of a planning section, a control section and committees for steel regions. With trade unions participating it will have authority to allocate credits for necessary investments thus alleviating the financial burden of enterprises. Overall commitments for a 5 year period will represent something like B. Fr. 100,000 million.

In response to the considerable financial aid given by the state private industry groups are expected to make corresponding efforts towards reconversion in the regions affected by the crisis. However, it is to be noted that results in this regard are still largely insufficient.

This entire restructuring effort is backed up by basic social measures. A major social breakthrough is the reduction of working time with full wage compensation by 2 steps of one hour each from a 40 to a 38 hour week during 1979.

This measure is understood to contribute to work share-out, thereby safeguarding employment in the steeel industry.

IN LUXEMBOURG: PAY GUARANTEE FUND, ANTI–CRISIS DEPARTMENTS AND TRIPARTITE CONFERENCE AGREEMENT

In Luxembourg where steel is the very backbone of the countries economy repercussions have also been severe due to almost complete dependence on steel exports.

Important social measures were taken as early as 1976 through collective bargaining and tripartite negotiations between trade unions, the steel companies and the government. A key point in the new collective contract at that time was the creation of a jointly administered pay guarantee fund. This constituted practical realization of a proposal submittet by IMF at the meeting of the Tripartite Iron and

Steel Committee of the ILO in September 1975 which called for the establishment of such funds. This fund was created in both Luxembourg steel companies, ARBED and MMR-A (Métallurgie Minière de Rodange-Artus) to finance an agreement for the workers protection against the negative effects of market fluctuations and restructuring. This agreement was signed by both parties as a supplement to the collective contract. It provided for joint administration of the funds.

In addition, the collective agreement increased the number of rest days for workers on continuous shifts from 20 to 23 days off work as a step for job saving.

Later, anti-crisis departments were created in both Luxembourg steel companies with the participation of a trade union representative. These departments have to provide employment for steelworkers not absorbed in the production process and fully guarantee their wages.

In March 1979 a new settlement, called the Tripartite Conference Agreement, was reached. It guarantees B. Fr. 22,000 million new investment until 1983 for restructuring and modernization of production facilities without expanding capacity. It also guarantees that there will be no dismissals until 1983 when new negotiations will be undertaken. A third guarantee given is that the workforce will not fall below a minimum of 16,500 workers as against 20,000 at present. A special committee will every 6 months examine the fulfilment of the Agreement.

Moreover, a new collective agreement of 1 January 1979 provided for two further rest days for shiftworkers apart from a 1,5 per cent real wage increase above the automatic adjustment of the cost of living.

UPHEAVAL IN THE FRENCH STEEL INDUSTRY DUE TO LACK
OF ACTION IN THE SOCIAL FIELD

In France where the steel industry has been in the throes of a crisis for some considerable time, severe social upheavels occurred in this sector in 1979 as a consequence of the announcement in December of the previous year of shut-downs of entire plants phased over a 2 year period, involving the loss of a further 21,700 jobs.

These decisions, coming on top of major cut-backs in the same regions in recent years, have had a disastrous psychological effect and have thrown workers and their families into disarray. The financial measures announced by the authorities to bring in replacement industries do not seem sufficient to create in time suitable jobs for the steelworkers affected.

Taking account of the worker's major concerns, as expressed within the plants and in mass demonstrations, IMF affiliated unions in France — FGM-CFDT and the Metalworkers Federation — Force Ouvrière — persuaded the authorities and the employers that more time was needed and that adequate provisions had to be taken to keep the areas concerned economically alive, enabling steelworkers affected to retire earlier and/or undergo appropriate retraining.

As a result of trade union pressure, projects were rushed through to provide some new jobs, 6,200 in the car industry in Lorraine by 1983 and 5,400 in the North.

As it is, these measures remain belated and grossly inadequate, so that urgent action in favor of further industrial development is required.

It is out of the question that trade union organizations stand surety in any way for industrial decisions taken on the basis of questionable calculations. However, authorities and firms must agree to regular checks together with the trade union organizations on the conditions for implementing or adapting such decisions.

178

Negotiations began at the end of May 1979 for renewing the social protection agreement for the steel industry which was to replace a previous agreement reached during the 1977 crisis. As far back as July 1966, a general agreement under the name of "Plan Professionnel" was concluded between state and employers, whilst accompanying social measures had been lacking for over 10 years.

The latest social protection agreement which was concluded on 24 July 1979 contains essential guarantees for workers whose employment is affected, with the following notable provisions:

— early retirement at 55 on 90 per cent of last gross earnings for one year, then 70 per cent up to age 65;
— the possibility of voluntary redundancy from 50 to 55, according to needs, on 79 per cent of last gross earnings;
— stronger guarantees in the case of changes within the industry or concerted restructuring outside the steel sector;
— retraining;
— continuation of the process of reducing working hours (1 hour for non-continuous operations);
— establishment of an improved minimum wage;
— extension of the agreement to cover virtually the whole French iron and steel industry.

FRENCH UNION PROPOSALS

In presenting arguments and proposals for job safeguards, conversion and new industry development, IMF French affiliates point to the low level of research and development in French industry in comparison with other advanced countries, the absence of a well-balanced and integrated regional development and the imbalance in the production structure of the French steel industry. Specific demands are put forward for:

— relaunching steel consumption in France to meet potential needs;
— re-arranging sufficient competitive capacity for liquid steel and rolled products;
— pursuance of a national coordinated investment production and research development programme;
— industrial planning for depressed steel regions and
— the improvement of employment conditions through social measures particularly the reduction of working time, namely shorter weekly working hours, introduction of a 5-shift system, earlier retirement at 55 years of age.

Strong emphasis is placed on the claim for a national tripartite set-up on matters of restructuring and industrial policy permitting trade unions to play an effective role.

SOCIAL DIFFICULTIES IN THE BRITISH STEEL INDUSTRY FACING A PARTICULARLY SEVERE SITUATION

The effects of restructuring have been particularly severe in Great Britain. In one year alone, 1978, more than 20,000 jobs were lost. British steel-workers are suffering the combined effects of a massive modernization programme, general problems associated with the restructuring of the world industry and the secular decline of British manufacturing industry and therefore of domestic metal consumption.

Although important features of the 1973 Ten Year Investment Strategy have now been cancelled, closures have continued apace. In each case the regions affected have been traditional heavy industrial areas, and the number of jobs dependent on steel extremely large. As a result, the prospects for alternative job creation seem severely limited.

The initial trade union response to rationalization associated with modernization was to accept job losses provided alternatives were made available. Under pressure of the recession on the one hand, and the increased appetite for redundancy created by Severance Payments on the other, the trade union position shifted to one of accepting closures where employees were not opposed to them. Many of those who took redundancy deceived themselves with a faint hope of a new job in the future, a hope which turned out to be a mirage in most cases.

In 1976 the British Steel Corporation founded BSC (Industry) Ltd, a job creation company with trade union directors as well as management representatives. BSC (Industry) Ltd has had some limited success, but its efforts are dwarfed by the scale of social problems. The particular difficulty in Great Britain is not the creation of infrastructure — these are easily made available under regional policy — but in motivating investment to use these infrastructures.

Under the Labour Government a number of measures were set on foot to create jobs by Work Experience, Temporary Employment Subsidy and other means. Unfortunately, a large part of this programme, particularly that affecting Youth Opportunities, has had its budget slashed by the present Conservative Government. It is also regrettable that even the limited amount of aid available to distressed areas has been curtailed, principally by changing the status of receiving areas. It must be said however that Regional Policy, especially since the withdrawal of the Regional Employment Premium in 1977, has been far more successful in creating capital intensive industry than jobs.

Where BSC employees are faced with redundancy advisory and retraining authorities work with the Manpower Services Commission to prepare them for new jobs in other industries. In Steel there have also been examples of cases where a young man in a closing works has changed placed with an older man in a viable works, permitting the first to continue to work and the second to take redundancy.

Finally there is a general reluctance to move away from established steel areas, even to jobs in other established steel areas, particularly when employees were built into a steel community as an escape from other depressed areas. Such was the case of Corby in Northamptonshire. Although there are some mobility incentives, they are unlikely to block down a natural reluctance to move and in any case the areas of "full" employment to move to are becoming increasingly rare.

POLICIES FOR EMPLOYMENT SAVING IN THE UNITED STATES STEEL INDUSTRY

In the United States, steelworkers employment is protected essentially through the union rights laid down in collective agreements and through plant level negotiations. There is extensive protection through the supplementary unemployment benefit system (SUB) for laid off steelworkers. Substantial severance payments are another major safeguard.

Coutinuous efforts have been undertaken by American unions, particularly the United Steelworkers of America-USA, for timely social adjustment measures through application and improvement of specific legislative measures in this field.

The USWA have launched many initiatives at various levels to safeguard jobs in an efficient steel industry. One of the main attempts for job protection and new employment possibilities lies in the struggle of the American trade union movement to bring about and maintain sound economic development in the framework of overall policies for substantial economic growth.

JAPANESE MEASURES FOR EMPLOYMENT GUARANTEES IN STEEL

In the Japanese steel industry a four stage measure was considered to regulate employment problems. The first measure is the elimination of all overtime, worked to a great extent prior to the steel slump. Once these possibilities will have been exhausted the next step is redeployment into other sections of the plant or establishment of the same concern. Only in the third phase, can a worker be temporarily laid off and, as a last resort, agreements are reached on voluntary fully compensated early retirement.

In accordance with the general adaptation pattern as described above, a majority of the involved enterprises have adapted themselves to the following measures in fact:

a) on a major premise that the employment should be retained and secured;

b) while redundant labour force has been adjusted mainly by no replacement of these employees, who are separated due to normal attrition;

c) the diversified rationalization programs were profoundly implemented, aiming at curtailment of the fixed and fluctuated expenditures, by virtue of elimination of excessive and/or obsolete facilities, reduction of stockpile by the intensive curtailment in production, introduction of continuous castings, energy savings, and improvement in the yield or extraction rate, and thus the enterprises were able to make profits even under the substantially curtailed operation, create a viable system to maintain each enterprise's economic balance, and finally overcome the crisis.

Meanwhile, the electric furnace makers of the small and medium scales were extremely beset with their surplus capacity and manpower, and coupled with their fragile business basis they were forced to implement the employment rearrangements including dismissals, applying the fourth step out of the afore-mentioned four adaptation methods in this case. At the same time, a number of the unemployment relief measures adopted by the State were also factually applied to the workers involved in these rationalization programs to the fullest extent, which had been gained through the extensive combat organized by the general, unified action among the various industrial trade unions to cater for the workers employed by the industries in structural depression. Precisely speaking, such relief measures covered the unemployment benefits payable for a period far longer than those applicable to the average workers unemployed, the incentive subsidies to promote possible re-employment for the workers concerned and the effective assistance in their job re-training.

There have also been various trade union initiatives with a view to stimulating steel consumption, for instance, through such projects as the construction of an off-shore airport on steel structures and scrap-and-build schemes in shipbuilding.

POSSIBLE INTERNATIONAL ACTION IN THE FRAMEWORK
OF THE EUROPEAN COAL AND STEEL COMMUNITY

In carrying out their national policies IMF affiliates indefatigably insist on coordinated international action in the steel industry and general economic development. In this context, European metal unions see a specific role for the European Community in seeking a healthy steel market through international negotiations.

This depends, however, on the political will of the member states of the European Community who, at the same time, should initiate dynamic measures for revival by adapting a wide ranging programme of investments for collective facilities on a community scale and within the framework of regional economic policies.

In fact, Article 56 of the European Coal and Steel Community (ECSC) Treaty gives the Commision the power to take action "if the introduction of new technical processes or equipment should lead to an exceptionally large reduction in labour requirements in the coal or steel industries making it particularly difficult in one or more areas to re-employ redundant workers" or " if fundamental changes in market conditions" for one of these two industries "should compel some undertakings permamently to discontinue, curtail or change their activities". It can facilitate "in any other industry the financing of such programmes as it may approve for the creation of new and economically sound activities capable of reabsorbing the redundant workers into productive employment".

The specific social measures that may be taken in connection with Article 56 of the Treaty concern non-repayable aid towards:

— the payment of tideover allowances to workers;
— payment of allowances to undertakings to enable them to continue paying such of their workers as may have to be temporarily laid off as a result of the undertaking's change of ability;
— the payment of re-settlement allowances to workers;
— financing of vocational retraining for workers having to change their employment.

These provisions of the ECSC Treaty are a unique and exemplary international instrument for practical measures to both initiate and substantiate national adaptation and re-employment programmes. As they involve an international authority in the carrying out of social action, they should provide for the coordination of social programmes to the mutual benefit of all nations.

This action can be all the more comprehensive as the European Community can facilitate the execution of investment programmes by individual mills while the European fund for regional development can provide funds for financing investments in the regions affected, besides similar measures of assistance through the European investment bank.

THE PRESSURE FOR EFFECTIVE SOCIAL MEASURES
WITHIN THE EUROPEAN COMMUNITY

Despite a number of action programmes and measures drawn up by the Commission to assist in solving problems in particular sectors like that of steel, all the above possibilities have been used insufficiently. Progress towards obtaining the Council of Ministers' approval for the necessary measures have often proved slow. No overall industrial strategy has been worked out and only a small fraction of the Community's overall budget is allocated to the redevelopment and restructuring of industry and in a field like the development of alternative energy resources which could provide additional employment. In view of these failures of European Community action the Consultative Committee of the ECSC explicitly stated that support measures taken solely at national level were wrong as this can distort competition and produce surplus capacity. It urged therefore the European Commission to draw up, in consultation with the relevant socio-economic interest groups, a more comprehensive programme of structural reform and industrial development, based on an overall strategy and designed to develop proper measures to stimulate new technologies and industries.

As to direct social action for affected workers, the European Commission in its report on "Measures to be taken by the Community in 1980 to combat the crisis in the iron and steel industry" points out that "the rationalization moves currently afoot, whether they concern the closure of plants or the improvement of their competitiveness, involve what is at times a wide discrepancy between lay-offs in the steel industry and the practicable temporary or permanent redeployment".

In the light of these circumstances the Commission has sent the Council a draft decision containing a proposal for financial participation by the ECSC over and above the conventional intervention measures, in temporary measures of an exceptional nature designed to benefit the steel industry. It is stated that such a participation would apply provided that the autonomy of decision of the social partners is not infringed and that a boost to employment could be achieved without harming the competitiveness of enterprises". These measures would have the following character:

— early retirement of workers who are at least 55 years of age, or of workers under 55 years of age if they have been engaged in work of an arduous nature;
— assumption of partial responsibility for the loss of wages resulting from measures aimed at phasing out jobs or overtime;
— payment of special allowances for the partial covering of additional wage costs incurred by reasons of re-organization of the labour cycle.

In the framework of further social measures to be carried out with the greatest possible international cooperation, the Sub-committee for Labour Problems of the ECSC Consultative Committee gave special attention to methods of implementing early retirement, the restructuring of shift or team work with a view to introducing an additional shift and proposals aimed at shortening the working week as well as restrictions on overtime.

PRINCIPLES AND MAIN DEMANDS FOR ACTION

The description of the various experiences that have been made in tackling manpower problems, adaptation and re-employment in the steel industry of different countries and possibilities of common efforts by the European Community is indicative of major items for action.

In this context, IMF and its affiliates submit the following main principles and demands for policies effectively protecting steelworkers in the restructuring of the industry, safeguarding their employment:

— employers, governments and trade unions must seek with greatest urgency practical policies to revive the steel industry and restore sound conditions, through measures ensuring full social protection during any required structural change;
— selective employment creation programmes must be carried out in affected steel regions, geared towards economic development, improving living standards through a higher quality of life and contributing to the satisfaction of the greatest needs;

- job creation and social progress require a dynamic economy in a context of pressure for better working and living conditions and based on fair expanding trade and management initiatives for sound and sensible investments;
- the provision of alternative employment must be a first-priority measure, and well in advance of any restructuring of steel facilities;
- in any case, and particularly where the opportunity of first-priority planning of alternative employment has been missed, guarantees must be obtained in free bargaining between employers and trade unions and with the aid of public authorities, to maintain intermediary job and income guarantees permitting the restructuring of the steel industry in affected regions;
- at the same time, rationalization must be accompanied by social progress in the whole economy to secure new market outlets so as to avoid mass unemployment. Because technical change is not tantamount to social progress, man-made technology must be subordinated to man. Similarly, market disruption caused by policy changes must be overcome by positive countermeasures in economic and social fields;
- to enable the steel industry to meet its social and economic obligations in a difficult period and to permit orderly production adjustments for the benefit of all, determined government intervention has become inescapable and is indispensable. It must be carried out without prejudice to the collective bargaining rights of the contracting industrial parties seeking progressive social and economic solutions;
- trade unions must be able to play an effective role in the establishment and implementation of programmes for the revival of the steel industry and in all connected measures for structural change and timely re-employment. Democratization systems in industry and in the economy are very important. Similarly, collective bargaining and trade union negotiations develop basic initiatives for effective measures aimed at overcoming the anticipated social effects of structural change;
- production facilities in a critical economic and social situation should be constantly reviewed and the threatening plant closures notified as soon as possible so that required measures can be taken in good time;
- in this framework when planning and timing new investments, full consideration must be given to existing facilities and their potential modernization, with a view to maintaining regional employment;
- temporary subsidies should be for employment security and social guarantees to bring about smooth industrial adaptation without loss of jobs. Over and above such indispensable aid to affected workers, there should be no subsidies that create market disruptions and maintain hopelessly obsolete manufacturing plants;
- the use of subsidies for conversion of production to preserve full employment should be carefully examined so as not to upset the market since they encourage the adjustment process with social safeguards. This positive social assistance should conform to the basic principles for international cooperation throughout the steel industry;
- basic principles should be worked out for a medium and long-term marketing concept allowing for constant international negotiations regarding national steel industry measures to find social and economic solutions compatible with fair competition;
- to this end, national measures for the re-organization of the steel industry from the outset should be taken after international consultations and in the light of factual conclusions drawn from the full awareness of worldwide problems;

— social plans should provide for the full protection of workers in all situations, in order to facilitate necessary adaptations. These plans include measures such as early retirement with a guarantee of income and full pension rights, additional rest days and special work programmes. Social planning commissions should be formed in each steelworks to permit farsighted measures that avoid any possible hardship;

— personal planning with effective trade union participation exists in all companies and must give the priority to manpower problems. It must be long-term and comprehensive to maintain a stable level of employment whilst improving the employment structure and safeguarding employed workers' personal and social needs;

— great emphasis is to be given to active manpower policy, with all its aspects of facilitating the availability of jobs, training and retraining, assistance for mobility, as vital measures for industrial adaptation;

— there should be an international exchange on social policy measures, particularly in connection with reorganization programmes for national steel industries and regional steel centres;

— an exchange on all relevant data on employment and social programmes is to bring about the necessary coordination to prevent social measures being taken to the detriment of other countries. Such cooperation is fundamental for mutual understanding of problems and the pursuance of policies that avoid social drawbacks in the expansion of new industry facilities whilst ensuring full social safeguards at production centres exposed to structural change. OECD is a unique platform through which such policies can be realized;

— an important measure for coping with reduced work availability arising from structural streamlining and other rationalization measures is the reduction of working time in its various forms. It is to be carried out on an international scale with the necessary flexibility consistent with national circumstances. International coordination of these measures will ensure similar social costs to the steel industry of the various countries and thus eliminate cost handicaps and facilitate such medium and long-term employment saving policies;

— a permanent link is to be established between the traditional steel industry in industrialized countries and the rapidly growing new steel industry in developing countries. Apprehensions on both sides, due to profound changes in the inter-relationship of world steel, should be rapidly overcome in the common interest for a sound overall development of this industry enhancing its social standard.

FURTHER WORK AREAS FOR OECD

Manpower problems, adaptation and re-employment should receive close attention with a view to effective international coordination of social policies in order to attain both the social and economic objectives of the mandate assigned to the OECD Steel Committee. To this end, the following work areas should be explored:

— policy elaboration, backed up by the continuous information system and surveys concerning production capacities, trade flows and forecasts, must concentrate on the right priorities to fully consider the human factor in this industry. This requires full consideration of all aspects of employment with regard to the state of the industry, its plans and development prospects;

— seek employment data so as to be able to assess the injurious impact of cyclical down turn, trade fluctuations and/or capacity retrenchment; and

185

the dimension of the social readjustment programmes. Such data collection would entail common efforts of government to provide substantial employment and social statistics on a common basis to show market and technological impact on employment, employment changes, resulting from Adjustment Programmes, natural attrition, dismissals in case of partial or total closure of facilities, training and retraining needs, state and chage of working time (work week, short time working, overtime, holidays) and plant and industry efficiency;

— reporting on investment projects should include quantitative and qualitative information on employment with regard to both newly planned facilities and its impact on the existing industry structure;

— there should be special studies on new important technological innovations in steel regarding their consequences on employment;

— through permanent contacts with governments, employers and trade unions in developing countries with a rapidly expanding steel industry, OECD should create a work basis for common enquires into steel needs and new steel production projects and prospects of steel trade between developing countries and traditional steel producers. This should be done with a view to sizing up the importance of geographical shifting of production and its impact in terms of employment.

There should be inventories for comparison of social policies in OECD member countries with a view to coordinating action in this field. Such inventories should be undertaken with regard to the following main programme areas:

— special programmes for maintaining as many steel jobs as possible in a viable industry under conditions of a fair steel market;

— activities carried out in member countries to successfully restructure affected steel regions through reconversion measures and active manpower policy;

— the role of collective bargaining and negotiations with government assistance to advance employment and income guarantees, projects for early retirement, provisions for relocation of the labourforce, retraining schemes, the guarantee and improvement of social standards, steelworkers protection through social plans and agreements for the reduction of working time in its various forms with full compensation;

— subsidy policies, their criteria and their social objectives and effects;

— conditions and government measures with respect to the work environment (health and safety) and the general environment in connection with the restructuring of steel facilities and the endeavour for reconversion.

This large field for OECD work on manpower problems and adaptation in steel industry and re-employment should merit the establishment of a committee of social experts with effective trade union representation.

Indications of practical experience already gained and concrete proposals outlined in this report show the scope of the problem and its basic importance. Action in this field is vital to secure better prospects for steelworkers.

CHAIRMAN:

Thank you very much Mr Sirs. In presenting you I forgot one of your titles — that of President of the International Metalworkers Federation's Iron, Steel and Non-Ferrous Metal Department.....

I would now like to ask Mr McBride, who is the President of the United Steelworkers of America, to present his paper on the steel industry in the international context.

THE STEEL INDUSTRY IN THE INTERNATIONAL CONTEXT

by Lloyd Mc BRIDE,
President United Steelworkers of America

The history of steelmaking has been fascinating, controversial and often volatile. Worldwide, about two million workers currently depend upon our industry for continued employment. Countless millions more, in hundreds of steel communities, have been witness to this history and have been subjected — sometimes hopelessly — to the adverse effects of market conditions which have precipitated corporate and/or governmental decisions on steel shutdowns or restructuring plans.

Steelworkers are engaged in hard and dangerous work; their safety and health require constant vigilence. Nevertheless, they are good workers, proud of their jobs, and have fought vigorously to improve the *conditions of work*. The industrial history of each country can well be documented by the social struggles of steelworkers. I refer to those struggles merely to indicate that the same determination by which steelworkers have expressed concern over the *conditions of work*, they are now expressing over the *existence of work*.

Job loss has always been a persistent fear of working people. Unemployment, occurring during economic downturns, has elicited unanimous demands from workers for government measures to overcome negative market conditions, thereby increasing the number of available jobs.

Unemployment is not just an economic issue. It is a major political problem, requiring appropriate and necessary political responses. Public policies have been developed to moderate the economic and social hardships of unemployment, and to stimulate further job creation. Such public initiatives relating to employment and income maintenance programs have addressed both short term problems of unemployment, and the longer term concerns for national manpower requirements.

As we all know, negative economic conditions are not uniquely national, either as to their causes or consequences. The existence of the OECD is a recognition of the international dimensions of domestic problems and of the need for international cooperation in alleviating national pressures.

Over the last decade, a peculiar set of problems have beset the steel industry, problems which are not just cyclical in nature. Steel unemployment has become more than just a temporary phenomenon. As plants shut down and steel companies implement retrenchment and diversification plans, unemployment has become permanent. These profound structural changes have affected not only individual steelworkers, but also the economic base of steeldependent communities, along with the texture of national manpower policies. It is for this reason that the United Steelworkers of America, the AFL-CIO, and indeed the International Metalworkers Federation, urged and endorsed the establishment of a permanent International Steel Committee within the OECD.

It is our firm conviction that the same special attention which has been directed to the macro-economic problems of OECD nations through this international forum, also be directed to the steel sector, especially since sharp trade fluctuations have forced a more rapid restructuring of our respective steel industries.

I. READAPTATION POLICIES

Unfortunately, the restructuring process is not occurring gradually, nor in a period of strong economic growth, so as to avoid social injury. Instead, a whole series of adverse factors have been compressed into a short time frame such as: recessions in steel markets, accumulations of obsolete facilities, under-utilization of productive capacity, shrinkage of export markets in developing countries, restrictive pricing policies, lack of capital formation for modernization and low rates of return on investments. Our respective governments have recognized that these factors result in unemployment of such massive proportions so as to threaten social tolerance levels. Thus, some well-defined social adjustment programs have been developed. I would certainly concur that one basic proposition on which there should be an effective consensus is that the social impact of large scale dislocations must be moderated by adequate transitional measures. While I am not in a position to judge how long the restructuring process will take before there is stability in the steel industry, experience indicates that turbulent years lie ahead.

A recent report to the OECD Council, *Positive Adjustment Policies* (April 26, 1979), emphasized the key role of such programs during industrial restructuring:

...effective selective manpower and employment policies is a necessary pre-condition for successful economic adjustment to (structural changes).

...Since governments have objectives other than economic efficiency, the choice of adjustment measures is never determined solely on strictly defined adjustment criteria, or on grounds of economic efficiency alone. Social and political trade-offs, which can not be judged only on economic grounds, must frequently be made...

From time to time, however, and in recent years with increasing frequency, resistances to adjustments are building up and economic dislocations of efficient significance or severity occur to require government support of the adjustment process which would not (or too slowly) be forthcoming.

There is now more awareness of the reality and inevitability of the steel restructuring. Therefore, there should be a full recognition on the part of both government and industry of the social obligation due affected steel workers and communities. I cannot agree — at least from my own experience — that such obligations and the need for adequate compensatory measures are fully accepted by industry and government.

Nevertheless, with or without adequate worker and community adaptation policies, steel industry restructuring will take place. The issue before us is how painful will it be. I believe that government policy makers — those who understand the structural changes that are occurring and who are encouraging a more rapid restructuring — have a clear responsibility to insure the development and implementation of policies and programs to absorb the shock. It is not necessary that such policies and programs be entirely government financed. However, they should be mandated as a matter of law.

There are many adjustment assistance provisions which can be negotiated by workers with their employers. But unions cannot be expected to negotiate contracts in which large shares of earnings be deferred in the expectation of a future shutdown. Early pension benefits, wage maintenance programs and severance pay plans have been developed. But there are limits to such options because wage increases are needed now and cannot be postponed. Moreover, these are individual benefits and are unrelated to the very serious community adjustment problems which must be confronted when a steel mill closes its gates.

Because of the realization that plant shutdowns will occur, I suggest that the OECD Steel Committee, as one of its principal tasks, undertake the development of firm recommendations on the content and implementation of adjustment assistance programs for workers and communities. Because of government involvement in rebuilding the steel industry, special attention should also be directed at the trade effects of any industrial adjustment assistance programs.

In this respect, I refer you to the original OECD decision to establish the Steel Committee which made specific reference to the understanding that the participants "agree to make every effort to provide effective programs for steelworker readaptation away from facilities affected by structural adjustments into alternative employment". The statement quite rightly warned that "domestic policies to sustain steel firms during crisis periods should not shift the burden of adjustment to other countries".

The uneven national responses to the social consequences of structural changes cannot be ignored. Fundamentally, our union has adopted a policy to encourage rapid investment in modernization. We are aware of the fact that an effective modernization program may result in reduced employment levels in our steel industry. A recent American Iron and Steel Institute publication, *Steel At The Crossroads*, predicts a drop in manpower requirements over the next ten years of some 22,000 jobs under the *best modernization scenario*. What this means is that there will be 22,000 less job apportunities in the American steel industry over the next decade. Nevertheless, our union has endorsed the need for an aggressive modernization program because the final result will be secure employment for American steelworkers. What we require now is some degree of economic security for workers so that the modernization process can proceed.

The point I wish to emphasize is that worker and community adjustment assistance in the US is minimal. Presently, the burden of adjustment falls mostly upon the worker himself and his community. Other countries have more effective means of easing adjustment and dislocation effects. The uneven quality and content of national adaptation policies are perhaps indicators of the political or governmental responses sought by workers with regard to import penetration during periods of intense restructuring. The human dimension of any profound structural industrial changes will elicit some kind of a policy response. Hopefully, the response will be positive and consistent with international economic and social principles. An OECD Steel Committee initiative in the area of structural adaptation problems would be most helpful — indeed, it is necessary. The OECD Council recognized the urgency of these concerns when it stated: "The Interrelationship of developments in the steel sectors from country to country and the potential that unilateral actions and policies can aggravate the problems of others have become clear". Failure to have an adequate adjustment policy carries with it the same disruptive potential.

II.   RESTRUCTURING POLICIES

I must admit, however, that worker adaptation policies receive very spotty acceptance both because of their inadequacy and because of steelworkers' concentration on structural policies. Other papers at this seminar have indicated the profound pressures which are propelling structural changes in the steel industry. Again, the OECD Council, in establishing the Steel Committee, acknowledged "in virtually all major steel producing nations, steel occupies a central place in the national economy. In a number of major areas, the magnitude of structural problems confronting the steel sector, and resultant social and economic indications of the necessary adjustment, are substantial".

The indicies of a growing crisis are well displayed. While world steel production levels increased 153 per cent between 1955 and 1976, ever greater percentages of that production have been committed to export markets. Thirty-seven per cent of Japanese steel production is dependent upon exports as is twenty-six per cent of the EEC's production. These are very large allocations and indicate a high degree of sensitivity to trade reliance.

The addition of new capacity in developing countries has put traditional steel exports under heavy pressure, especially during periods of slack or slow growth in steel demand. Excess capacity and lower utilization rates negatively affect exporting and importing countries alike. Pricing weakness and increasing loss of domestic markets to imports cause great uncertainty over the viability of restructuring programs. Since 1969 there has been a decrease of 100,000 steel industry jobs in the US and imports have fluctuated between fourteen and eighteen per cent of available steel supply. Negative restructuring has already eliminated some 5 per cent of our steel capacity and further retionalization is expected, especially as our economy becomes more recession-oriented.

Since the closing of major US Steel facilities in 1977 and again in 1979, there has been an intense focus on the role which government policies play in steel restructuring strategies. This has been a particularly sensitive period both in terms of national policies and international cooperation. As we all know, a number of steel trade actions have already occurred: The Voluntary Restraint Agreements of 1969 and 1974: the US specialty steel quotas of 1977 through 1980; the EEC Davignon Anticrisis Plan of 1978; and the US Trigger Price Mechanism of 1978. There have also been bilateral arrangements and, of course, anti-dumping actions. Monitoring of such trade actions by the Steel Committee is essential so that trade transparency can be achieved and injury avoided.

As it develops its program for work for 1980, the Steel Committee must give priority consideration to the global restructuring efforts. An attitude of sensitivity to rapid fluctuations in trade flows is necessary if domestic steel industries are to initiate and maintain restructuring and rationalization programs. As we are painfully aware, higher levels of import penetration during recessionary slowdowns cause unemployment. It is difficult to accept steel imports at prices below the cost of production.

There is international recognition in the GATT agreements that such trading practices are not a sound and fair basis for stable trade relationships. In terms of unemployment, we have noticed that US steel import levels are high when domestic capacity utilization rates are low and workers are laid off as a consequence.

However, I believe there has been a shift in national concern away from only the short term adverse effects caused by a decline in steel demand to one of uncertainty over the long term prospects of being able to implement structural improvements. This is an extremely critical issue.

A steel industry is vital to each industrial nation. Developing countries also recognize the role that a domestic steel industry plays in satisfying domestic demand as they move toward greater self-reliance. In order to cope with the multitude of adjustment factors facing the steel industry, government policies are becoming more specific and definitive. Each country must develop a national industrial steel sector policy thereby creating a proper environment for structural reforms. As workers we are additionally concerned that such industrial policies insure that steel investment funds are not diverted into non-steel activities or away from traditional steelmaking communities. Difficult policy assessments will have to be made as to the relationship between capacity levels and domestic demand, as well as on the reliability of and access to export markets. Current global excess steel capacity, in spite of prospects for favorable future demands, is forcing such a review.

Along with the emergence of national steel policies, there must also be an international steel sectoral approach. International coordination is needed so as to assure that domestic steel measures neither distort trade relationships nor disrupt restructuring programs. A more complete international dialogue and understanding is required because of the large commitments of resources necessary to achieve stability for the industry and protection for steelworkers and their families. Plans for modernization and/or rationalization must be allowed to unfold so that abrupt dislocations can be avoided. Recognition must also be given to the fact that undue import levels, resulting from unfair or predatory pricing, will put heavy strain on capital formation requirements, restructuring plans and employment levels.

Temporary national measures to restrain dumping, like the US Trigger Price Mechanism, do not assure stable trade relationships. Neither do threats to engage in adversary confrontations under the GATT codes. The main challenge to the steel industry is whether the necessary steel modernization programs can commence and a reasonable reationalization process be undertaken. Such a prospect can be fulfilled only if the OECD Steel Committee can develop policy measures for orderly restructuring activities.

An OECD Steel Committee draft work proposal identifies the positive aspect of such a policy possibility:

"Consideration would also need to be given to readaptation measures that would assist in the modernization and structural improvement of the plant in the participating countries as part of the effort to improve the competitive position of the different industries. The results so far achieved by the measures and their likely future period of application would be considered in order to ensure that measures taken promote adjustment in the steel industry in as positive a way as possible. It will also be necessary to assess their compatibility with this objective of returning to a situation of full competition and their relationship to measures aimed at achieving fully competitive enterprises in the longer term".

CONCLUSION

By encouraging structural changes in our respective steel industries we can avoid defensive domestic policies. We can, through a positive policy of modernization, maintain high standards of living for steelworkers and provide economic security to steel-dependent communities. To accomplish this goal, we need an international framework so that difficult decisions can be made in an atmosphere conducive to the avoidance of market disruptions. While the steel restructuring process is not dependent on trade actions, a potential for trade-related damage exists. The OECD Steel Committee, however, can play a positive role in facilitating necessary policy adjustments.

There are pressures on our steel industry to meet other social, occupational and environmental health obligations. Our union maintains that these commitments can be met even during the period of revitalization and modernization. The option to abandon these social requirements is nevertheless attractive, particularly if there is no assurance of an acceptable steel market once the modernization program is completed.

Our governments have the responsibility to develop industrial policies for steel. Through the OECD forum, they have the opportunity to coordinate national efforts at modernization to overcome any trade disruptions that might occur. As workers, we hope that the opportunity will be seized so that all of our nations can achieve stable, productive, and profitable steel industries.

CHAIRMAN:

... I would now like to call on Mr. Sanden who comes from the Norwegian Trade Union for Iron and Steel. And he will be talking to us on the restructuring of the industry and the readaptation of manpower in the industry.

RESTRUCTURING OF THE STEEL INDUSTRY AND THE REEMPLOYMENT AND READJUSTMENT OF LABOUR

by Knut Arne SANDEN,
Editor,
The Norwegian Iron and Metalworkers' Union

1. INTRODUCTION

The Norwegian Iron and Metallworkers' Union considers it to be of great importance that this Symposium is taking place. We also believe it to be important to have such discussions at a high international level, as the problems facing the Trade Union Movement in connection with the difficulties in the steel industry are based on international causes and therefore they have to be solved internationally.

Even though we do not expect that this conference will arrive at unambigious guidelines as to how the problems in the steel industry should and will be solved, the exchange of views taking place at such a conference will have its effects and importance.

FUTURE DEVELOPMENTS

The steel industry will also in the 1980's be confronted with problems of low-growth in consumption and production. One uncertainty in this situation is the possibility that the difficult international political situation could lead to what one could call a "war economy" in some countries, i.e. increased armaments and defence budgets. This could lead to better conditions for the Western steel industry, but in our opinion this is not the way to solve the industry's problems.

In light of the low growth prospects for the economies of the OECD countries in general, and the limited number of possible growth areas for steel consumption, especially in light of the energy situation, we will not be able to avoid an extensive restructuring of the steel industry. We admit that in a number of fields the capacity is too high, and both for production and consumption of steel the growth that will take place will mainly occur in countries that are not members of the OECD.

This will put certain limits on future developments, which should be met in two ways. In the first place, by stimulating the economic growth in the OECD area and by stimulating the general consumption of steel. Secondly, through finding acceptable social arrangements when we restructure the steel industry and make the necessary reductions in the industry's production capacity.

We consider it to be important and the only correct solution to prepare within international frameworks what we would call national plans for the development of the steel industry. By national plans we mean plans covering the whole nation, with the active participation of the public authorities, employers and trade unions. This is necessary in order to obtain a managed development which does not result in unemployment. In this area, market forces must not have their free play.

3. THE NORVEGIAN SITUATION

The Norwegian steel industry is very small on a world scale, and the problems facing the workers in this sector in Norway are far less severe than what is experienced in certain other countries. Nevertheless, we also feel the effects of the international steel crisis, and one of our steelworks has gone bankrupt. Therefore we are very much concerned about safeguarding the conditions for the remaining part of the Norwegian steel industry in order that it may be profitable and provide the necessary secure jobs.

As part of this work our Union has taken the initiative for a structural analysis of the Norwegian steel industry in order to study the conditions of competition which this industry will face in the years to come, and how we in Norway should go about it in order to maintain a national steel industry, which we consider to be very important.

The Norwegian authorities have replied in the affirmative to the undertaking of such a study and our Union will be represented on the committee in charge of this work.

This shows that the Trade Union Movement is willing to go along with a necessary restructuring of the industry, but we insist on participating in the restructuring process. It is our aim that the structural analysis should lead to a national strategy, which should, inter alia, include a closer cooperation between the firms of the industry, and a more rational division of production between the firms, also on a Nordic basis. With a small home market it is necessary to concentrate on some types of products and in co-operation with companies in other countries to cover the Norwegian consumption of steel. This is important to secure the jobs on a long term basis.

4. PRINCIPLES ON FULL EMPLOYMENT

It is the primary aim of the Norwegian Trade Union Movement to maintain full employment, for two reasons: Firstly—unemployment is socially unacceptable. It leads to such great social problems that unless unemployment is low our societies would not survive and function as real democracies, and furthermore, unemployment is humiliating to the individual affected. This is a situation which we in a welfare society just could not sit back and accept. Secondly—the power and influence of the Trade Union Movement is directly dependant on the unemployment figures. When these figures become too high it will not be possible for the trade unions to obtain a breakthrough for their policy in society. This is also something which we cannot accept.

5. WHY ENGAGEMENT ?

In recent years the Trade Union Movement, not only in Norway, but in most countries, has engaged itself very strongly in the shaping of industrial and economic policies. The reason is simple—it concerns our jobs which must be secured in order that we may have the remuneration and living standards which we consider reasonable. The trade unions therefore have a deeper reason for engaging in the future of companies and industries than the owner interests which are concerned first and foremost with the profits to be earned in the various industries and companies.

6. THE STEEL INDUSTRY

It is our point of departure that we must maintain a national steel industry with as many jobs as possible. The increased competition in the years to come will force through an increasing restructuring of the sector. This is a development which the Trade Union Movement must recognize. But we cannot accept that this restructuring is forced on the workers without them being brought into the decision making process at any point or at any level. Nor can we accept that such a restructuring creates unemployment and lower living standards in society and for our members. In other words, the question is not to say yes or no to a restructuring, but in which way it should take place and over which period of time. The time factor is decisive. At the outset the employers and the trade unions will have different desires and interests, a wish for a rapid readjustment versus safe jobs. Therefore, the rate of the readjustment will depend on how solutions to the social problems created by the restructuring are found, either by employment guaranties, the development of new jobs, or by social arrangements preventing a reduction in the standard of living when loosing the job.

Pure capitalist interests cannot be made the basis for the development in the steel industry. It should not be for the employers and owners to set the pace and extent of the readjustment process. The Trade Union Movement, and not least the public authorities of the various countries must be brought into the decision-making directly in order to obtain a controlled and managed development.

7. SPECIAL LABOUR MARKET MEASURES

I mentioned that one Norwegian steelworks has had to close down. This happened two years ago and a group of scientists have collected data on those who then became unemployed. The steelworks that had to close was the dominant employer in a small community. One hours' travel from this small community, however, there is an area with a growing number of jobs connected with the oil activities in the North Sea.

The report shows that two years after the closure 73 per cent have found new work, while 21 per cent is still out of work. The interesting thing about those still unemployed is that they were the weakest groups that were hit by the closure, i. e. those with the least education, those over 45 years of age, those on hourly wages, and the women. I have a number of figures I could mention, but my conclusion is that even in places where one manages to create alternative employment, special labour market measures are necessary to assist those most affected and the least competitive workers.

The other conclusion is that half of those who found work by travelling two hours every day were not satisfied even though they got higher wages, and they wanted to work closer to their homes. This underlines the necessity to find alternative employment in those place where people become unemployed. Job mobility is something that can be accpted, but geographical mobility is far more difficult for the workers and this must be taken into account in a restructuring process.

8. SUMMARY

On the part of the Norwegian Trade Union Movement we would point out the following elements: The general economic growth should be stimulated, trade between countries should increase and not be cushioned by protectionist measures, consumption of steel should be stimulated as far as this is possible, by amongst other things giving the shipbuilding industry new orders. Research for new usages of steel should be undertaken.

Trade with the developing countries should increase. Growth in their domestic markets will help alleviate the situation in the steel industries in our countries.

A restructuring of the steel industry is necessary; this readjustment should be planned and managed with the full participation of the trade unions on all levels. The restructuring should not take place at such a pace that it will not be possible to solve the problems of a readjustment of labour. More should be put into the development of alternative jobs. Where this is not possible at the same rate as the restructuring which is taking place, the workers should be given a guarantee of employment within a certain period of time. More should be put into active labour market policies and the retraining of manpower in order to adapt it to the jobs that are vacant.

Regional development programmes should be extended in order to prevent social catastrophies in the regions with the highest dependency on steel industry. Social standards should be secured for those who become redundant in order that living standards in society as a whole do not decline. In connection with the above mentioned points reduced working hours may play a certain role, but it will not alone solve the problems we are facing.

Direct price subsidies and assistance to production will not solve the long term problems, but support will be necessary in the adjustment period.

The supply of capital to this industry is a problem, and with the low level of profits, and the big investments that are needed, necessary capital has to be made available when the private stock market is not able to do so. The workers and their trade unions should be given increased influence over their own jobs and their development, by participation on boards of directors and other decision making bodies.

9. CONCLUSION

Our views are of a principal and general nature. On our part we hope that the OECD in its further work on steel problems will place greater emphasis on the social aspects and on arriving at certain general guidelines on how the work should be done. The Norwegian Iron and Metal-workers' Union has followed the work of the Working Party of the Steel Committee with great interest.

The information and evaluations which we have received have been important for our work in Norway and they were the releasing factor for our request for a structural analysis of the Norwegian steel industry. We believe that the Steel Committee for its part has benefited from the consultations that have taken place with TUAC.

It has been a point of departure for our Union that these problems cannot be solved in isolation, by public authorities, either at the national or international level. It is the cooperation between the social partners and the authorities both at the national and international level, which will bring us further in developing the secured jobs we wish to maintain and in improving our standards of living. Only in this way shall we be able to secure a further harmonious development of our societies.

CHAIRMAN:

I would now like to call on Mr. Sambrook, who will be taking over for Sir Charles Villiers, to present the paper concerning the British Steel Corporation's programme of job creation in areas affected by restructuring.

Mr. SAMBROOK:

First of all, may I apologise for Sir Charles Villiers' absence. You know that we have serious problems, and I think it is perhaps no surprise to you that he felt that he had to remain in the United Kingdom at this time. But it was a particular regret to him that he could not come because he is very interested to hear how the forecasting was discussed yesterday and how we perceive the situation ahead. Also, he knows of the disruption which is caused socially by the various retrenchment programmes which every company has to undertake as we see that situation facing us in the future. He has taken a particular part in that because he is Chairman of BSC (Industry) Ltd., I shall try to describe that company and indicate how it works.

Yesterday we spent our time on the analysis. I wish that what we had all decided, at least as concerns the direction of demand, had been rather different. Certainly, it was a matter of disappointment to me that my own views and those of my colleagues were confirmed — that the direction is not one that is going to let us grow out of our troubles and, therefore, we have to take some action. The action which yesterday's Chairman, Mr. Hodges, summarised was that we had to get on and match capacity to demand. Now it is not surprising that in the papers we have heard this morning a great deal of concern is being expressed at the social disruption which will result. None of us disagrees about that. None of us would willingly wish to face it. But, in the light of the facts as we perceive them at this time — probably they are not going to be facts at all since somebody said yesterday that the one thing you could be sure about was that the forecasts would be wrong — if one assumes that the direction is right, we would argue that delaying action only makes matters worse. We do believe that it is up to governments, to managements and unions to try to ameliorate the social impact and to try — if we cannot find the employment — to set up the right conditions which will provide new businesses, new opportunities, new men, using new technology. Only in this way is it possible to give people new hope.

We all look back on the decade of the 70s as sad and difficult. It was a bad decade for many people around the world. The first signs for the 80s are that life is not going to be all that different, especially in steel. So we have to encourage what is new. We have to get people to look ahead to find new opportunities and then to provide fulfilment for people, jobs which will satisfy and jobs which will motivate. We have looked at BSC (Industry) Limited as a link between what was and what will have to be. Sir Charles often calls this "the hinge" between the 70s and the 80s.

But the trouble is that these are only words. It is easy to say that employers must act, they must have responsibility, they must look after their displaced employees. It is easy to define the responsibilities of government, but the problem is what actually should be done, It was in 1975 that we moved in a very small way by forming a company wholly owned as a subsidiary British Steel, which we call BSC (Industry) Ltd. Its purpose was to join others in trying to attract new businesses to areas where obsolete plants were being closed. It has always been the intention to keep it as a very small company doing a very big job.

But since 1975 life has become more difficult. Now we have to close more capacity more quickly, with all that this entails. A very large number of jobs are going to be involved. Some of those jobs are concentrated in areas of chronic unemployment and economic difficulty; some are not. We have put our efforts into those areas where there is the greatest difficulty.

We set up a special job-creation team. We looked for certain characteristics. First of all, we tried to be sure that we looked to other than big corporation men as leaders. We felt that people who are used to a deeply constructed hierarchy, where it takes many floors from the bottom to reach the Chairman's office, are not for this organisation, which we styled in a very flat way. We were willing to accept any proposition which appeared able to create jobs. There was only one restriction: that the whole affair had to be lawful.

We also set up very effective links with national and local governments, with development agencies, with the Community and, in fact, with anybody with money to help us out. But our money was made available not as a large sum but essentially for lubrication. We were very surprised when we talked to quite large companies who were wishing to expand or to relocate, and certainly when we talked to small companies, how few knew anything about the cash which was available to assist them. They did not know about regional aid which they could get in the UK. They did not know about money from the Commission and the European Investment Bank. They certainly seized the chance when we built speculatively factories in areas where there was unemployment because they could move in very quickly and get ahead with whatever ideas they had. They also took advantage of retraining funds which were available if they were to take on workers from coal and steel. We concentrated our own money on the provision of land at a relatively cheap price (land which we already had); on buildings (either new or converted), on the leasing of equipment; and on consultancy, feasibility, technical studies and so on.

It will come as no surprise to you to hear me say that the last thing the British Steel Corporation could do was to act as banker to any interested new employer in the UK. We did advertise our services very widely and were told that it would be pointless. Everybody advertises but nobody takes notice of the advertisements. But, that was not our experience. We were smothered by replies on each of the three occasions we decided to advertise. We were able to explain to those who approached us that we could offer a package — a package of people with a variety of skills who we knew and could recommend, as well as land and access, mainly to other people's money.

We have had some success. We currently have about 250 firm projects being established at a time of enormous economic difficulty. I am not saying that BSC (Industry) is entirely responsible — that doesn't matter — but these companies are there. We have 300 more firm projects which we feel sure we can carry through to fruition. We achieved about 3,500 new job commitments last year, 5,000 this year, and we are confident we can get 10,000 next year as we gear up our operations.

All right, these don't match the total need. But they do answer some of the points raised by the previous speakers, that we have to find jobs for people who we displace from their present place of work in the steel industry. Of course, the pressure is always for redundancies to happen only at the time when new jobs become available. The British Steel Corporation cannot promise this. We do not think that any company can make such a promise. But it is a responsibility that we have as an employer to work together with local and national governments.

I thought that I might just draw attention to one particular novel idea which has been introduced by BSC (Industry): the neighbourhood workshop. The whole thing was not so much a physical matter, but very much the psychology of letting people in localities see a place of work. If two, three, four or a dozen men have an idea, there is a workshop to which they can go and try out the idea. They are not burdened with legalities. They are not required to find large sums of capital. We have used our redundant buildings. We have serviced those buildings and have let them at low but economic rents. We have introduced a minimum of formality, as I said, when they were taken over.

198

In this idea lies the mainspring for future small but also some medium and large size businesses. And that is where we believe new jobs will come from in steel-making areas. Of course this is costly. Of course we get a lot of assistance from elsewhere. For the British Steel Corporation alone, it has cost in recent years about one thousand pounds for each new job that we have found. Our total budget this year for this activity — and some question that — is about ten million pounds. Quite a bit of that we shall spend in the purchase of assets which later we shall be able to sell.

Certainly, we feel that in those areas where we are having to run down our traditional steel businesses, we have to do something to give people hope. We know that in the period up to closure there is terrible disappointment and a great deal of bitterness. We know equally that at the point of closure if people can see that there is opportunity for new jobs, bitterness quickly disappears and people are encouraged to seek a new life. That is everything that British Steel (Industry) Ltd. is about. We believe that the employer does have a responsibility, and it is a moral one. It can never be a statutory one, for a bankrupt outfit cannot find the money. Nevertheless, an employer does have a responsibility for trying to find new employment for his workforce. We shall never be the banker. We do possess a great deal of knowhow and we shall supply some of our own resources for the lubrication.

I hope that other people will think that this is an area where they might do more to ease some of the difficulties of change. I do not profess that I can answer all the details, but I did take the opportunity of having with me Mr. Patrick Naylor who is the Chief Executive of this new company and who can talk to people privately or in a dialogue at the full assembly.

(Following is text of Sir Charles Villiers' prepared statement)

JOB CREATION BY THE BRITISH STEEL CORPORATION IN MAJOR STEEL CLOSURE AREAS

by
Sir Charles VILLIERS,
Chairman, British Steel Corporation

INTRODUCTION — THE TASK

The major changes that have affected the world economy in the 1970s have forced major adjustment on the steel industries of the advanced nations. Adjustment would have been required in any case to bring about the benefits of technological progress, and the concentration of production which this involved. But the collapse of the economic expectations of the early 1970s has meant that this adjustment has to be harsher and deeper if the mature steel industries are to stay in the business in the 1980s.

In Britain the need to change direction radically was fully realised in 1977/78.

Once the fundamental changes which have taken place in the market were recognised, the Corporation then began to bring capacity more closely into line with the lower demand. A none too easy task in a mature industrial society such as that of Britain. It has meant a radical restructuring programme with far reaching consequences. One of these consequences being the shedding of many thousands of jobs. Under the programme the intention is to deman to a level of 100,000 by August 1980. This is to be compared with a current figure of 152,000 and a 1975 level of 190,000. Clearly, the Corporation has by any standards embarked on a dramatic course of action which has far reaching social implications. Most of these job losses are in the less prosperous parts of the country where the harmful effects, in the absence of remedial action, would be considerable and lasting. Often they affect communities where for decades there has been almost total dependence on just one or two industries such as steel and coal. This means they are extremely vulnerable areas in the current crisis.

As a Corporation the BSC is extremely concerned over the social consequences and hardships which will result from its restructuring plan, as are the Government; one can no longer decimate a community such as the steel town of Corby and walk away. Indeed, special consideration has been given by successive Administrations to steel closure areas in respect of improvements to infrastructure, factory building programmes and the rigorous use of legislative powers to assist new job creation. Even considering these initiatives by Government, the Board of BSC felt that more had to be done to put back the job opportunities being lost as a direct result of the steel closure programme. So in 1975 the BSC took the unprecedented step of setting up its own job creation agency, BSC (Industry) Limited.

BSC (INDUSTRY) — THE JOB CREATORS

BSC (Industry) is a wholly owned subsidiary of the British Steel Corporation and its goal is not just simply to bring new job opportunities to steel closure areas but to create a climate conducive to job creation, The company is controlled by its own Board of Directors which includes 6 leading trade unionists, certain members of the Main Board of the Corporation and myself as Chairman.

It was made clear from the outset that the new company would cease to operate in an area once its job target there had been reached. Initially the company's role was essentially to advise on the assistance available to businessmen from the various sources including the British Government and the European Coal and Steel Community. However, over the past two years the company has been given more executive muscle, considerably more power and access to substantial resources. This has led to a growth in company size to a point where today there are 36 people employed. Of those employed by the company approximately one-third are based in its London Headquarters and the remainder are located in its 9 operational regions. In the main the central staff provide services and specialist aid to their colleagues in the regions. By design, the management structure is a flat one with each executive enjoying considerable freedom of action and decision making power. Such a system means that the group are able to respond quickly and effectively to the needs of the businessman who has a promising project.

It should be said that job creation is a far cry from producing steel and it needs a special breed of executive if it is to be done successfully. Therefore, one must not leave the activity to redundant company managers but rather consider the best people you currently have and also be prepared to go outside. In particular, it is important that those chosen can bring marketing flair, dynamism, and the ability to deal with and understand the problems of management at all levels of business. The team is effectively there to create an executive bridge between the sterility of Government incentives and the real needs and problems of the entrepreneur. It is a fact that substantial British Government and European aid had been available for several years prior to the creation of BSC (Industry) Limited. But they were still a mystery to many businessmen and often the businessman was totally unaware of the beneficial effects such aid could have on his operations. So a considerable amount of time and effort is put into promoting the package of incentives available. The current package may include the following major benefits:

1. *Cash grant*

   The British Government, under its regional aid policy, gives cash grants of up to 22 per cent of the cost of new buildings, plant and equipment. In addition, certain projects would qualify for further discretionary Government cash grants.

2. *Medium term loans*

   A major advantage offered to projects locating in a BSC steel closure area are low cost medium term loans of up to 50 per cent of the cost of the projects fixed assets. These key loans are provided by the European Coal and Steel Community and the European Investment Bank.

3. *Land and buildings*

   The British Government has pursued a very enlightened factory construction programme. Under this programme factories are built on a speculative basis in advance of there being a suitable tenant. The team's ability to offer an industrialist a ready made factory is often a deciding factor, and cannot be stressed too strongly.

201

4. *Recruitment benefits*

   Ex-steelworkers, and their new employers, can benefit from a compre-hensive retraining scheme. Under this scheme there is financial assistance for steelworkers undergoing retraining and BSC (Industry) is pioneering ways of helping employers offset the cost of in-house on the job training through various avenues including the EEC Social Fund.

5. *BSC (Industry) help*

   BSC (Industry) does not itself provide direct financial assistance to its clients but it does enjoy close-working relationships with all the relevant British and European organisations and so is in a good position to advise on the financing of a project. In addition, it can and often does provide specialist aid and services covering such things as market research, technical studies, general and strategic consultancy advice and detailed location studies, all at no cost to the client. It can also provide critical help with the provision of land, buildings, plant and machinery.

The company's promotional strategy has been crucial in its success to date and is interesting in that it goes against most of the advice given by regional economists and consultants. The usual argument is that one should set out to attract projects from those industrial sectors enjoying high growth rates and from those companies which are particularly compatible with the geographic areas of concern. The BSC approach is simply to attract all the projects which are currently active regardless of their type, that is to say regardless of whether they are in the service or productive sectors of industry. The major advantages of the BSC (Industry) approach are:

— The wide spectrum of companies attracted allows one to create a diversified economy and so avoid the single industry dependence of the past.
— There are many more projects to choose from with a consequential increase in number of jobs created.
— There is potentially a much wider choice of job opportunity so one has the ability to cater for the special needs of groups such as school leavers.
— The promotional message can appeal to all, be simple and direct, e.g. "convince us you have a viable project and we will do our best to make it happen", rather than "if you are in micro-electronics and have a project we will help you go to town Y".

In addition to advertising companies and individuals with business proposi-tions are also found through referral by interested agencies including government departments, direct enquiries to BSC (Industry) Limited and promotional seminars held in the major cities of Britain.

There has over the past year and a half been three major advertising campaigns conducted in the U.K. press. On each occasion the response was so great that the campaign was prematurely terminated simply because the flood of enquiries threatened to overwhelm the company's staff. This ability to generate significant numbers of good projects has meant that now the company is able, to a large extent, to control its workload and be quite critical when assessing any new venture. This last point means that the team are now in a position to separate out "the wheat from the chaff" at a very early stage and so minimise the time spent on fruitless projects. Nonetheless, processing enquiries is still a very labour intensive matter as in order to bring them to a successful conclusion staff work very closely with the client through all stages of the negotiations. They offer him a choice of location, visit sites with him, and advise him on the salient features of each location. In parallel with these visits a systematic examination of all his needs is made, covering areas such as finance, labour, buildings and marketing. So that once a site has been found the team can then pursue how the Corporation and other agencies, both public and private, can contribute towards the businessman's needs.

The business propositions being dealt with fall into four broad categories; the mobile international project which could be located in a number of countries, the mobile national project with several domestic sites to choose from, the indigenous company with an expansion project for his current location, and the start-up situation. As might be expected each of these four types demand a different kind of action from the team if they are to be won for a steel closure area. Of particular interest is the approach we are taking towards the encouragement of new small business.

There is a growing concensus of opinion among politicians and informed observers that it is the small business which offers the best opportunity for job and national wealth creation over the next decade or so.

In an effort to encourage the formation of small businesses in the steel closure areas, BSC (Industry) introduced an exciting new concept in industrial regeneration—the neighbourhood workshop. The first of these was on a six hectare site at Tollcross, a closed iron foundry, where we saved certain buildings from demolition. These buildings attracted over 54 small companies employing close to 400 people within 4 months of the doors being opened for business. An interesting and very encouraging fact is that 43 of these companies were in their first business premises. The critical issue here was to create the physical and psychological environment which would be conducive to the birth of new businesses. What should be noted here is that the workshop concept can provide new jobs and businesses quickly and cost effectively on old steelworks sites.

Under the Tollcross system the entrepreneur takes space via an informal licence with an absolute minimum of legal and bureaucratic red tape. If due to a need for further expansion, or indeed for any reason, the tenant wishes to move, he merely gives three months notice and leaves without financial penalty. The usual need for financial guarantors has been waived and provided he operates as a 'good neighbour' he will enjoy all the security of a long-term lease. On site catering for the tenants and their employees is provided by a licencee, operating like all other tenants on a fully commercial basis. There are also offices and conference facilities provided for the use of tenants on the site. It should be said that each tenant pays a commercial rent and is in no way subsidised. The success of the Tollcross experiment has led us to make a decision to set up similar ventures in all our closure areas and this contribution to new business creation is now a major element of the BSC (Industry) effort. This success has meant considerable exposure on national television and radio and in the press and it has also done a lot to raise the morale of the local business community. Clearly the idea is not just applicable to steel closure areas but does, we believe, offer hope to many depressed regions throughout the industrialised world.

As with any properly run company, BSC (Industry) operates within an agreed budget, but unlike the conventional business its concern is not with the generation of profits but rather with the creation of jobs as cost effectively as possible. So the team are pre-occupied with an unusual parameter—the cost per job. Historically, the cost of BSC (Industry) has averaged less than £1,000 per new job commitment, a small amount in relation to the overall national cost of job creation. Currently, the staff costs, establishment charges, advertising and other revenue expenses are running at nearly £1m per annum. Additionally, in the current year, we have voted "funds for job creation" totalling £10m.

RECENT RESULTS

Since stepping up the level of operation of BSC (industry) in March 1978 over 3,500 organisations or individuals have been attracted by the package on offer. At any one time the team are dealing with 300 or 400 potential projects and since March 1978 they have secured in excess of 250 job creating ventures. It is interesting to note that despite the onset of recession, it is still possible to attract good, viable projects. Indeed, it would seem from recent evidence that the flow is increasing rather than diminishing. This is in part due to a growing awareness among entrepreneurs of just what is available to them through BSC (Industry). For the year beginning March 1978 the target of 3,000 new jobs was set and was just achieved by March 1979. In the current year which ends March 1980 our target is 5,000 new jobs and we are presently on course to exceed this. Unfortunately, in the light of the recent worsening of our commercial situation, and the financial consequences of this, we are having to accelerate our closures, so that while these figures are impressive we must do significantly better in the future. Our target for next year is 10,000 new jobs and we recognise that this is a considerable task by any standards but it is one we are confident of accomplishing.

Whether or not the greatly increased targets for the future can be met is not solely in the hands of BSC (Industry) but is critically dependent on the continuing support of a number of bodies. These include the British Government, the Local Authorities, British Trades Unions and the European Economic Community. This dependence underlines the fact that BSC (Industry) cannot and does not operate alone in a vacuum but is part of a team. All the members of which strive for a common goal—the creation of new viable long-term jobs in steel closure areas, The task is clear, but the problems in achieving it are formidable. For example to illustrate just one aspect, that of providing factories; taking 4 people per 100 sq.m. as the norm for industrial buildings, to create space for 4,000 new jobs, the cost for buildings alone could be as much as £20m.

In Table 1 are summarised the new job and project commitments, made between April 1978 and November 1979 by companies who have decided to locate or expand in BSC (Industry) areas and have received help and support from BSC(Industry). For those interested in a more rigorous analysis of the results to date and our future task I refer them to a recent document published by BSC (Industry) entitled "Job Creation—An accountability report from BSC (Industry)".

SUMMARY

I note that one of the fundamental aims of the founders when creating the OECD in 1960 was "to achieve the highest sustainable economic growth and employment and a rising standard of living in member countries, while maintaining financial stability, and thus contribute to the development of the world economy".

A statement which I am sure we all endorse and one which many others pay lip service to but do little about. I firmly believe that over the past couple of years we have shown that even in these days of depressed economic activity such an aim is not a futile one. It is clear that given the will we can cope with the traumatic changes facing our industry while at the same time contributing in a substantial way to the creation of a healthier industrial base. Such a transformation does, however, require the complete co-operation and understanding of many powerful factions within our society. While it is the steel industry which is today in crisis it is bound to be followed by other mature industries tomorrow and the sooner we learn how to build new enterprises to replace the old the better for all.

One thing is certain, the more quickly and efficiently nations respond to the challenge, the greater very considerable rewards in terms of future prosperity and higher standards of living for their workers.

In conclusion I should say that we are moving through a process of major industrial changes. This requires major adjustments both social and economic. I believe that in BSC (Industry) we are demonstrating ways to assist industrialised nations to cope with these adjustments.

Table 1   SUMMARY OF FIRM NEW JOB AND
PROJECT COMMITMENTS, MADE BETWEEN
APRIL 1978 AND NOVEMBER 1979 BY COMPANIES
WHO HAVE DECIDED TO LOCATE OR EXPAND IN
BSC (INDUSTRY) AREAS AND WHO HAVE RECEIVED
HELP AND SUPPORT FROM BSC (INDUSTRY)

Location	Projects already operating or committed	Estimated total of all job commitments to March 1982	Estimated employment from projects as at November 1979
Cambuslang/ Motherwell	89[1]	1,550	650
Garnock Valley	31[1]	700	200
Derwentside	5	400	150
Hartlepool	25	1,100	600
Deeside	27	1,200	200
Blaenau Gwent	20	1,500	450
Cardiff	10	600	150
Totals	207	7,050	2,400

[1] Includes Neighbourhood Workshop Projects.

CHAIRMAN:

Thank you Mr. Sambrook...

Ladies and Gentlemen, at this stage I think that we should reflect on what has been said by the four previous speakers relating to the employment consequences of restructuring and the whole strategy that relates to this. I think that it would be appropriate before calling on other speakers that we would pause a minute and see if there are questions from the audience, contributions to be made, criticisms to be presented.

The idea of a symposium, as I understand it, is not that there is a continuous flow of presentations — however important those might be — but that we try to structure the discussion. And so, I would now like to pause and call on other participants either to present an opinion or to ask questions from the various speakers or from any other party concerned

Mr. ELLIOTT:

I would just like to express a few words of caution, not about the problems that have been identified this morning by the various speakers, but about some of the trade policy prescriptions for the world steel industry, particularly as expressed or implied in Mr. Florkoski's paper.

I should say at the outset that my perspective is that of a relatively small economic power which instinctively starts to get nervous when the big traders get together on any issue in which we have an interest. When the discussion turns to possible changes in the rules of the game, or mutual agreements that the rules need not be observed, we start to get even more nervous since our market can become exposed to the disruptive diversion of exports seeking a new destination when some of their markets are closed.

My concern about the paper, therefore, is the implication throughout that there is a need for new rules. For example, the summary section calls for solutions in five issues areas. I am particularly uneasy about the fourth suggestion, that it is unrealistic to expect the international trading system to ever become devoid of some form of restraints on steel trade and that it would be reasonable to permit trade policy measures to remedy market disruption.

What the paper seems to ignore is that there are existing rules which already permit such actions under certain and controlled conditions. The GATT, for example, provides for measures to deal with problems of injurious foreign dumping. It provides escape clause action for disruptive imports, and it permits countervailing action against injurious subsidization of exports. The best safeguard for the system is not, in my opinion, any agreement that would permit a departure from these rules. Rather, we would prefer affirmation by all countries that they will stick to the rules and accept the obligations they have already undertaken. In this way, the legitimate interests of both exporters and importers will best be served.

The other issue identified, that of bringing developing countries into the dialogue, is also facilitated by maintaining observance of the GATT rules. Many of these countries which are not members of the OECD and do not participate in the Steel Committee are already members of the GATT.

The same theme of the need for new rules appears again in the paper which identifies four sets of policy areas. I have no quarrel with the argument in favour of stable government policy framework which provides a reasonably consistent policy environment within which industry can make its decisions based on market considerations. However, the reference to trade measures designed to deal with market disruption again implies that no such internationally agreed rules currently exist. As to adjustment programmes, the best assurance that adjustment measures will be vigorously pursued is probably the strict adherence to the existing GATT rule that safeguard measures must be temporary and must include provision for compensation to those countries whose rights are impaired by such measures.

Mr. Florkoski's paper addresses institutional arrangements. Here, the paper acknowledges that there are some existing rules, for example in the GATT. It is interesting though that the references are made only to the new GATT codes agreed in the MTN; no mention is made of the existing body of GATT trade rules, including provisions related to impairment of previously negotiated concessions and to the settlement of disputes in a way that protects the interests of both exporters and importers.

My inclination is to approach with extreme caution the suggestion that the existing body of trade rules should be supplemented with special measures on steel or any other sector for that matter. It is not clear from the paper what kind of special sectoral rules the author has in mind which would go beyond existing rules. Again, as a small country which attaches great importance to the way in which GATT rules have been carefully drafted to protect the interests of small countries through the operation of the most-favoured-nation principle, I am most concerned about suggestions that different rules or some derogation from the rules should apply in a particular sector.

The Steel Committee was established in response to a perceived crisis in the steel industry. Indeed, it does very useful and important work in providing a forum in which a number of major steel producing countries can discuss and explain their respective steel industry problems and their policies with respect to these problems.

It would be a pity if the discussion in the Steel Committee were to be used as an alternative to adherence to the rules. In my opinion the expansion of the horizon of Steel Committee activities to embody new trade rules has two important dangers:

— It might well institutionalise the crisis and provide longer term justification for certain member countries to avoid adherence to existing trade rules; and
— Secondly, and equally important, it might provide implicit cover for the major steel producers to agree among themselves to new rules in isolation from other countries which have a legitimate interest in international steel trade.

CHAIRMAN:

Thank you very much Mr. Elliott, I notice that your nervousness was of a subjective kind, but didn't relate to your statement. I would simply like to say that what I was calling for, so that we could have some more order in our debate, is reactions on the employment consequences. If we immediately go into the trade question, which we will certainly go into through the day, then I think our discussion will be less clear. I have a number of speakers who have asked to talk, but I ask them if it is on this particular subject to which Mr. Sirs, Mr. McBride, Mr. Sanden and Mr. Sambrook addressed themselves, which is employment, structural adjustment and so on, and not yet the overall questions of what we should do in the Steel Committee...

Mr. Judith who is the Chairman of the Consultative Committee of the European Coal and Steel Community...

Mr. JUDITH:

From Mr. Florkoski's report you might get the impression that co-determination is preventing restructuring. I would like to contradict this. In the Federal Republic of Germany since 1962 we have had constant resort to rationalisation measures to reduce capacity. These measures in each case were adopted in consultation with the organisation concerned, including the trade unions and their representatives in the companies concerned. Since the time I mentioned, we have eliminated 100,000 jobs without incurring social need.

Since 1975 we have been adjusting the paths that we follow, and up to the present period a further 40,000 jobs have been eliminated from the iron and steel industry. This took place with the acceptance of the trade union organisations and also involved the acceptance of those working in the steel companies.

Something that was put very clearly yesterday was that we have considerable over-capacity at the world level. Anyone who accepts this, and everyone must, has to be convinced of the necessity for restructuring throughout the world steel industry. But to present the situation as though it were only a matter of economics is to put things in terms that are too simple. Restructuring is not just a matter of economic policy nor of the policy of the industry. It is also a social problem and for this reason we need the co-operation of all those who are concerned in the restructuring process; we need their co-operation to make restructuring meaningful.

I myself have participated actively in the adoption of certain measures, in particular for the Saarland where it was not just a question of restructuring at the national level, but also involved co-operation across frontiers. Let me therefore try to give some indications as to the way in which one must proceed with restructuring.

First of all, I think it is right and important to bring together a small group of people who can think things over. They will produce certain ideas and there are various combinations possible in this field. These ideas have to be presented to a group of experts representing the social partners. This I think is the starting point, but this is not the time when one chooses final objectives. At this initial stage of the discussion, we can determine what is feasible, politically feasible or not. And then from this discussion certain concepts have to be worked out and these from the financial, technical and personnel point of view have to be justifiable.

It is important in connection with these three components to add a fourth component and that is the factor of time. No restructuring concept can be rushed through. It is not a question of slamming on the emergency brakes; anyone who has been on a train when the emergency brakes have been put on knows how things fly through the air and get all mixed up with each other, including the passengers. This process has to take account of the differences in various regions and countries.

Once the objective is defined, we come to the most important part which is informing the people concerned so that they know exactly where they are going. I think that people working in a company have a right to know in what direction their company is going. But it is not just a question of informing the people who work in the company, there is also the matter of informing public opinion. Public opinion too has to be informed so that at the various levels we can think in terms of creating new jobs to replace those that are to be lost. Only when everyone has become convinced that these measures are necessary can the government begin to act.

I can tell you that restructuring measures that have been carefullly prepared can be implemented. But it is up to all those responsible for the process to make sure that there is no turning back. Corrections can be made in the course taken once things are underway, but there can be no radical turnabout. This would give rise to problems for the people working in the company, and it would also upset public opinion. People who are responsible for restructuring measures must have a firm grasp on the rudder. For this reason, I would like to again stress the fact that it is necessary to carry out the restructuring, but in all cases we must make sure that all those concerned are convinced of the necessity for this; we must also make sure that technical and social measures accompany each other.

CHAIRMAN:

I should now like to ask Mr. Rebhan, member of the OECD Consultative Trade Union Committee, to speak on the same topic.

Mr. REBHAN:

I would like to return to the social aspects of the steel crisis. I would like to draw your attention to the catalogue of trade union action that is proposed in Mr. Sirs' paper to mitigate the effects of restructuring on the workers of our industry; and I would like to underline what Mr. Davignon has said, that steel is important. Whether we are employers, governments or unions we have a vested interest in the fortunes of the steel industry, and as Mr. Judith has said, the vested interest is very well understood by the workers on the shop floor.

One thing is abundantly clear, and this was pointed out in Mr. McBride's paper: there has to be international co-operation to find socially and economically acceptable solutions to these problems, and this is essential. Let me repeat again that the trade unions are prepared to play their part but, in return, want to be assured of certain minimum guarantees.

We cannot accept the situation, where jobs in the industry are lost without any alternative employment being made available. I think it is ridiculous to listen to some people saying we are training for jobs when we are training for jobs that are non-existent. We cannot accept a reduction in the living standards of workers. Therefore, we have to have what is known in the jargon as active manpower policies and general social-economic measures which entail these guarantees.

All of us I think here in this room agree we must get the steel industry out of the doldrums and get it back into a profitable position. Jobs must be created through selective employment, creation programmes, especially in those regions where the steel industry is the major employer. Whether we like it or not — and as Mr. Florkoski says, industry is threatened sometimes because of too much government involvement — government intervention has become inescapable and indispensable in this industry, as Dr. Marshall pointed out yesterday. Only by the full commitment and active participation of the trade unions can these structural changes be made relatively painless.

Mr. Florkoski speaks in his documents on subsidies and he quotes the IMF and its original position that we presented here in 1978. But, he also says that labour must become more dynamic. We disagree with his conclusion because unfortunately workers do not have tenure like university professors and others. The argument he quotes and the government statistics about workers not wanting to go back to work because unemployment compensation is too high remind me of the argument which I recently read in the proceedings of that great organisation in my own country, the National Association of Manufacturers, who were arguing in the year 1924 whether the work week should be reduced to 50 hours. One of the arguments presented by certain steel companies in those days stated that, if the work week was reduced to 50 hours, it would result in workers spending too much time in taverns and having too much time to abuse their wives. I think that has been proven wrong.

Let me say that a reduction of working time is an essential and an inevitable component of any policy to resolve job scarcity brought by structural change and technological progress. And we have to have international co-ordination that will ensure that the burden of social costs to the steel industry is spread equally to the various countries thereby eliminating financial handicaps, and facilitating medium and long term employment saving policies. This is the area where the future work of the OECD should be concentrated and where we would like to have it concentrated.

We urge that in any considerations with regard to the state of the industry consideration be given to its plants and development prospects and that priority be given to the human factor in this industry. I appeal to governments and employers in this forum to co-operate in the key task of gathering adequate and comprehensive employment statistics, capable of being compared internationally, which will give a clear picture of the market and technological impact on employment.

When we are reporting on investment, which is so important, we should include qualitative and quantitative information on employment both with regard to new plant facilities as well as the impact on existing industrial structures. We need studies on technical innovation and its consequence on employment. I think this is obvious. The further area should be a compilation of what we call inventories like those that were listed in our document, which include the role of collective bargaining, the relocating projects, early retirement, training schemes, subsidy policies and health and safety, among others.

Steel workers may not understand all the complications that are discussed in this Symposium, but they are dead right, let me tell you, in their gut feeling that the managers of the steel industry and the managers of the economy have a great deal to answer for.

In conclusion, let me say that I agree with Mr. Florkoski that the Steel Committee should play a more active role. I would like to plead for the establishment of a committee of social experts, mentioned by Mr. Judith, which would include those from the trade unions whose major task would be to concentrate on these problems.

Mr. Foy:

Mr. Chairman, is this the appropriate time to raise a question with respect to your opening remarks, or would you prefer to hold that for later?

Chairman:

In all objectivity I would suggest that we keep the suspense open a little longer. So I would like to call now on Mr. Rydh who will put a question on the employment side.

Mr. Olof Rydh:

I would like to give you a brief account of some of our experiences in restructuring the Swedish steel industry. It is a small industry but the problems are enormous to us. And I can tell you it is not a glamourous job; it is hard and awfully frustrating.

The first thing which I would like to stress is what Mr. Judith has already said, that it is necessary that the actions taken are accepted by the employees. This is not only a justified union demand. If you wish, you may look at it from a pure economic point of view. Because if you are to manage to get all the results you hope for from adjustment and modernisation in efficiency and competitiveness, you must have a wholehearted participation from the employees and you cannot get that unless you have an acceptance of the actions taken.

Sometimes you might say that when shutting down you do not need any enthusiasm from the employees—they are out of business. I would say that this is rather short-sighted because there are steel workers left in other factories; they know that the next time it will be their turn and they will be very well aware of whether the actions taken in the works being closed are accurate.

To get this acceptance I think you need co-determination and I will not comment on Mr. Florkoski's paper; that has already been done. But I will say that the employees and their representatives must take part in the decision-making process from the beginning. It is not appropriate if they are brought in when, in fact, some crucial decisions have already been taken. This decision process must also be allowed to take its time, perhaps sometimes more time than management is ready to give. It takes time because the unions are democratic institutions and, as you know, democracy is a time-consuming thing.

Secondly, our experience tells us that adjustment process itself must be carefully planned and must be allowed to take its time. We must have the time to create new jobs, to do the retraining of those being laid off, and during this period, as Mr. Sirs has pointed out, employees must have a guarantee of their income. Mr. Sirs has also shown us that such agreements for income guarantees have been reached in some countries. We have one in Sweden. I think it is very necessary to have a smooth restructuring process because we cannot throw out thousands of people in the street and just hope that they will get a new job sometime in the future. The effects of such a policy for the individuals and for society as a whole have been described by other speakers here today.

But we still have the biggest problem left, and this is my third point, the creation of new jobs. We have heard from the British Steel Corporation about their efforts to create new jobs, and I am not the one to judge their success. I can say from our experience in Sweden that we have noted two things.

The first is that we can never place a responsibility for creating new jobs entirely on the industry alone. This is sometimes done by politicians who think that it is an easy way to handle the problem. Our experience has shown us that the companies are mostly not very successful in trying to create new jobs; they are in fact raiding the country for new employment for the old steel districts, and they are not successful. They have to concentrate their interests more wholly on steel production. In our experience the responsibility must be on society and on public authorities. Placing responsibility on companies in the first place will only postpone good solutions for the employment problem. I think we must from both sides, industry and labour, put pressure on our governments to take this responsibility.

The other experience we have had is that we can never solve this problem in the steel industry and in the steel districts, unless we have a growing economy, especially a growing industry sector. It is no use talking about mobility or dynamic labour, occupationally or geographically, unless you have alternative jobs. By saying this I have brought myself back to where we started yesterday morning, that is the question of general economic growth. In the session yesterday we talked mostly about economic growth and steel demand. Here we have another example of why we need faster economic growth in our economy, as was pointed out by Mr. Rebhan yesterday.

I would like to repeat what my fellow countryman Mr. Höök said very correctly yesterday, that economic growth is in the hands of the politicians. I think we ought to make very firm demands on our politicians in this matter, from the industry as well as from the labour unions. There is a lot at stake in the steel industry for society as a whole.

These conclusions are drawn from our Swedish experiences. We share the problems and I think we can learn a great deal from one another. I will therefore stop by once again repeating the proposal from the labour unions to establish a working group on manpower and social problems within the framework of the Steel Committee of the OECD.

CHAIRMAN:

I still have three speakers on my list which I will then close. I would call now on Mr. Møller, then on Mr. Marshall and then on Mr. Doyen. After that we will return to presentations and I will answer Mr. Foy's question.

Mr. MØLLER:

I shall be bried and precise, ask a question and   deliver the answer. First of all, yesterday we got dimensions on the past and this morning you opened by saying that we lost 120,000 jobs in the EEC. Second, yesterday we got forecasts on the future on capacities and on demand. But we have not got any answers on the question of future increases in productivity. Until we get this answer, we cannot find out how many people will be employed in steel making in the coming years. We know that we will have a shift in our production processes and our technology; and we know that the more we restructure, the higher the productivity increases.

So when I did not receive any answer to my question yesterday, I produced an answer myself. Let me give an example. Let us say that we have 1.5 million employed in steel making today in the OECD area. Secondly, let us assume that we have, as we discussed yesterday, about a 2 per cent increase in demand for steel in the OECD in the coming years. Let us say that we really restructure so that we have increases in productivity of about 6 per cent a year. This will leave us as a result a net reduction in numbers of jobs of between 60,000 and 80,000 people a year in the coming years. This is a lot of people.

So to find out what kind of financing we should need to create new jobs, I made a very simple calculation. If a new job in a new industry would cost about $ 50,000 to build up, this leaves a financial burden of about $ 4000 million a year to create about 70 to 80,000 new jobs a year. I am very happy to be able here in this forum to give the answer to my own question and to let you find the money. I will deliver the people to work there. Thank you.

Mr. Michael MARSHALL:

There have been a number of calls this morning upon governments and the actions which governments might take. It may perhaps be appropriate as a Minister in Her Majesty's Government to give one or two short comments.

I think that recognising the realities of the problems facing the steel industry, it is evident to all that in our country we have seen a shift away from support for the modernisation of the industry to adaptation and problems of change, indeed to the wholehearted acceptance, as we would see it, of the social responsibility which is necessary when we talk about the practical effects of matching supply and demand, as we did yesterday.

The three, perhaps obvious, instruments which we have adopted have been generous redundancy payments, the concentration of regional aid to areas most affected and, in terms of the work which is being done within the Community, the assurance that the schemes for readaptation and training which the Community has organised on a very effective basis are deployed as quickly as possible. If I may say so, Mr. Chairman, we regard this attempt to find co-operation within the Community as perhaps the key factor, this is because quite apart from these immediate remedial and social matters, we think that perhaps the most significant move within the Community and the area where we wish to give greatest support in discussing these matters is on the question of transparency. Mr. Florkoski in his opening paper mentioned this and you yourself, Mr. Chairman, also referred to it. We think that in terms of the understanding, the dialogue, the commitment, it is self evident that transparency is an important factor, but it is also, we believe, the way to go on trying to get away from the mutual suspicion which hinders international co-operation on the widest basis.

While recognising the point made by the Canadian delegate this morning, it seems to us that if we can have clear understanding within the Community and expand that to the OECD's Steel Committee, this is making progress and perhaps we can move into stage 3, the wider fields that he suggested.

Could I say in conclusion, that while we all recognise the problems, the degree of emphasis contained within the broad picture which I am painting, the fact that we have had this morning contributions from both the management and the trade unions from within my own country, the fact that our steel consumers are represented by Sir Richard Marsh behind me and the fact that I as a government spokesman have this opportunity to say a brief word, I hope augurs well for international co-operation in the way that I have suggested.

Mr. J. DOYEN:

I should like to continue with what I was saying yesterday. Since I am still not satisfied with the replies, I should like to reiterate my concern.

If I have understood what the economists were saying yesterday, we are to have moderate growth this coming decade in the steel industry, a major industrial sector. This needs to be put in the context of the overall growth of our economies. Consequently, for the developed countries the steel industry will become a kind of wasting asset because jobs will be reduced every time it is restructured and the general economy will not be able to master the problem of employment and unemployment.

Therefore, now that we are present here, I should like to ask the following: what measures could be envisaged in order to ensure better economic growth? How can we begin to see more clearly with regard to steel industry investments and how can they be properly planned? How can we, here in the OECD—in other words the developed countries—take into account the huge changes that are taking place all over the world as a result of the political wishes and plans of developing countries? How can we deal with this situation? What kind of reaction is possible?

Yesterday I received no answer to my query. That is why I am repeating it today. As workers, we subscribe to these objectives. We wish to have an influence on economics because it has implications for employment. In my view, employment is a major aspect of the general economic problem. Economics is not just a laboratory experiment. It also has human implication; it concerns human beings.

My second point is the following: Do the economic decision-makers and political authorities represented here today agree that before making any decision, they will consult employers and workers at the national level and also take account of the cost of social measures? This refers to the cost of social measures for workers staying on in the steel industry as well as for those who will have to leave. We cannot possibly say to thousands of workers: "thank you for your hard work; we have paid you, and now we owe you nothing." This would be a very shortsighted argument. We have to think about the social consequences. Do we agree to include this social cost in the debate on the steel industry and to consult employers and workers on the subject?

My third point is that I was glad to hear the representative of British Steel talk about the moral responsibility of decision-makers as regards industrial restructuring. I share his concern. The only way to solve employment problems consists not in providing social measures but in giving new jobs to people. It is therefore necessary to summon all energies, explore all possibilities and seek imaginative ways of re-employing workers. For a worker, nothing is more crippling than to lose his job and become a beggar with social benefits. Workers want work. Consequently, are all those in this room resolved to co-operate and meet the objective of creating new jobs and to formulating ideas for industrial restructuring in the countries facing problems in this area?

Finally, we are now at the OECD, where yesterday Dr. Peco expressed a wish and a hope in his paper to which I subscribe: today's Symposium should not be an end but a beginning. On the basis of all the problems discussed, and with the goodwill of all concerned, the Steel Committee should continue to consider the situation, to contemplate measures, and to formulate proposals, recommendations and a code of conduct where solidarity will be a main element.

Dr. MARSHALL:

I seem to be playing a role of calling attention to what I think are some realities. I did that yesterday. I would like to do it this morning. I believe that one of the early speakers said that the condition of the industry is not the fault of the workers but of the government and the management. I would disagree with that. If you look at the statistics in the United States of America in 1967 the average steel worker was compensated about 28 per cent more than the average manufacturing worker. That premium was probably justified due to the risk and the skills involved in the type of work which you find in the steel industry compared with the rest of manufacturing.

By 1977, ten years later, the average steel worker made 64 per cent more than the average manufacturing employee in the United States of America. The premium almost doubled—in fact it did double. That premium is a result, I think, of the very effective bargaining power of Mr. McBride and his organisation, upon which in other circumstances I think he would be complimented. But in the situation in which we are now, I think that is one of the causes—not the only cause but one of the causes—which has weakened the American steel industry and a cause that has to be faced in any matter that has to do with restructuring.

Someone else said that the steel workers might not understand much of what we talked about. But I believe that they would understand the concept that there is a fixed economic pie at any point in time, and that if one group gets more of it, other groups get less of it. I would argue that one of the problems that the steel industry faces, not only in the United States but in European countries, is that the steel workers have been successful at gaining a larger share of economic wealth than other sectors of society. In fact, what is happening now is they are having to pay for that in having fewer jobs, or the potential of fewer jobs, to go round.

This is an issue which cannot be ignored. I repeat, I do not think that it is the only cause but for the representatives of the workers to insist that this is a problem of only someone else's making is I think incorrect.

Let me go on and say one more thing about one of my own experiences in the last year which relates to the issue of restructuring. Many of you are aware that a significant amount of job loss has taken place in Yougstown, Ohio. There have been two plants closed there by the Jones and Laughlin Steel Company and a third plant closure has been announced by the United States Steel Corporation. The net amount of unemployment is probably somewhere between 10 and 20,000 in that city.

When the first closure was announced in September 1977 there was a great amount of concern but almost no planning had taken place for what to do, at least on the part of the local people. There was no capital to revitalise the Campbell works, the mill that was closed. The work practices were not competitive; labour costs were standard U.S. steel worker contract costs; there was no management; and most people figured that there was also no hope.

However, the local clergy—the Archbishop of the Catholic diocese of Youngstown, and the Archbishop of the Episcopal diocese of Cleveland which is a surrounding area—started a group called the Ecumenical Coalition to try and find some way to salvage some steel working jobs out of that closure. Much to the dismay of my business partners I got involved in that activity. One of the great experiences that emerged—although it was a failure in getting any more jobs created—was a realisation that workers at the local level when confronted with the reality of steelmaking economics find many creative ways to get actively involved in solving their own problems.

We have talked about co-determination—I am not sure that I would subscribe to that because it has a lot of political ramifications, but the joint working between workers and management was clearly productive in this case. We were able, through discussions with the local organisation which represented unemployed workers and through the co-operation of Mr. McBride's union in Pittsburg, to make agreements with those workers to reduce manning, to take reduction in vacation pay and some fringe benefits and more important to take their incentive compensation, which is a not insignificant portion of total pay, as part of a profit-sharing plan.

There was also an agreement that the Companies would not reduce the base take-home pay which had been negotiated by the steel workers. That represented about 60 per cent of the total compensation costs. On the other 40 per cent there was a tremendous amount of flexibility after several long and arduous meetings. The net result of all that was a reduction in labour costs projected to be 24 per cent. As Dr. Crandall mentioned yesterday—he and I agree on one aspect—steel making is labour-intensive when you look at the unit costs of making a tonne of steel.

The net result that we felt after putting it into some financial models, was that this mill could have been reopened on a reduced scale creating some one thousand jobs in steelmaking, not in some other type of activity, if we could find the necessary capital.

Here ends what I consider the successful part of my story. The United States Government refused to guarantee loans that would have allowed the hoped-for negotiation and purchase of the assets from the owning company and an attempt to introduce modern steel making technology of the electric furnace type. The

Government chose not to guarantee those loans, for various reasons—some good, some I think not so good, and that project failed. The lesson I would draw from that, though, is twofold. First, there needs to be some long-term constructive planning in these kinds of situations; and secondly, workers and people who plan to run whatever organisation is re-opened have to have a large amount of trust and a willingness really to understand the economics of steel making to see if they cannot work their way out of this problem.

Mr. FOY:

My colleagues and I are glad to be here to participate in this first Symposium of the OECD Steel Committee. My purpose today is to provide a brief American steel industry perspective on issues of concern to our industry and our Government.

I appear today as chairman of the American Iron and Steel Institute, and you must bear in mind that the Institute is made up of many diverse companies.

One month ago, the American Iron and Steel Institute released a publication entitled, *Steel at the Crossroads: The American Steel Industry in the 1980s.* It is the most comprehensive analysis of the industry's problems and prospects that we have ever issued.

But it is more than a "publication"; it is more than a "study"; it is more than an "analysis". It is a *petition to the Government of the United States.*

The basic message was presented to key officials of the Carter Administration a month earlier, in December, 1979. I can therefore assure you that our message, including the industry's prescriptions for policy changes, is being discussed *currently* and *seriously* at high levels of our Government.

Copies of the booklet are available to you, but permit me to summarize its primary message very briefly.

The American steel industry is at a crossroads, a time for decision by Government on the kind of economic and regulatory climate it will provide for the domestic steel industry.

Continuation of existing policies will encourage American steel companies to allow their facilities to become over-aged and non-competitive. The industry's ability to supply the United States would necessarily decline, and steel consumers would have to rely increasingly on foreign sources of supply.

The other alternative would be the adoption of new policies that would permit the rejuvenation of the industry through creation of an economic climate permitting modernization and revitalization.

It is not difficult to choose the more desirable alternative, but putting it into effect — more specifically, providing the necessary financing for modernization — is a matter of profound importance.

Therefore, we have urged our Government to formulate an American steel sector policy based on three key elements:

First: Achieving a reduction in the average age of facilities, and improving the overall efficiency of American producers, will require that recent levels of capital investment be more than doubled, to about $7 billion per year.

Financing this unprecedented level of capital investment will require much greater cash flow from faster capital recovery and improved profit margins. This requires that existing antiquated and non-competitive capital recovery laws be scrapped in favor of more realistic schedules designed for these inflationary times, and more in line with prevailing world standards.

Second: Environmental laws and regulations in the United States are more demanding than can be justified by the protection of public health. If these laws and regulations are modified, in order to reduce the mandated capital outlays, that will help American steel companies to modernize and revitalize.

215

Third: The capital commitments required for such a revitalization program will require firm assurances from the Government that dumped and subsidized imports will not be permitted to disrupt the domestic market, particularly during the period of modernization.

Those three elements must be considered together. Actions dealing with each element independently would not be sufficient. Co-ordinated action related to all elements is necessary if the American steel companies are to launch and sustain this revitalization effort.

The program I advocate would not impose undue hardship on present trading partners, and would establish a framework within which domestic producers could rationally invest in the future.

Only with assurances such as these would the capital market system in the United States be willing to commit sufficient funds from private sources on the scale required to modernize certain needed segments of the domestic steel industry to meet world standards.

We believe that this is the only way a market economy can, or indeed should, provide capital to its industries. I want to emphasize that the program I advocate does not involve Government subsidization or financial assistance.

It must be understood that our steel industry is presently suffering from the cumulative injury resulting from years of dumped and subsidized steel imports. Our program requires assurance that American producers will be able to recover from those injuries.

I an confident that the Government of the United States will respond positively to our proposal for the revitalization of the American steel industry. The alternative, continued loss of competitive position and increased net imports, would be a most regrettable situation. I can assure you that the American steel companies will not acquiesce placidly to gradual liquidation. They will take whatever steps are necessary to defend their legitimate interests.

I fully realize that some steel producers represented here today have endured five consecutive years without profits. Many others haven't done much better. In relative terms, it might appear that the steel companies in the United States have been prospering. That is not the case.

Over the last decade our return on investment has been among the lowest of all American industries. Our companies have not had sufficient funds to modernize many of their facilities, which therefore on average have become relatively older and less competitive by world standards. The ability of American steel companies to serve the United States market is being threatened. American steel companies cannot continue along the present course and maintain the world-class facilities that the American market demands.

We recognize and appreciate the adjustment efforts of other steel industries and their governments. Such efforts should continue. We cannot, however, accept that they be carried out to the detriment of the American industry and its future propects. The charter of the OECD Steel Committee requires that member nations refrain from shifting the burden of adjustment to other countries. We intend to live up to this commitment, and we expect others to do likewise.

I would be remiss if I did not comment on the participation of the steel producers of developing nations in the world steel market. These countries are currently experiencing the most rapid relative growth in demand for steel. They may choose to add capacity in order to satisfy some or all of their own needs, but they should be prudent about planning capacity for export trade. As is the case with the more developed countries, their own production objectives must not be achieved at the expense of the United States.

The United States Marketplace, which everyone here recognizes as the most open and transparent market in the world, is already suffering from the largest net tonnage deficit in steel trade of any major steel-producing region. American steel companies can and will provide efficient capacity that should mitigate the problem.

I hope my message has been clear. The United States must maintain a strong and healthy steel industry. This requires that the American steel companies revitalize their facilities and improve their efficiency. We have urged our Government to limit the disruption of our domestic market by dumped and subsidized imports by aggressively enforcing our trade laws. My company and others have stated that they are prepared to take whatever steps are necessary to achieve this objective.

The Chairman stated that at the conclusion of my comments I could raise the question that I had indicated earlier and my question goes to this point. I was somewhat concerned and a little confused by the numbers as I understood them, Mr. Chairman, when you spoke in your opening remarks about Europe having reduced its capability by something in the range of 2 to 3 per cent, and if I understood you correctly, the United States has increased by 4 per cent. Now, numbers can be used in many ways. But the numbers for the American steel industry—and I am talking now of capability—indicate that in 1970 the American steel industry had capability in the range of 157 to 160 million short tons. We presently estimate that capability to be about 153 million short tons. I think some confirmation of that can be got from the statement Mr. McBride made that since 1969 there has been a reduction of approximately 100,000 jobs in the steel industry. Now, it is true that some of that 100,000 was due to more efficient operations. But a large part of it was lost in cutbacks which have been made by the American steel industry.

CHAIRMAN:

Mr. Foy, as you said, one of the difficulties that we are all faced with is to compare and assess data on the same basis. I was referring to a table which existed in the document which I read coming here on the train this morning. So I quoted from that paper.

Mr. FOY:

Well, I do not think it is accurate and I think that the mere fact that Mr. McBride's own numbers of employment show that he is down 100,000 men in the last ten years indicates that there has been a reduction in the American steel industry.

CHAIRMAN:

That is why we set up a Steel Committee... so that one day at our 157th symposium where our grand-children will be, we will agree on the figures and disagree on the rest. Maybe today we could agree on the rest and remit the figures to the Steel Committee. That would be my suggestion for my conclusions later.

But thank you for putting the question and I answered it in the spirit in which it was put.

I would now like to give Mr. McBride the floor... Then I will suspend the meeting for this morning and I will ask Ms. O'Reilly if she accepts to make the presentation as the first point of our afternoon agenda...

Mr. McBBRIDE:

With an eye on the clock, it would appear that I have three minutes to respond to Mr. Paul Marshall's comments.

First, I don't think any of us can avoid a degree of responsibility for our being at this particular point. All of us have contributed to that. Certainly government policies have had a major effect on the ability of the steel industry and other industries in our country to function. In particular, government trade policies have resulted in the disappearance of some of our industries, have resulted in the flourishing of others and have resulted in limitations on some of those that still remain, the steel industry for one. Much as government may elect to place the blame on management or the labour movement, the fact is that government has a share of responsibility for whatever position we are in. And particularly with respect to trade policies.

Reference has been made to the fact that the union of steel workers has bargained hard and has done a reasonably good job in carrying out its obligation to its members. I think that is correct. I do not think that we have been over-enthusiastic in that respect. We have probably not done as well as some other sectors of our population have done—probably not as well as management consultants have done in improving their standard of living! I happen to believe we have not done as well as some people in management have done. I know damned well we have not done as well as the oil industry in our country and their management.

We have attempted to carry out our responsibility in a very responsible way. We have attempted to co-operate and encourage the use of new technology and labour-saving devices. We have agreed to the use of incentive systems to promote productivity, and we have avoided, despite the charges that some have made, the tendency to be guilty of feather-bedding. Our productivity operations I think are notable. The productivity in the steel industry in the United States has been one that has been improving along the line.

In 1978 as a result of a reasonably good level of operations there was a productivity increase in excess of 5 per cent. That is remarkable and is certainly one which demonstrates the value of full capacity utilisation.

I notice now that my three minutes are up. In carrying out my obligation to the Chairman for letting me have the floor, I would conclude in this fashion. We are geared to adapt ourselves to gradual change. We are geared through collective bargaining or through government action to make and accommodate gradual change that could result from the flow of new technology into the industry. Increasing of productivity, reduction of man-hours per ton, are the things that we are geared to do. We must do those things though cognizant of the influences and responsibility of the separate entities that enter into running companies. Of course management has a heavy responsibility for the success or failure of any enterprise and management guards in our country this responsibility very carefully. There is a very fine line beyond which the labour movement is not permitted to cross—to get into certain areas that management considers their own responsibility. They represent the owners. They do not own the operations but they represent the owners and they guard this area of their responsibility very carefully.

Government no less looks upon its area of responsibility as its exclusive prerogative and guards it very carefully. They listen to us; we attempt to persuade them and influence them. Sometimes we do, but in the final analysis we find that our system of government has a very heavy influence on the ultimate result of our efforts.

And of course, the union has a particular responsibility under our system and looks upon it as a very heavy responsibility and will continue to carry out that responsibility in the best way.

I have infringed upon my privilege and I am sorry, Mr. Chairman. I would like to go on but I can see that I should not.

CHAIRMAN:

Thank you Mr. McBride for your statement.

CHAIRMAN:
As we decided this morning, Ms. O'Reilly has been kind enough to begin this afternoon's meeting and give us the point of view on the consumer side.

A CONSUMER ADVOCATE'S PERSPECTIVE
by
Kathleen F. O'REILLY
Attorney at Law
Executive Director,
Consumer Federation of America

All too often policymakers and consumers alike have viewed consumer issues in their narrowest sphere—issues only as they directly relate to finished consumer goods. Obviously the typical consumer's shopping list of food, appliances, furniture, clothing, etc. does not include the notation, "Don't forget to pick up 1/10 ton of steel." Thus there is a temptation, or perhaps merely a tradition, of overlooking the very significant stake which consumers have in the steel industry. The OECD is to be applauded for recognizing the importance of considering the viewpoint of the individual ultimate consumer rather than identifying only the impact on the direct industrial purchaser who uses steel to manufacture automobiles, airplanes, etc. Hopefully the consumer viewpoint will also inject into the discussion a more objective examination of the totality of factors to be considered and will serve as a balance to the relatively narrow view understandably embraced by those whose livelihood is at stake.

Consumers have a four-fold interest in steel.

1. INDIRECT CONSUMERS OF STEEL PRODUCTS

As consumers of products which contain steel, we deserve quality, durability and price which is set in a vigorously competitive market.

2. UNWILLING CONSUMERS OF STEEL WASTE PRODUCTS

Against our will, we are also the consumers of steel industry *waste products*—the enormous quantities of particulates, sulfur oxides, and various hydrocarbons which steel plants emit into the air. Millions of consumers are daily exposed to carcinogenic coke oven emissions. In 1975 twenty per cent of all man-made particulate pollution in the U.S. came from the steel industry.

We are the consumers affected by the solids, acids, heavy metals, arsenic, cyanide, phenols, ammonia, oil, grease, and heat which steel plants discharge into the water.

We are consumers who endure pain, the cost and the emotional disruption of asthma, bronchitis, emphysema and cancer caused by air pollutants.

We are the consumers who must pay the medical insurance and legal bills which result from industrial waste. Consumers are not willing and should not be forced to tolerate dirty air, contaminated water and their associated health burdens, in order to accommodate the steel industry's desire a larger profit margin for their shareholders. As consistently reaffirmed by the United States Supreme Court, a corporation is a creature of the State. It does not have an inherent right to exist. It is a privilege conveyed by the public. If a conflict exists between ultimate profit maximization and the public health, corporate activity must yield to the larger public good.

3. CONSUMERS PAY THE COST OF GOVERNMENT AND STEEL INDUSTRY DECISION-MAKING

As consumers/taxpayers, we pick up a lion's share of the tab for the *cost* of government and steel industry decision-making. In this connection, industry unhesitatingly reminds us that consumers pay the price for environmental and worker safety standards. It is a cost well worthwhile. To the surprise of some, consumers do not advocate achieving the lowest possible product price regardless of the consequences.

The American consumer movement has a proud history of recognizing that the cost of a product should reflect the cost necessary to assure a safe, quality product; a decent wage and suitable work environment for the labor force; and a profit for the manufacturer. Indeed the organized American consumer movement had its origins in fighting the sweat shop conditions of the last century, which when exposed in such works as Upton Sinclair's *The Jungle* shocked consumers into action. We are encouraged that consumers in so many countries share those goals of economic and social justice.

As to the cost of environmental controls, when a few years ago, Proposition 13 (calling for sharply reduced state taxes) was approved by the voters of the state of California, a wave of rhetoric was triggered about taxpayer revolt against government spending. Yet a simplistic reading of that trend ignores the fact that voters have been more discerning about which government cutbacks should be approved than one might expect from the headlines. Do not forget that the California supporters of Proposition 13 voted *that same day* to *increase* significantly state environmental protection activities at taxpayer expense.

Polls show consistently strong public support for the cost of environmental and worker safety standards.

1. *Cost versus benefits of regulation*

(Survey by Opinion Research Corp., 1978)
"Are the Benefits Worth the Cost Necessary to:"

	Yes	No
Ensure safety, dependability of products or services	47%	15%
Ensure equal employment opportunities	42%	21%
Protect the environment	42%	19%

2. The 1976 Harris polls asked respondents whether or not they favored federal "legislation or regulation" in a number of specific areas. The most popular areas in which people *favored* federal regulation were the following:

— Product safety and quality standards (85 and 83 per cent favored regulation in these areas, respectively)
— Pollution controls (82 per cent)
— Corruption (bribes, payoffs, illegal contributions) (75 per cent)
— Equal employment opportunities for women and minorities (72 and 69 per cent, respectively)

3. "The Added Costs of Ensuring Product Safety, Providing Employment Opportunities, and Protecting the Environment are Regarded as Worthwhile by Significant Proportion of the Public."

Costs Added by Regulations are Worth it to...	No	Yes
Ensure safety, dependability of products or services	15%	47%
Ensure equal employment opportunities	21%	42%
Protect the environment	19%	42%
(Council of Better Business Bureaus) October 14 - 17, 1979.		

Fortunately, the United States steel industry has recently increased its efforts toward meeting federal environmental regulations. Clearly consumer support for strong environmental standards does not mean, however, that we favor inconsistent, overlapping, unduly vague, or inefficiently enforced regulations. Yet, as stated by an Environmental Protection Agency official this month:

"... the iron and steel industry is among those industries with the largest number of facilities out of compliance with air pollution control regulations. For example, as of December, 1979, 45 per cent of the iron and steel facilities were out of compliance compared to 19 per cent of the power plants, 55 per cent of the primary smelters, 15 per cent of the pulp and paper mills and 18 per cent of the petroleum refineries."

Testimony of Frans J. Kok, Director Economic Analysis Division, Office of Planning and Evaluation, Environmental Protection Agency Before the Ohio House Steel Task Force February 14, 1980.

Although anticipated expenditures for future compliance translate into increases in the cost and price of steel, the increased costs will, according to the Federal Trade Commission, be less than ½ of the increase Japan's steel industry will incur because of their even tougher pollution control standards. It is, therefore, unlikely that the US steel industry will face an economic disadvantage against its toughest major competitor in the world market because of pollution controls. And of course EEC countries are also experiencing higher environmental control costs. Numerous studies have concluded that compared to the societal cost of protecting the environment, controls are much less expensive than the medical, insurance and legal costs of contaminants. The lessons of Love Canal, PCB's, PBB's, etc. were hard learned.

As to the implications of plant closings necessitated by pollution control standards, it might be argued that environmental regulations played an important part in the closure of some steel industry facilities during the last decade. The number of actual job dislocations according to EPA is:

"estimated to be 2,907 with 300 of those occuring in Ohio since 1971 and 513 nationally in 1979. This compares with total employment of 600,000 in the US steel industry—about 450,000 of these are engaged in steel production. In addition, EPA has listed one plant nationwide that threatens to close because of an inability to meet environmental regulation; US Steel's Geneva Works in Utah. Regarding these plant closures, EPA recognizes that older, less efficient plants may have particular problems complying with pollution control requirements. The steel industry's problem with such plants recently came to national attention when US Steel announced plans to close a numer of facilities. Many factors influence a decision to close a plant. In these cases, the environmental requirements were only a minor factor, as most of the affected plants were in compliance with applicable air and water requirements."

(Kok testimony, page 6 and 7).

Workers displaced because of a public policy which sets environmental standards, are entitled to receive from that public, payment for job re-training and relocation programs.

When analyzing any set of "costs" to the steel industry, it is important to remember that we are all prisoners of our own bookkeeping system. While industry alerts us to the costs of worker safety programs, do its ledgers also calculate the savings which accrue when overall worker productivity has increased as a result of fewer missed work days because of ill health?

When industry tells consumers about its costs, does it alert consumers that we are also paying the tab for industry lobbyists and litigators who spend millions of dollars in the halls of Congress and in the courts resisting environmental standards.

Does industry tell consumers about the costs of its violations of the antitrust laws? For example, how many consumers are aware that representatives of the nine major steel companies in Texas repeatedly met in hotel rooms to add up the tonnage of anticipated rebar steel projects (hospitals, schools, etc.) and then prorated the assignments among themselves according to a predetermined arrangement. It is estimated that between 1969–1972, this Texas scheme alone, resulted in $10–30 million in overcharges (including treble damages).

And when industry tries to justify relocating a plant (closer to a body of water, or in a location with cheaper labor) how many consumers appreciate how cost inefficient such a policy might be. As recent years have demonstrated so dramatically, such corporate stripmining of urban areas wreaks an unconscionable havoc on those abandoned communities and exacts an enormous tax-payer expense. Industry should be forced to sustain a heavy burden of proving the ultimate public efficiency of removing a significant industry from a community which eliminates not only the jobs of the plant workers, but also damages the support and related service industries in that community, and forces the citizens of that community to wrestle with the consequences of the un or under utilized schools, roads, hospitals, etc., which the company left in its wake. In many instances rail/truck transportation of the steel would be cheaper than the cost of upheaval in abandoning one community and starting over in another. Where senseless government regulations impede efficient, cost effective ground transportation, they should be reformed.

Looking toward the future of the steel industry (domestically and internationally), certain considerations should be part of the public debate.

As students of the industry have suggested, it is unlikely that there will be a significant increase in the overall demand for steel. Nostalgic remembrances of the days when aluminium, plastic, fiberglass, etc, did not exist will not change that reality. Consumers are sympathetic to the steel industry's crisis and willingly endorse principles of compassionate assistance to the worker victims of the industry's inevitably changing complexion. But no industry can expect the public to prop it up against those developments which are but one more chapter in the history of the industrial revolution. We are witnessing an ongoing cycle. First the US and EEC watched with frustration as Japan had the advantage of cheap labor. That gap is closing. Now Japan sees itself at a disadvantage because of the even cheaper labor in Korea and LDC's. Absent a regressive return to the days of colonialism, history suggests that the cost of labor in these countries will eventually become part of the game of "catch up." Consumers in every country should support demands for adequate wages and safe working conditions for steelworkers.

One of the best hopes for any sector experiencing industrial decline is to support comprehensive federal economic policy which recognizes the natural link between any single industry and the economy's overall growth and direction. Solutions which rely on separate treatment for separate industries (protectionism, subsidies, relaxed environmental/safety standards) are unlikely to have long term impact and instead may simply postpone the inevitable.

In the US the frightening erosion of our economy is more directly related to our energy crisis than any other factor. It is easy (and misleading) to attribute the problem to what some perceive as our overdependence on OPEC. Yet Germany and Japan (both *more* dependent on imported oil) have stronger GNP's, lower inflation and unemployment rates, than the United States. The decision to decontrol crude oil is already exacerbating a serious economic situation, will cost American consumers more than $1 trillion dollars over the next decade and is already having negative impacts on industrial sectors including steel. The steel industry should join with consumers in fighting for the reimposition of energy price controls. Ironically there has been a scandalous lack of meaningful commitment by our government to comprehensive conservation programs (weatherization, retrofit, solar, etc.) which according to last year's Harvard Business School's *Energy Future,* could reduce domestic consumption by 40 per cent *without* imposing human hardship, *without* increasing inflation and unemployment and *without* increasing the already excessive economic power of the energy industry.

Likewise the economy, including the steel industry, would undoubtedly suffer if our nation proceeds with a crash effort to establish a massive synthetic fuel industry. In addition to the grave environmental and worker safety problems it poses, a synthetic fuel industry will demand enormous capital, create relatively few jobs and have no impact on consumption for at least ten years. Additionally, it is questionable whether we have sufficient water to meet current agriculture needs and provide the voluminous amount of water necessary for a synthetic fuel industry. Finally, the large reliance on subsidized rail transportation necessary because of the coal involved in synthetic fuels, will exert an upward pressure on the transportation cost of other commodities—again affecting the steel industry. In short, this is one more example, where bad government policy will undermine an already weak economy, with a ripple effect on the steel industry at a time when its future depends in large part on a strengthened overall economy.

The steel industry craves increased (or at least sustained) demand. Undoubtedly the industry has profitted in the past from the stubborn insistence of our auto industry to build large, heavy, gasguzzling autos. Those days are coming to an end. We urge the steel industry to use its clout to promote government policies which replace the auto industry's excess demand for steel, with a commitment to achieve our nation's serious need for expanded mass transit (including rail). Be at the forefront of pushing for this conversion and you'll have the active support of consumers.

Help us face the reality that in most areas of the country we *don't* need more hospitals, more schools. What we do need is more low and moderate priced housing.

Steel industry demands that the US government foot the bill for the large amount of capital needed for modernization must be scrutinized closely. Recognizing a relative inelastic supply of such funds, one might reasonably ask why that money should be made available to steel rather than high technology industries which produce a product more attractive to the export market and thus with higher prospects for long term growth.

"Dumping" practices are just as offensive to consumerism as predatory pricing, restraints of trade, unfair trade practices and similar incidents which inject instability and blatant inequity into the marketplace. Consumers are not misled into encouraging dumping because of some arguably short-term price relief it may offer. Dumping should not be tolerated, yet we warn against what could develop as an excessive backlash which responds to dumping by establishing unjustifiably stringent trade barriers. Recognize that industrialized nations are each others' best customers. If we insulate ourselves from international competition in steel, what nations will buy our computers, airplanes, timber, farm products, etc? Those industries, their workers, and consumers of their products have rights, too.

4. CONSUMERS' STAKE IN COOPERATIVE INTERNATIONAL TRADE

As citizens, consumers have a serious 'stake in world peace and the role international trade cooperation plays in maintaining peace.

History has taught us that economic autarky (the equivalent of the "me-first" syndrome) destroys world trade, weakens the economies of all concerned, and ripens nations for repressive political takeover and international conflict. Consumers in every country depend on their governments to abide by commitments to maximize international cooperation as symbolized by this OECD symposium. Thus in the final analysis, the question "What about the steel crisis?" must be answered "Compared to what?" Let us not be seduced by so-called solutions which may well be worse than the crisis—solutions such as:

— strident protectionism which ultimately threatens world peace
— policies based on the delusion that the steel industry will see a healthy expansion in the near future
— relaxed environmental/worker safety standards
— relaxed antitrust laws, a consumer's best assurance of competition
— corporate strip-mining of urban areas in pursuit of the greenfield approach
— excessive reliance on government funding, which funding then becomes unavailable for other investments which might be more cost-efficient
— parochial demands for tax relief, etc. rather than collective efforts to an overall sounder economy through such measures as reimposition of energy controls, commitment to mass transit and opposition to a costly crash synthetic fuel industry.

In the long run we must remember that important as steel is to you because it is your livelihood, it must take an inferior role to the larger good of world peace. And so we hope that not just symbolically but substantively this OECD Symposium is a collective and an individual commitment to ensure that there will be vigorous, co-operative international trade because it is one of the best weapons we have to ensure that world peace and harmony will remain.

* Consumer Federation of America (CFA) is the nation's largest consumer membership organization composed of more than 200 national, state and local groups commited to the goals of consumer advocacy and education. CFA includes grassroots consumer groups, senior citizen, rural, labor, cooperatives and other organizations. Established in 1968, CFA

— advances pro-consumer policy before Congress, the Administration, regulatory agencies, and the courts
— assists state and local consumer groups
— increases public and media awareness of consumer needs.

A 1979 University of Chicago Graduate School Study ranked CFA as one of the top ten most effective lobbying organizations in Washington.

CHAIRMAN:

I should now like to ask Mr. Tesch to present his paper on the steel industry in the 1980s.

THE STEEL INDUSTRY IN THE 1980s

by
Emmanuel TESCH
President of ARBED
President of EUROFER

From the outset, I must make it clear that I have not sought the opinion of my EUROFER colleagues about the thoughts set out in this paper. They are entirely my own and the responsibility for them is therefore mine alone.

The subject proposed—"Policies for adjustment, modernisation and adaptation of the steel industry in the light of expected world developments"—raises two questions:

— The preliminary question of expected world developments by which I understand developments in the steel sector;
— The main question of adjustment, modernisation and adaptation policies to cope with these developments.

The motive forces operating in the steel sector are of three kinds:

1. Normal economic forces,
2. Events of extra-economic origin,
3. Entrepreneurial action.

By normal economic forces is meant, first and foremost, the changes taking place in the two essential parameters of steel supply and demand. In every free economy country and at the level of all international trade, prices and trade flows are always conditioned by the confrontation of supply and demand. Forecasting what is going to happen in the world steel sector means first of all arriving at certain assumptions about possible developments in the two parameters of supply and demand, or more exactly supply potential, which is another word for capacity in relation to effective demand.

— What are the dynamics working in this field?
— The crisis that began in 1974 and whose full effect hit the steel industry in 1975 slowed down demand more than expected, whereas it was more difficult to apply the brakes in the case of the capacity increases decided upon, or initiated, in the buoyant years of 1973 and 1974 when things were almost "too good".
— As a result, overcapacity at the world level for 1979 is estimated at 87 million tonnes.
— A study that looks at steel prospects to 1973 estimates that there will still be some 56 million tonnes overcapacity in terms of excess deliverable supply potential.

The answer seems simple: an adjustment policy to reduce overcapacity either by shutting down plant with the poorest performance or calling a halt to investment in capacity increases.

225

The dynamics of the steel industry are much more complicated if we leave the general world level and look at what is happening at the level of the major world areas, countries and regions.

There is one geographical structure specific to production capacities and another specific to steel demand.

In the old days, steel industries firmly established in the industrial countries with a bulging file of customers, met the demand from far-off countries by long-range exports.

The supply structure long remained anchored in the countries where the industry grew up, whereas demand's geographical structure, highly mobile and dependent on the general economy, shifted its centres of gravity.

That the location of production centres should ultimately change to match trends in the geographical structure of demand seemed inevitable.

Either the old-established steel industries in the industrialised countries could chase demand by setting up subsidiaries elsewhere in the world or else local and independent steel industries could be set up to meet the new and increasing demand.

For very many reasons, the latter process took the upper hand.

In the period before the crisis, when potential supply and effective demand were never far out of step with oversupply alternating with overdemand in a four-five year cycle, trade flows fluctuated to a certain extent without changing too violently or too suddenly or displaying any irreversible trends. The parallel alternation of high and low prices allowed companies with higher production costs to continue in business.

Since the date of the general economic crisis, persistent overcapacity at world level and the new situation in geographical supply and demand structures and in the cost factors of production have inevitably placed trade flows under such pressure that they have changed considerably.

Producers not prepared to give away any of their markets, including their own domestic market in some cases, to which they felt they had an external and inalienable right, fought with all the weapons they had and although, in terms of tonnage, trade flows were not too severely disrupted, the effects on selling prices are familiar to us all.

This describes the disarray of the early years of the crisis. For more than a year now, however, abnormal turbulence in trade flows would seem to have ceased under the vigilant watch of all the authorities concerned.

The policy that seems to be prevailing is that the exporting countries should exercise some measure of restraint and not be too aggressive in increasing their footholds in markets which otherwise would not fail to react with all kinds of protective measure.

This brings us to the second type of force affecting the dynamics of the steel sector.

Events of non-economic origin

The purely business factors such as supply, demand, price, costs and trade flows, represent only one aspect of things and although this is important it describes only one part of the real picture of the steel sector as it is today.

In fact, an economic world obeying no other law than this field of commercial forces and in which the enterprise could act without restraint in a market open in every respect and in competition with other enterprises all treated in the same way is today, in this pure state, not much more than a model of theoretical thought or perhaps an objective for the distant future.

Although, prior to the structural crisis we are now passing through, the steel sector still had some affinities with such a model, the suddenness, scale and length of the crisis have provoked and are continuing to provoke defensive action which has considerably increased its remoteness.

In Europe particularly, the crisis has taught us an object lesson and established certain facts that must first of all be identified before they are criticised or appraised in relation to an economic ideology. In the past there used to be a large number of small steel companies. A slightly severer short-term crisis than usual could cause one or other of these companies to close down, thus re-establishing market equilibrium, without too serious a social disaster or economic and financial upset.

Nowadays, these small units have grown and merged into big groups. If one of these came to an end it would mean that a whole region would be reduced to an economic desert or that a country which was previously a steel producing country would lose the whole of its industry in this sector with the tragic social consequences this would entail. This is no longer acceptable, at least not in all its suddenness and without a period of adaptation to allow the human hardship that accompanies such changes to be cushioned.

Whereas earlier, any excess supply was removed by the disappearance from the market of a small and unprofitable company, nowadays the only way is to introduce restructuring measures in big steel groups which have to shut down their least-viable plants and works.

This is often more than the financial resources of the big companies can cope with and the inevitable result is public intervention in all kinds of ways.

This is stated as an observed fact: on the evidence it has to be recognised that government aid has become part of the steel scenario and merits thought about the new dynamics it has triggered off.

One particular question is whether aid measures in one country are not likely to provoke a similar reaction in other countries, and thus start off a process of escalation in public support and hence a compartmentation of markets, in spite of all the efforts made to open up world trade.

It has to be said that, in its principle, it would be contradictory and intellectually dishonest for an entrepreneur to want the permanent advantages of protectionism and public grants in his own country and unobstructed access to, and equal treatment on, markets elsewhere.

Steelmakers are therefore caught between two constraints:

— acceptance of certain protective public support,
— the need to keep foreign markets open.

These two constraints are not contradictory provided public support measures comply with certain criteria and obey a code, in which the code of subsidies would be only one part.

The criteria justifying government support could be as follows:

— it must be a purely defensive measure aimed at survival or preventing too disastrous an upheaval—not an aggressive measure aimed at increasing capacity or grabbing a larger market share;
— this defensive support must be geared to the volume, duration and specific nature of the problem and therefore conditional on internal restructuring with the shutdown of non-viable plants and the co-ordination of efficient units, and must guarantee a return to adequate competitive capability;
— public support must also be as non-discriminatory as possible, in other words it must cover as many companies as possible in the same situation.

Failing adequate transparency as regards public support, and if it is not restricted in volume and time by the application of well-defined criteria, trouble could well develop in relations between steelmakers.

Every steelmaker claims that the crisis warrants special measures in his particular case. It is important to realise that instead of there being one, general crisis, regions or countries each have their own crisis, often with very different characteristics from others, which does not facilitate reciprocal understanding. For example, if steelmaking locations are classified by three parameters, namely:

— intensity of demand,
— intensity of supply,
— intensity of production costs,

areas can be rated on a scale with the following graduations:

— the worst off are the steelmakers in regions where there is more or less stagnant demand, a big margin of excess capacity and, on top of this, escalating production costs. This is true of Europe which is being forced to reduce its capacity and in the meantime, to accept too large a proportion of exports at undercut prices. The European crisis is threefold: under-utilisation of capacity, inadequate prices and excessive costs;

— somewhat better off are the steelmakers in regions where there is more or less stagnant demand, undercapacity and production costs that are still moderate. This applies to the United States which wants to retain its satisfactory capacity utilisation factor and at the same time protect itself against foreign competition. Here the crisis is primarily that of prices to fund plant modernisation;

— in an even better position are the steelmakers in areas with considerable overcapacity but showing a reasonable increase in demand and with moderate production costs. This is the case of Japan, capable of a high proportion of exports on terms that are still remunerative because of its low production costs. Here the crisis is one of overcapacity although the rate at which this can be absorbed seems to be higher than elsewhere;

— best placed of all are the steel industries in countries or regions where demand is growing steeply and where production capacity is developing and production costs are very low. This applies to the new industrial countries like Brazil, Korea, Taiwan, South Africa, etc. It is important in these countries that capacity expansion should keep in step with growth in demand if international prices are to continue to be remunerative for everyone and if these countries are to avoid, in the long term, running into the same difficulties as today's big exporters.

This classification of developments in the steel sector is purely intended to show how difficult it is to establish a bridge of understanding from one country to another, each seeing the crisis idea in terms of its own and devising its own measures to ward it off.

The steel industry policy needed to deal with this problem is to keep public support measures to the essential minimum, to give them sufficient transparency in order to prevent harmful accusations of ill-intent and, if possible, to draw up a code of assistance, on pre-established criteria, to prevent aid escalation and the disastrous compartmentation of markets.

The third and last field determining the dynamics of the steel sector is *entrepreneurial action.*

The performance of a steelmaking company depends on:

— the quality of its commercial environment proper;

— its relative position in the field of extra-economic action;
— its internal efficiency.

The entrepreneur's specific field of action is the latter.

First and foremost, in my view, the entrepreneur must make a right judgement of the commercial environment in the locations of his plants. This environment may have changed appreciably since the crisis and still be in a process of irreversible structural change. The "commercial environment" notion has already been defined by three parameters: demand structures, supply structures and production cost structures.

In other words, if the "commercial environment" of a company or a steel plant is running down, the steelmaker has to have the courage to face up to this fact. He cannot, as in the past, reckon on being able to offset this structural recession in his immediate environment by a lasting increase in exports.

Whereas, as in the past, he was able to invade areas where there were no steel producers, he will, from now on, come up against a large number of local producers (or at a closer range than his own) in the new industrial countries fully competitive both in quality and price. Fortunately, the new resistance to direct exports of rolled products is partly offset by indirect exports of steel in many, more sophisticated, end products.

These are the high-technology steels on which producers in the highly industrialised countries should set their sights.

The entrepreneur should therefore begin by arriving at a clear definition of his policy to invest or disinvest and to increase, or not, his capacity.

As a general rule, there is a considerable potential for productivity improvement investment both in Europe and in the United States and, in my view, steelmakers should give precedence to this kind of spending.

These comments on the general strategy of steelmakers having been made, each of them will have to face up to the need to have his own specific restructuring policy. At this level, general recommendations are less relevant than specific measures to be put into effect.

Mergers and groupings of all kinds offer the entrepreneur a wide and scattered range of production tools which have grown up at various places according to individual strategies.

If this has not already been done, these plants have to be looked at from the angle of their rationalisation potential.

Then begins the down-to-earth study of what has to be done to dispose of the most obsolete plant, cut out any functional and operational duplication by a specialisation programme, shortern in-firm transports distances, etc.

That then, in reply to the question posed at the outset, is the adjustment, modernisation and adaption strategy that is incumbent upon, and the responsibility of, the steel entrepreneur, who must carry it through in conditions causing as little social distress as possible.

CHAIRMAN:

Thank you Mr. Tesch. I should now like to give the floor to Mr. Amaya, Deputy Minister of International Relations.

Mr. AMAYA:

Mr. Chairman, I will make a brief observation on present and future steel market situations as I see them, and frank remarks on the policies to be followed by steel-producing countries.

As one of the government officials who engaged in operations to tide over the steel crisis of 1977 along with M. Davignon, I recall the scenes in which we discussed seriously with Mr. Strauss and Mr. Solomon of the U.S. Government how to modernise the steel industries and still maintain fair and open trade. The principles for implementing anti-crisis measures were that no actions be inconsistent with GATT provisions; that fair and open trade be maintained; and that domestic policies to sustain steel firms during crisis periods should not shift the burden of adjustment to other countries. But in the subsequent three years, I regret to say that the fundamental situation surrounding the steel industries of most countries has remained almost unchanged.

The steel situation in 1978 and 1979 was not unfavourable for the steel industry. 1980 is also not expected to be a bad year on a world scale, although there are many factors which will increase the cost of production, such as oil and other raw material prices. I would like to share the view expressed by Mr. Kono yesterday that demand for steel as a basic material may grow steadily without a steel shortage in the course of the first half of the 1980s. But the plant and equipment for steelmaking are getting older every year and will not be able to produce as much as expected without constant maintenance.

You will recall that the Solomon Report and the Davignon Plan stressed the importance of the adoption of domestic policies as well as fair trade. These reports are the reasons why Japan made its mind up to be co-operative on trade with the steel industries of the U.S. and Europe to cope with the difficulties of 1977. Yet, particularly in the United States, I have seen few significant changes in domestic policies on revitalisation, such as capital formation, in order to bring plants and equipment up to date and keep them there.

We have difficulty in understanding to what extent this reflects lack of industrial planning and to what extent lack of government policy. There seems to be little anxiety in the U.S. steel industry to make new investments for modernisation, although we were informed many times that plans for new plants were ready. Instead, the U.S. steel industry imported approximately 5 million tons of coal from Europe in 1978 and the same amount in 1979 without replacing coke ovens which are the main and, indeed, critical equipment for steelmaking.

We have noted that the major U.S. companies have been taking steps to close obsolete facilities and concentrate on improving selected existing facilities. Perhaps, given the already depreciated state of much existing capital equipment and the cost of green field plants, this is the only economic course. We wonder, however, as outsiders, whether an essential question is not being left unanswered, namely, the existence of a commitment by enterprises and government to maintain an efficient industry. Perhaps you may feel that I have spoken too bluntly, but I think that the advantage of the Symposium is to exchange views freely and frankly, my intention is to make full use of this opportunity.

Obviously, each country must decide its policies for itself, including the relationship of its government and basic industries. But we are concerned that, if what I have just called the essential question is not faced, then there is temptation and political pressure to try to deal through trade restrictions with what in reality are problems of domestic policy.

All participants in this Symposium should recall the initial commitments to the decision of the Council establishing a Steel Committee and positive adjustment policies adopted as part of the communiqué of the June, 1978 Ministerial Level Council Meeting.

Developing countries, especially newly-industrialised countries, are paying great attention to whether the revitalisation plans of advanced countries' steel industries can be accomplished without protectionist trade measures. In this regard, we have to bear in mind that the measures taken by the U.S. and the EEC will play a role of great importance. The important point is that we should not be short-sighted. We should establish long-term restructuring plans. The only way to overcome these difficulties in compliance with positive adjustment policies and the "initial commitments", is to arrive at the optimum mix, technically and economically, of building new plants and equipment and applying modern technology to old plants.

A study of the Japanese steel industry in the 1980s is being undertaken by the Japanese steel industry itself. The fundamental concept of the study is a responsibility to supply steel products sufficient for our customers' demand inside and outside Japan. The industry, in establishing long-term plans, is determined to install new replacement equipment even though it has comparatively more modern equipment than other countries. This is being done to maintain competitiveness through higher continuous casting ratios, oil-less operation of blast furnaces, so on and so forth. The Japanese Government highly appreciates the forward-looking attitude of the Japanese steel industry.

I emphasize the Japanese steel industry because it is they who make and execute plans not the government. There exists, I think, a prevailing myth that in Japan, government and companies are incorporated into one entity. But, this is just a myth, not a scientific observation.

In light of the foregoing comments, we are naturally disturbed at unconfirmed reports that a United States steel company will shortly file dumping complaints against the EEC and Japan, and that the Trigger Price Mechanism will thereupon be suspended. I hope this is not true. From the outset up to now, we have maintained close contact and co-operation with the United States Government in making the Trigger Price Mechanism run smoothly and successfully. Our exporters have meticulously abided by the rule of the mechanism. Therefore, we would be astounded if the anti-dumping suit were filed against Japanese steel exports to the United States. We have had our own complaints about the Trigger Price Mechanism, but it has served a useful purpose. To have it succeeded by a series of complex litigious preceedings creating great uncertainty in the trade, could be little less than tragic.

For our part we remain ready to be co-operative with other governments in implementing whatever measures, including the Trigger Price Mechanism, to maintain proper order and stability in the international trade.

In conclusion, I do not sympathise with the view that there may be steel shortages in some countries assuming imports do not increase unless governments take the initiative to stimulate new investment. We regard it as the responsibility of industrialists to take the initiative to make new investments with long term prospects in order to support sound, free economies. In this connection I appreciate the effort for restructuring taken by some companies in certain countries. On the other hand, government must help create a suitable environment for this, but should not intervene in the free market for irrational political reasons.

As explained many times, we are sympathetic towards the difficulties which Americans and Europeans face. We are determined to be co-operative in the solution of these difficulties. Without criticising each other and in view of the long term prospects, we confirm our determination in this Symposium to achieve the revitalisation of the steel industries in the light of positive adjustment policies.

CHAIRMAN:

Thank you very much, Mr. Amaya. Now I would like to give the floor to Mr. Vanik. We look forward to listening to his statement on an industrial policy for steel in the United States during a period of trade friction.

THE DEVELOPMENT OF AN INDUSTRIAL POLICY FOR STEEL IN THE UNITED STATES DURING A PERIOD OF TRADE FRICTIONS

by
Congressman Charles A. VANIK (Ohio)
Chairman, Subcommittee on Trade,
Committee on Ways and Means
US House of Representatives

It is an honor to be here with this distinguished gathering of world leaders in steel production and labor.

I am not a steel expert. But I am entering my 26th year of service in the United States Congress, and I believe I can contribute some insights on American trends in the politics of steel and trade as we enter this new decade.

There are, of course, some exciting, controversial rumors regarding steel trade during the past several weeks, and I will want to discuss these. But I believe it is the purpose of this Symposium to look at some of the longer range issues so that we may be better guided by an industrial vision or plan. Therefore, I would like to take up these longer range, structural adjustment issues first.

RESTRUCTURING AND MODERNIZATION

The abysmal profit levels of recent years are forcing a restructuring and modernization of steel operations throughout the world.

Japan, totally dependent on energy imports, is making great gains in reducing the energy requirements of its already modern mills.

In Europe, some are to be congratulated for making very difficult decisions to close obsolete and over-capacity plants. Britain, in particular, in cutting a third, has taken an action which I doubt that the private enterprise companies of America would ever dare take at one time — and if they do, I do not plan to be in the Congress to take the heat! (Still, in Europe, where 50 per cent of the steel capacity is state-owned, it is not clear in America what is the master plan of your restructuring. I hope through the Steel Symposium that your overall blueprint will become clearer.)

In the United States, the issue of modernization and adjustment is perhaps more difficult and complex than elsewhere. American industry probably has the largest proportion of overage and inefficient steel capacity in the world — and yet its need to be understood and assisted must overcome a basic cultural bias[1] against big business, an adversarial position (what might be called "bad press") that goes back for decades and that must be changed.[2]

*See notes 1-2 on following page.

232

STATUS OF AMERICAN STEEL INDUSTRY

I believe that, in general, the American industry is competitive in its home markets and, in light of the low return on investment, has been doing a reasonably decent rational job in attempting to modernize.

The problem is that with profits probably continuing below average and the increasing modernization abroad, coupled with continued domestic inflation, the domestic industry is losing competitiveness rapidly. The American Iron and Steel Institute has recently issued a major report (the so-called Orange Paper) entitled "Steel at the Crossroads: The American Steel Industry in the 1980s", which I believe shows rather uncontestable evidence that the domestic industry is entering a period of rather rapid decline unless countermeasures are taken.[3]

1. The term "cultural patterns" was suggested to me by a speech by Nippon Steel Corporation's Tadayoshi Yamada, at a seminar sponsored by the Japan Society, Inc., November 29, 1979. In his remarks Mr. Yamada summarized the problems facing the domestic industry — and even more, the problems facing the American industry when it attempts to compete against steel production in countries which have a totally different "cultural pattern":

"We admit that the Japanese views on profit and return on investments, the comparative benefits of high debt leverage, the much closer Japanese government-industry dialogue, and much greater freedom of action to cooperate for the overall good of industry than our American counterparts enjoy — all these are difficult concepts to accept and to fit into the American culture. Such cultural differences post problems not previously identified with sufficient specificity and importance as to merit new forms of legal institutions and cooperation, but such cultural patterns now have emerged to full visibility. We feel that existing US dumping and other unfair competition laws do not take these cultural differences into account, and they offer inadequate opportunities to discuss such unjustified charges and to settle genuine disputes expeditiously. Perhaps new thinking may be needed and some new mechanisms must be designed to resolve these conflicts."

2. The American government must also learn how to be more effective in helping workers and communities adjust to steel mill closings. My Trade Subcommittee held a hearing in Youngstown, Ohio on December 27, 1979, which showed a woeful failure by the Executive Branch to help localities attract new lines of business or train workers for different types of jobs. My Subcommittee would be interested in hearing from others how their nations have responded to large mill closings in medium-sized communities.

3. As the Orange Paper states:

"In the last ten years, productive capital expenditures in steel (total capital expenditures of steel companies, excluding non-steel and environmental expenditures) have averaged just over $2 billion per year (1978 $) – far from sufficient to maintain a competitive industry. If productive capital expenditures continue at a $2 billion rate for the next ten years, the average age of equipment (assuming current productive capability is maintained) would increase from 17.5 years to almost 20 years — completely unacceptable for even borderline cost competitiveness. This would mean that approximately 25 per cent of equipment capacity would be over 25 years of age by 1988, as compared to the current 20 per cent shown in Chapter III.

"While one could theoretically imagine the industry operating with older and older equipment, the implied inherent loss of competitive position makes this a practical impossibility. Given current inadequate rates of return, domestic producers are not in a position to absorb further losses of competitive position. Thus, if capital expenditures continue at their current inadequate levels, the inevitable result will be even older equipment and a loss of competitiveness that will force plant closures. American steel producers have reached the point where rates of return and capital expenditures have been too low for too long. Both must increase significantly from current levels, or production capability will sharply decline.

"If $2 billion per year (or less) is an insufficient level of modernization and replacement capital expenditures, what is required? To maintain current production capability and the same average equipment age, expenditures of $3.0 billion per year will be required over the next decade, but merely preserving the current replacement cycle (35 years) would be a losing game, because foreign competitors are modernizing rapidly. To remain competitive, American steel producers must do likewise."

The AISI paper goes on to argue that to effectively modernize, some $7 billion a year in new investment will be needed. I personally question whether that is not a "wish list" figure, and whether a level of $5 to $5.5 billion annually would be adequate. The exact amount of additional investment is not important at this time; what is important is a national awareness that substantial additional investment is indeed needed.

CAUSE OF PROBLEMS

How did the United States, once the world's steel leader, get into this situation? There is blame for all.

Wage settlements have been excessive and labor has not always co-operated with innovations. Management has had a neanderthal approach to social concerns for safety and pollution control. Management has failed to see its industry as a dynamic one that justified strong levels of R & D and innovation, and it has been more concerned with making tonnages rather than quality. Government has been involved heavily — on steel's back more often than as a helper, multiplying regulations, ignoring the enforcement of trade laws, and forgetting the research needs of the industry.

Basically, each element of society was doing what it perceived as immediately best for itself: labor was maximizing wages, industry was maximizing short-term profits, and government was seeking to prevent inflation. Put together, the three policies have brought the steel industry to the edge of disaster. Capacity is shrinking — perhaps 10 per cent or more of our existing facilities need to be closed down and completely rebuilt.

American needs immediately to make some fundamental changes in its approach to steel.

EVOLUTION OF STRUCTURAL CHANGES

But this is not a forum for the casting of blame. What of the future? What of the new decade?

Without any conscious direction (and with little understanding) there is a substantial change under way in the American steel industry.

It has become increasingly clear that the American steel industry is evolving into three distinct sectors: integrated, mini-mill (or non-integrated), and specialty.

Mini-mills are expanding. The market share of non-integrated steel mill shipments is expected to grow from about 13 per cent in 1978 to about 25 per cent in 1989. These mini-mills and specialty shops tend to be modern and profitable. About 52 per cent of their 1978 production was through continuous casters (compared to 12 per cent for the big integrated mills).

It is reported that in 1978, the average return on investment for the dozen largest integrated companies (accounting for 63 per cent of domestic shipments) was 6.2 per cent — and even this is overstated because it includes more profitable non-steel investments. But for the six largest non-integrated producers which accounted for 61 per cent of their industry's shipments, the average return was 12.3 per cent. Nine of the major specialty steel companies showed an average return of 11.1 per cent.

In other words, as big steel divests, local markets are being captured and well-served by a growing number of electric furnace mini-mills.

In the specialty steel sector, the industry appears to have used wisely for adjustment the temporary import relief provided by three years of quotas (which ended just two weeks ago). The specialty steel sector appears to be modern and ready to compete in world markets with high technology products. The fact that several US specialty firms are major exporters attests to their competitiveness.

But frankly, these evolutionary changes are not enough and they do not fulfill the needs of the United States for a world class steel industry in the decades ahead.

Mini-mill products are of limited nature and quality. The United States, as a keystone in the defense of the Western democracies, must have a steel industry capable of supplying a major share of its basic steel needs — and that means modern, integrated steel mills.[4]

In addition, a modernization of the integrated mills in the next five to eight years is absolutely essential. By the mid-1980's the American industry will be competing with a more energy-efficient Japanese industry and a substantially more modern European industry. The early 1980's are critical. At least 10 per cent of existing capacity must be replaced and additional capacity added to share in market growth.

THE NEED FOR A NATIONAL CONSENSUS IN SUPPORT OF A MODERN STEEL INDUSTRY

The first and most important step which is required for the restructuring of the American steel industry is a national consensus that a modern steel industry is a national priority and that steel can and should be a dynamic part of the economy.

I believe such a national consensus or (to borrow a phrase from MITI) national industrial vision is rapidly developing. If nurtured, this new approach to the industry in America will help make a fundamental difference in the years ahead. This is the good news in my message of today.

By definition, this national consensus must be a partnership. The role that industry and labor must play can best be defined by others. Let me say to my colleagues from American steel management and labor that:

— industry must develop a long range strategy and build public support by making its commitments for modernization better known;
— industry must realize that the American people, like the people in Japan, want clean air and water: the industry has wasted more good will capital in opposing pollution clean-up than it will ever save in financial capital;
— labor must show wage restraint; merely to agree "not to oppose" innovation is not enough; labor must help bring innovation about;
— labor and industry should study the management/worker relationships which have worked so well in many Japanese firms operating in America: perhaps we can learn how to pull together rather than just spar together.

But the major tasks for restructuring the American industry lie in the government sector.

THE EVOLVING AMERICAN INDUSTRIAL POLICY

I see an industrial policy beginning to evolve in America — and this is a dramatic development, a major change in American politics. Hopefully, it can be one which will draw on the best of the lessons from Japan and Europe — and avoid the mistakes made elsewhere.

---

4. As domestic capacity shrinks, the national security question is becoming more urgent. Foreign steel supplies 50 per cent of the steel used west of the Rocky Mountains. In this one region, at least, there must be a serious question as to whether the nation's national security needs can be met.

There are cycles in the social histories of nations and in the tone of Congresses. I have spent 40 years in elected office in America. I was first elected to office as a young Councilman in a steel-working ward in Cleveland, Ohio. I count much of my first success at the polls to documenting the diseases which the air pollution from the steel mills inflicted upon the people of my ward. I was elected to demand action by the government to force the mills to clean up. I first served in the Great Depression, in an age which demanded and got government income security programs and regulation of big business — escape valves in lieu of riot and revolution. For 40 years, politics and government consisted of confrontation between big government and big business, especially big steel.

I mention my past, adversarial role toward steel only to highlight the change I see in myself and many members of the majority party in the American Congress.

In the last few years, there has been a noticeable trend in government and in business to move both closer together: not a marriage, but an occasional "date and kiss", with both eyes open. Certainly I would never have dreamed a decade ago that I would be urging a joint industry-labor-government policy for steel — and in the Congress I am still considered one of the hardest critics of the industry.

Is American government getting more "conservative"? Perhaps. Perhaps business has gotten a little more liberal and has a bit more social conscience.

But I believe it is primarily foreign competition, the success of United States efforts to see the rapid recovery of the world economy after World War II, that now is pushing American business and government together. There is an acute awareness that our productivity trends are abysmal, that our innovation level seems to be sagging, that we are not doing well in trade and therefore the dollar is in trouble. There is an awareness that these problems are due in part to the wonderful successes in Japan and Europe and that it is often difficult for U.S. firms to compete against the government aid available in Europe or the working together spirit of the Japanese.

To respond to these problems of international competition, I believe that the United States is groping towards an industrial policy that will, in time, make a major improvement in our nation's productivity and contribution to the world.

Today, conditions are similar to those in the summer of 1977: there have been large plant closings, massive layoffs, declining profits, and lower mill utilization. Just as the summer of 1977 spurred the creation of the Solomon Task Force and its Report, so today's conditions will spur a new, and I hope more successful effort to develop a comprehensive adjustment plan for the American steel industry.[5]

The industrial policy which is evolving will seek to:

— lessen the contradictory and unnecessarily expensive aspects of various social regulations (worker safety, environmental protection, anti-trust);

5. The Solomon Plan of December 1977 has not made a decisive difference for the American steel industry, because it was not comprehensive enough. It attempted some important, but relatively costless (to the Treasury) changes. Even the few recommendations which it made carrying a price tag (slightly faster plant write-off and $500 million in loan guarantees) took several years to implement. There is no "costless" way to revitalize an industry as capital expensive and vital as the integrated steel industry.

My Trade Subcommittee is currently working on ideas for a "new Solomon Report. Also in the Congress, "steel caucuses" have been formed in both the House and Senate. Composed largely of representatives from steel producing areas, these caucuses are on the lookout for ways to make legislation more compatible with the modernization needs of the American steel industry.

In addition, the Administration has helped form a new Tripartite Committee of labor, industry, and government to examine ways in which the three sectors can work together for their joint gain. It is expected to issue a major report this summer.

— be more sympathetic to the industry's international needs;
— spur research and development efforts and the level of innovation throughout the economy (i.e., R & D efforts probably will not be concentrated just on the steel industry); and most importantly,
— provide additional capital for the steel industry.

The need for capital *committed to* steel plant modernization and replacement is the most important part of the new partnership. The exact nature of how the capital flow can be increased is now and, for at least the next year, will be a major topic of debate in the Congress. I hope that regardless of the form of assistance, the debate will always center on how to make the industry more efficient at the least cost to the rest of the economy.

Let me comment for a minute concerning the internal American debate on how to provide additional capital.

The domestic industry is pushing legislation (sponsored by a large majority of the House of Representatives) which would provide for substantially more rapid depreciation write-offs for *all* business, not just steel:

— ten years for structures;
— five years for machinery and capital equipment;
— three years for autos.

Unfortunately, this legislation will have a cumulative cost to the United States Treasury of $122 billion in its first five years. The Congress seldom provides corporate tax relief without providing at least as much (and usually twice as much) to individuals. Using this traditional "formula" for tax cuts, I believe the proposed depreciation bill is simply too expensive. The Federal government, in an era of budget deficits and inflation, is not rich enough to pay for this type of "buchshot incentive".

Further, I do not believe that providing faster cash flow for *all* capital investments will make investment in *steel* any more attractive, relatively, than it is today. What is the guarantee that the investments will be made in the vital steel sector rather than gold coins or sugar futures?

I urge the American steel industry and its workers to break away from support of this overly broad depreciation scheme and support relief targeted to the special needs of its industry. If there is depreciation relief, it should be targeted to certain core industries. Like the Japanese Ministry of International Trade and Industry, the United States should establish a National Priorities Board, perhaps with combined public-private membership, to designate or target certain "core" industries (such as steel, other materials industries, and energy saving/mass transit endeavors) and new technology/high R & D industries (such as robotics and semiconductors) for depreciation relief.

In an age of federal deficits, I do not believe we need to provide depreciation relief to oil the roulette wheels of Las Vegas. If the core industries (such as steel) are healthy, the roulette wheels will be kept busy.

In providing capital formation, we must establish priorities — and I believe the Congress will agree that steel modernization is a priority.

I have also proposed a new loan guarantee program, modeled after the Reconstruction Finance Corporation which was used so successfully during the Great Depression and early World War II to keep core industries afloat and to construct whole new industries on a crash basis. I believe loan guarantees are unrated in the American economy. There are some signs now that Chrysler may not need to call on the federal government for its $1.5 billion loan guarantee. The very act of federal backing has restored private sector confidence in the company. A similar program for the steel industry, designed to accelerate the installation of energy-saving devices (continuous casters, new coke ovens, etc.) and to meet the costs of pollution control devices, could be an important new source of capital, with minimal government interference and probable profit to the public.

Regardless of the details over form and amount, I believe that there is a consensus in the Congress that the steel industry must obtain additional capital for modernization. I believe that the capital will be found to rebuild and to grow with the market.

I am excited by the prospects presented by a joint-industry-labor-government cooperation. It has been said that the American steel industry is the most profitable in the world except for the much smaller Canadian industry, and it has continued to be so in the face of adversarial relations with its workers and its government. *Think how successful the industry can be if all three elements work together!*

TRADE POLICY DURING THE PERIOD OF MODERNIZATION

A nation's steel modernization cannot be carried out in a vacuum. We live in a world of trade. The United States, as the only large industrialized nation which must depend on imports for a portion of its domestic supply, is particularly sensitive to the ebb and flow of trade and the impact that surges of imports can have on price and on economies of scale. The American industry realizes its modernization must be carried out in the face of import competition. The US industry has given up being all things to all people and does not expect to supply all domestic demand.

Let me be very clear. As we close down old facilities — and before new capacity is available — some may plan to capture a larger share of the American market. No one should plan on permanently capturing that larger share. The American industry will be rebuilding — not for exports, but to meet a substantial portion of the domestic market — and to supply the basic national security needs of the United States.

I do not want to throw any bricks, but as the United States looks out over the world and sees the efforts of others to modernize, it sees that others are very carefully controlling access to their markets as part of the effort to assist in domestic steel modernization.

Thanks to the Davignon Plan, imports into Europe habe been held to about 10 per cent of consumption — compared to 15.2 per cent of consumption into the United States market — and this despite general agreement that US costs of production are still lower than European costs.[6]

The Plan, which seeks to control internal trade and establish "hot house" prices free from world competition, even seems inconsistent in many ways with the laws governing the Community. The control of imports is nearly absolute, and has no basis in foreign costs of production.

In the Pacific, I find it hard to understand why more low-cost Korean steel was shipped across the Pacific to the United States and sold below Japanese TPM prices in 1979, than entered Japan, across the narrow Korea Strait.

Trade, and import penetration levels are particularly sensitive now, because of the OPEC-inspired recession facing the world economy. OPEC keeps moving back the date of world recovery and the long-predicted "steel boom", and by keeping

6. "Financial Times", December 17, 1979:

"The Davignon anti-crisis plan is proving its worth, however, as a control upon steel imports into Europe. In the first half of 1979 imports of European Coal and Steel Community (ECSC) products represented just over 10 per cent of the Community's steel requirements. That is considered a respectably low figure at a time when so many developing nations are expanding their steel industries rapidly and are anxious to widen their export trade."

many nations' utilization levels low, it encourages dumping and subsidization, thus fueling trade friction among the nations. In the United States, the economic downturn and the dramatic downsizing of autos[7] (OPEC again), makes the pressure for trade war in the Congress very real. The State Department keeps telling us that every unemployed European steelworker votes Socialist or Communist. Let me tell you that every unemployed American steelworker demands a rabid, protectionist Congressman!

Thus, recognizing the mutual social problems which trade competition can create, it is more essential than ever that we reach agreement among ourselves on how trade frictions should be handled.

We can, of course, continue to rely on the traditional responses to discriminatory pricing practices, i.e., the application of anti-dumping duties, or its offspring, the TPM.

When it was first proposed, I questioned not only the need for the TPM, but also certain of its legal and political aspects. However, despite its many defects, I believe that the TPM has served and can continue to serve, as a stabilizing force in the U.S. market. The U.S. General Accouting Office will soon issue a report critical of the TPM. The press accounts of this report seem misleading. I believe that the TPM has done a good job in ending price discounting and reducing the level of steel dumping in the U.S. Before rushing to judgment, I would ask the American industry to consider what the past two years would have been like if there has been no TPM. The GAO report does make clear, however, that both Customs and now the Department of Commerce could and should do a better job of administering the program. Certainly, in equity to the Japanese who have cooperated so thoroughly in this program, there should be better enforcement. We may also want to consider a second TPM level for European producers.

Rumors that a large number of sweeping antidumping petitions are about to be filed in the U.S. (primarily against Europe) have focused debate on whether such an action would or should result in the termination of the TPM. There is no legal obstacle to the continued operation of the TPM in such circumstances. Therefore, I am urging the Administration to continue the TPM so as to provide needed certainty during a time of rebuilding. It is particularly important to continue the TPM for the sake of the smaller producers, who do not have the staff or resources to pursue a wide range of andidumping complaints.[8]

It is difficult to chastise the steel companies for resorting to antidumping cases. Throughout the long and arduous negotiation of the MTN, the American Executive Branch (and to some extent the Congress) urged the steel industry not to oppose the new Agreement or the tariff cuts, on the grounds that the industry stood to gain from the new Codes governing dumping and subsidies. To have created this expectation (and support for the MTN) and now to decry its use is hypocritical — and destructive to the overall efforts to build an American industrial policy.

At the same time, for one or two companies to purposely set out to render the TPM redundant raises questions as to their motives: Is there a desire to force some sudden bankruptcies and thus reduce competition? What will happen in the Western US market? Will they use the resulting confusion to precipitate massive layoffs at older plants, rather than seeking to modernize and adjust in a more orderly and creative manner?

---

7. The mood of workers in the United States is particularly sensitive because of the crisis our auto industry is undergoing, and the spiraling level of imports.

8. Failure to continue the quotas on specialty steel for another two years was justified by the Administration in large part by the fact that by February, 1980, the MTN codes would be in place and would serve as adequate safeguard for the industry. To discourage the specialty — or carbon — steel industry from using those Codes would now be inappropriate.

For these reasons the industry might wish to consider the advantages of filing a petition on a limited number of products to serve as a test case. Such an action would test the adequacy of the new law as a remedy against unfair imports while continuing TPM stability.

In my opinion, the debate over the relative merits of the TPM versus the antidumping law can be counterproductive because, by perpetuating the myth that the industry's problems are predominantly trade related, it narrows the area within which remedies are sought. It should be clear from the previous discussion that the problems faced by the industry, e.g., capital formation, environmental regulations, etc., are due to many factors in addition to other countries' trade policies. Thus, remedies limited to the trade area will not address all the industry's problems. The TPM and the Davignon Plan are, after all, temporary measures. I hasten to add, however, that while vigorous enforcement of the trade laws will not alone ensure the modernization of the American industry, lax enforcement will ensure the gradual withering of the industry.

Nevertheless I believe we must broaden the context within we seek a remedy to the plight of the steel industry. At the same time in developing our trade policy in this area, we must be sensitive to the intense pressures experienced by each of us as we seek to rationalize our steel industries.

LONG-RANGE TRADE COOPERATION

I have said that in time the TPM should be terminated. For the past fifteen years, world steel trade has been marked by a patchwork series of ad hoc schemes designed to prevent "excessive" and undue trade disruption. The United States has been guilty along with the rest. Our VRAs and TPMs have contributed to the confusion — but appear relatively "ineffectual" and simple compared to the "guidance", VRAs, and minimum prices operated by others.

As we move through the difficult period of adjustment and modernization, we should work toward a system of trade that "graduates" from all of these bilateral, confusing, and indefinitely promulgated devices.

These gimmicks sprout up primarily when there is an economic downturn — and then are slow to wither away. Recognizing the terrible political pressures which occur during downturns, perhaps it would be best to develop some clearly understood indicators of such steel downturns in various nations. As signaled by the indicators, the nations would understand that economic conditions in one or more countries called for a special awareness among all trading parties to the dangers of market disruption and to the need to avoid increasing market share or lowering price in a manner which would exacerbate injury. Steel is such a vital sector and such a cyclical sector of the world's economy, that some form of almost automatic and clearly understood Article XIX-type action should be agreed on and placed on stand-by until called for. Such a system of minimizing — or at least not compounding each others steel cycles — would provide a clearer, more transparent form of trade than the current system of schemes and gimmicks. In normal times, the trade rules of the new Tokyo Round Agreement would govern and provide economic discipline. In times of distress, a clearly understood form of temporary escape clause import relief could be triggered on and off by predetermined international agreement. Some system such as this is needed to remove some of the unnecessarily destructive features of temporary overcapacity in the steel cycle.

I offer this idea as one for further discussion and debate, realizing that it may have many defects. I believe those defects are less than those of the present system.

PARTICIPATION OF NON-OECD MEMBERS

Before closing, I want to commend the delegations from certain non-OECD countries for participating, and I hope that in the coming months the Steel Committee membership will be expanded to include all nations with the potential or desire to enter the steel export markets.

American steel imports from newly industrializing nations are increasing rapidly. In some cases their cost of production is lower than even Japan's, due to new technologies and low wages. In others, the cost of production is a mystery, due to a maze of government subsidies.

I suspect the continuing OPEC price rises will continue to postpone the development of significant new capacity in the developing countries. Still, if only a portion of the Lima Declaration goals are achieved, the impact on the pattern of world steel trade will be awesome, and carries the most important implications for Japan's exporters.

As LDC's build plants to achieve economies of scale, they will often have temporary capacity far above home demand. They should not casually assume others will absorb that export. Full and frank discussions at these steel meetings will provide invaluable information to new steel exporters on how to maximize the benefit of their steel investments. Failure to understand or participate in these meetings could lead to costly investment errors and the squandering of precious capital.

FINANCING AND ENERGY ISSUES

The work of this Committee is also important to two other committees here in Paris: it is important that we establish a policy on export financing credits for the construction of new steel plants and to encourage maximum cooperation among the nations on ways to save energy in the steel-making process.

On the first point, increased steel development and import substitution among the world's developing nations is essential. No one should deny or hinder that development. But it does seem that many LDC plants come on stream with considerable temporary capacity, which then enters the export markets and compounds the pressure on the older steel mills of the world. Export credits which create new capacity for the export markets make no sense whatsoever. We should agree among ourselves to end this doubly destructive financing competition.

In the United States, steel consumes about 4.5 per cent of the total energy demand of the Nation. Throughout the world, it is one of the principal energy gluttons. Cooperation among the nations on the latest in energy-saving ideas is essential. Japan's leadership in this area is particularly to be commended. The advances she and others are making will serve as an aid to the International Energy Agency in its efforts to respond to the oil price and supply crisis.

OECD SHOULD CONSIDER ADDITIONAL SPECIAL COMMITTEE IN HIGH TECHNOLOGY SECTORS

I believe that the Committee has proven useful for exploring and beginning to find ideas for avoiding trade problems in a fundamental core industry. In the future, I foresee trade controversies of new and special bitterness in the area of high technology products and subsidized research and development. Trade and investment in semi-conductors, for example, is rapidly becoming a cause of friction between the United States and Japan (and to some extent, Europe). I propose that a Committee, similar to the Steel Committee, be created within the OECD to begin to understand the nature of the developed world's growing trade, R & D, and investment problems in *very high technology industries.* A forum in this general subject area can help trade disputes on the crucial industries of the future from becoming explosive issues.

IMPORTANCE OF OECD STEEL COMMITTEE

The Steel Committee and Symposium such as this offer us the chance to periodically step back from the daily turmoil and look for the long-range trends. Hopefully, this forum will be a place where issues can be researched and explored — issues such as the future of the iron age in an era of new products, the impact of energy changes on steel-making, the demand for steel in autos and other products, the shrinking world supply of cheap materials and alloys. Questions such as these need a world consensus and a world input. The Steel Committee is uniquely situated to contribute to preventing problems in these areas from becoming crises.

There is only one certainty in the new decade — and that is uncertainty. Much of the world's steel industry has not responded well to change in the past. The Steel Committee is an opportunity to make the responses of the industry more flexible and more dynamic — to understand the direction of events.

I view this Steel Committee as a most important forum, fulfilling one of the major objectives of the American Congress when it called for sectoral negotiations as part of the Tokyo Round MTN.

In the Congress, we occasionally receive the impression that this Committee is still not taken seriously or as an important forum. I believe the delegations here today belie that impression. Let me say very clearly, that the Congress believes, probably more strongly than the Executive Branch, that a healthy American steel industry supplying a significant portion of domestic demand must and will be maintained. I believe that the Congress is prepared to move unilaterally to protect — in the rawest sense of the word — basic, core American industries.

As far as I can gather, the United States does not plan a steel industry designed for export, we expect a considerable portion of our market to be supplied by imports, now and in the foreseeable future. We also expect everyone to understand the importance we attach to having a domestic industry which supplies the substantial portion of our market and which can fulfill our national security needs. Everyone in this room should support a strong American steel industry as an essential part of the credible defense of the Western democracies. I will do everything in my power — and I believe a majority of the Congress would support this — to guarantee that imports do not destroy the domestic industry's efforts to replace capacity and to modernize.

The Steel Committee is a place where we can all appreciate the terrible political and social burdens of correcting the problems of overcapacity and obsolete facilities. The mutual appreciation of these burdens should help to prevent uncoordinated responses which unfairly increase the burden on any one participant. Our mutual goals should be stable, but competitive markets, based on efficient production, which ensure the rights of producers, consumers, and international partners. Open and frank discussion will permit all the democracies to work out the best possible adjustment to the future and to achieve our mutual goals.

CHAIRMAN:

...I would now like to give the floor to Mr. Alan Wolff who was the first Chairman of the Steel Committee, and therefore has some experience with that new body. All his other titles are well known.

Mr. Alan WOLFF:

I will not read my paper which has been distributed, but make a few comments on the subjects that we have been discussing today. Being the last speaker on the formal programme gives me some advantage and some disadvantages. I will now be somewhat redundant with regard to those things that I agree with.

I am very pleased to be here. I was very devoted to the work of the Steel Committee and the OECD activities in the steel sector. I would say for one thing we would not be here today had it not been for the dedication of a few of the people in this room, including and most notably Viscount Davignon, to the creation of an international process through which greater understanding could be obtained in the steel area. The idea of the symposium was to get broad representation of governments, management, labour, consuming interests, both developed countries and less developed countries into a single room and to try to examine in common the outlook in steel and what those here might do to influence events in a positive direction.

It is extraordinarily timely that this session has been called right now. The continuation of the status quo with respect to the United States as the world's largest open market is, according to newspaper accounts, subject to some doubts. This presents something of a major challenge to trade policy makers as well as a number of the other things that have been pointed out in these two days.

The assumptions upon which we all act are quite important and they vary somewhat considerably, as we have seen yesterday. My assumptions tend to be on the negative side. I don't think we can assume anything but slow growth in the coming years. This morning's newspaper accounts indicate that there is a competition among the OECD Member governments in seeing who can impose the highest interest rates. This is the opposite of the investment incentives to renew capital stock that the steel industry participants have been calling for in these two days. I don't think that it is safe to plan any policies on an assumption of a deceleration of energy price increases; that doesn't seem to be within our control yet. I don't think that we can rely on high net imports by developing countries to save the developed countries with respect to demand, in view of the enormous and growing debt of the developing countries.

So we face a number of questions. Who has to restructure and by how much? In terms of reduction of capacity or of trade flows or of expectations, to what extent will this subject be pursued within an agreed international framework? To what extent will there be conscious, co-ordinated action? And how quickly will the adjustment process take place, both of industrial structures and trade patterns?

Trade policy alone provides an inadequate response because without restructuring in steel, without renewal of our capital stock, we are condemned to trade action after trade action, for the next decade.

Now I might make just a comment at this point about how one characterises trade action – although I don't want to get into this in any detail. There has been some implication in some of the papers that have been provided and in some of the discussion that we should all avoid protectionism. Everyone is agreed that we should all avoid protectionism, but lumped into the idea of protectionism for some reason are anti-dumping and countervailing duties.

We went through a very long negotiation, for about six and a half years in Geneva, on the subject of how anti-dumping and countervailing duties should be brought and how they should be applied. It should not be, in my view, thought that to countervail or to bring an anti-dumping action causes somehow a disequilibrium. To me it does not. It restores a balance, and any United States countervailing or anti-dumping action I assume would now be wholly consistent with the international agreements that we have just spent so much time negotiating. It is to me like saying that Western opposition to the Russian invasion of Afghanistan is an unwarranted provocation by the West.

243

The purpose of that digression was solely to defend the legitimacy of what we did in Geneva. Even though there are unilateral trade actions which have the full sanction of international agreements, I do not believe invocation of those remedies would be a preferred route. We need to adopt a package of policies, industrial policy and trade policy, domestic and international.

I do not believe that we have the time to negotiate an international safeguard arrangement for steel, without an understanding of the conditions of trade that will apply in the interim with an agreement on what we shall each have as domestic policy. We missed the opportunity for engaging in future planning of what international institutions should be like at some point. Now our governments must deal with the present and they must deal with it now. Nor am I convinced that what we necessarily need is an international safeguard arrangement for steel, although I have an open mind on that question.

The narrow purpose of my paper was to talk about trade policy. I think one could say a lot in it in terms of trade policy about government interventions in the markets, subsidies and dumping that has taken place. I don't think it is useful to concentrate in any way on recriminations or a diatribe against government intervention. What we need are solutions. We have perhaps enough now in the way of discussion of the problems.

So the responsibility lies on many of the government representatives in this room and private sector representatives for the future of amicable economic and political relations; the responsibility lies on perhaps everyone here but myself, as I have opted out. We have the principles to live by; the OECD ministers have artfully drafted them; they are called "positive adjustment". Now we need to have something more specific in terms, I believe, of an international understanding and not just an exhortation to moral behaviour. I believe we need a specific agreed approach on how to accomplish the transition to a world market shaped by international competitive factors. The apparent alternative, and it seems to be rather stark and staring us right in the face, is to investigate and litigate endlessly. The main employment increase will be for lawyers and cost accountants, to which I am not opposed in principle, but there are far better objectives to which these talents might be directed.

What is needed in terms of international co-operation is unprecedented. Some of the groundwork has been prepared, as I say, by the OECD ministerial declarations on positive adjustment, and the Steel Committee's charter. But no government entity has been willing today to sit down with another government to do what is needed now, to take the difficult decisions that have to be made.

The United States perhaps will have to move the furthest of any government in terms of changing its approach to industrial policy problems. We need badly, it seems to me, change in our system of promoting capital investment. We need badly a revision of our macro-economic approach and our sectoral policy and analysis.

Others will have very difficult adjustments to make as well. The major exporters of steel will have to accept changed patterns of trade.

There is no mandate for this conference to reach conclusions; it is entirely appropriate that it should not attempt to do so. But what I would hope is that we would not all go our separate ways after this afternoon and that this would become like so many other international conferences having increased common understanding of the problems and then allowing a general deterioration of the overall situation with ensuing confrontations which may well be avoidable.

It may not be particularly evident from the Cassandra-like statements that I have just made about overall outlook that, in fact, I still remain basically an optimist. I think that our governments can jointly manage the problems ahead and now is the time for them to do so.

THE TRADE POLICY CONTEXT OF STEEL SECTOR PROBLEMS: DEVELOPING AN INTERNATIONAL APPROACH

by
Alan Wm. WOLFF
Former Chairman, OECD Steel Committee
Partner, Verner, Liipfert,
Bernhard & McPherson

BACKGROUND

We live in a world in which there is no shortage of challenges for economic policy-makers. In the next few days and weeks, trade policy decisions will be taken on a number of issues affecting large volumes of trade, involving synthetic fibers, steel and automobiles. There will be others. These are only the beginning. Unfortunately, these are not unrelated, isolated issues, coming together by coincidence.

Vast adjustments are taking place in the world economy, many more will still be required. These adjustments would have occured in any event, but the sharp oil price increase and its broad economic impact have accelerated the need for major adjustment in many basic industries. One of the most obvious visible signs that adjustment is needed is the build-up of pressures to impose trade restrictions. However, trade flows which generate complaints are not generally more than a part of the problem. They are a symptom that other factors affecting an industry require attention.

The outlook for the immediate future offers no general relief. The next few years are likely to be characterized by slow growth, a continuing escalation of the price of energy, and from this and other sources, continuation of an undesirably high rate of inflation. Eventually, if not as soon as had earlier been predicted, added to these factors adversely affecting the basic industries of the developed countries will be the challenge of new sources of competition from the newly industrializing countries.

This is not to paint a picture of unremitting gloom. It does mean that there will continue to be a series of substantial challenges to official economic policymakers over the next few years. The ultimate results are not likely to be improved by an absence of early strenuous efforts to work together to find mutually acceptable solutions to common problems.

Of course, the economic conditions cited above have been present for several years. Why then should there be serious cause for even greater concern for the next five years, given our survival reasonably intact through the last five. Several answers are possible: First, the overall economic crisis of the 1970's proved manageable. Aggregate demand, was sustained to a greater degree than had been expected

245

through a recycling of funds, among other measures. A sharp decline in non-OPEC developing country resources was forestalled by massive private lending. Trade patterns were maintained, in part through government support. More importantly, existing trade policies favoring nonintervention which had been adopted during more favorable economic conditions were continued. In part, this was due to a strong philosophical commitment to these policies. In part it was because the need for broad structural changes only became fully apparent slowly.

THE INTERNATIONAL POLITICAL CONTEXT

There was another, less tangible, factor that was perhaps an even more important reason why economic confrontations were kept to a minimum among the three largest trading entities (the U.S., the EEC, and Japan). This was the process of the multilateral trade negotiations and the close working relationships that it developed among the key participants.

It is easy to forget the serious confrontations that were avoided successfully through solutions found during the last half of the 1970's. Between the United States and Japan, enormous hostility pre-dated the Strauss-Ushiba understanding of January, 1978, and this could have easily resulted in across-the-board trade restrictions. Between the European Community and the United States, a cheese war (including canned hams) nearly occurred that could have widened into a full trade war. Numerous other issues remained just beneath the flash point: two U.S. countervailing duty cases against the most basic features of other countries systems of taxing goods (the VAT and the Japanese commodities tax); a confrontation with Brazil over its export subsidies; a myriad of complaints over other countries' product standards, government procurement, customs valuation and tariff issues.

A number of these issues could have resulted in the unraveling of the existing system of international trade commitments. The Tokyo Round of trade negotiations provided a vehicle and a motive for finding solutions. It also provided a close daily working relationship between Strauss, Ushiba, Haferkamp, Davignon, and a large number of others. The fact was that each participant in the negotiations was very reluctant to take any precipitous or excessive action adversely affecting another in the midst of a great cooperative effort. This structure for the joint management of problems is no longer present to the same degree.

THE DOMESTIC POLITICAL CONTEXT

Another factor supporting the management of international trade problems was the fact that there was considerable unity in the United States over the conduct of trade policy. This has usually not been the case. The U.S. constitution divides trade authorities in two, in a way that has generally not worked well: The Congress has complete power over the regulation of foreign commerce, but only the Executive can negotiate agreements with other governments.

The result has been an uneasy association in which every multilateral trade agreement since World War Two that required Congressional consent (all but tariff agreements) has either completely failed to receive approval (i.e. the International Trade Organization, the World Trade Organization, the American Selling Price Agreement) or arguably was seriously impaired by the Congress (i.e. the International Antidumping Code). Against this background the unprecedented vote of Congressional approval of the MTN agreements, 395-7 in the House of Representatives, and 90-4 in the Senate, is particularly remarkable. The Executive Branch (the President's negotiators), the Congress, and the private sector were generally united in support of the negotiated results and the accompanying implementing legislation, the Trade Agreements Act of 1979.

The roles that individuals played in this event can best be told elsewhere. It is the institutional arrangements which deserve some brief attention here. In every country, there are two systems of law and politics in which trade policy officials are held accountable, the domestic and the international. An elaborate domestic support process for negotiating the codes, in existence to varying degrees in each major MTN participant, provided the domestic basis for reaching an international accord, which would then be acceptable at home.

The domestic system was perhaps most formal and elaborate in the United States, but analogous systems also existed in Canada, the European Community and Japan. In principle, these systems could be utilized to obtain domestic support for negotiated solutions going beyond the MTN. In the United States, the same procedures that assure prompt and direct Congressional consideration of the MTN still remain available in law, as does the private sector advisory system. Moreover, U.S. law has a specific provision authorizing negotiation of an agreement of the type that would be required for steel.

The domestic political support, built in the OECD member countries in support of the improved international trading system emerging from the Tokyo Round of multinational trade negotiations is a great asset, both for continuing general progress toward strengthening the framework for international trade (e.g. for completing work on unfinished issues, such as a safeguarde code, an agricultural consultative mechanism, a wheat reserves agreement, and for work on new issues such as the trade barriers which affect international trade in services). It also can serve as political support for negotiating solutions to individual sectoral problems.

Conversely, however, this domestic political support will erode quickly in the United States if concrete progress is not made in two areas – full implementation of the MTN codes of conduct and, this is particularly important, if there is not vigorous enforcement of the trade remedies (anti-dumping, countervailing and invocation of code remedies) contained in U.S. statutes. These remedies are designed to provide prompt relief. They are simpler to invoke, and, for the first time they are explicitly in conformity with international agreements. The anti-dumping and countervailing remedies have also been designed to exclude, in so far as possible, any exercise of discretion by their administrators.

UNITED STATES TRADE POLICY

Given the fact that the United States is the largest market for imported steel, and that most of those present at the International Steel Symposium attach some importance to continued access to that market, it might be useful to describe briefly the assumptions underlying U.S. trade policy to indicate to the extent possible what considerations will guide U.S. officials in deciding what measures should be taken at the present time.

To understand U.S. trade policy, it is useful to examine the philosophy and theory on which it is based. It flows directly from the absence of a coherent or comprehensive U.S. industrial policy. As a general rule, Americans do not see the role of government as foreseeing where U.S. international competitive strengths lie, or to support the production of one commodity over another. The U.S. government does not intervene at the border pursuant to an industrial plan to protect any industry, nor does it attempt to stimulate or retard investment or production in particular industries. It is equally neutral with respect to foreign investment in the

United States, or U.S. investment abroad, or with respect to producing for export or for domestic consumption.

In short, the theoretical basis of U.S. trade policy is not mercantilist. It is a laissez faire policy, altered on a piecemeal basis by a series of other policies, such as acting against unfair or injurious trade practices, including dumping or subsidization or other forms of injurious disruption of the domestic market.

To reinforce the overall laissez faire policy and ideology, the U.S. government operates not only on a "hands-off" basis but also with "eyes-closed" as well. Within the government, there are few resources devoted to examining, much less predicting, what is happening in the various sectors of the economy.

United States policy is often criticized abroad as being excessively legalistic. This accusation is made particularly with respect to the administration of U.S. trade remedies. Where a trade policy official in a country with a more managed economic system is probably looking at the health of a particular industry and its need to export, as well as at the bottom line of the country's international payments position, his American counterpart is most often examining the question of whether the trade is "fair". Some see a test of "fairness" as a mask for protectionism in the United States. This is too simplistic and a largely inaccurate conclusion.

U.S. officials have a bias against intervention in trade. They, and more importantly the U.S. private sector and its elected representatives in Congress, do differentiate between "fair" and "unfair" foreign competition, however. It is a rough test, perhaps, and often requires excessively fine distinctions to be made. At its extreme, severe results seem possible, threatening to plunge the United States into a major economic confrontation, for example over whether the rebate of the value added tax or Japanese commodities tax was a countervailable subsidy. In this matter, U.S. law required an examination not of what the optimum size and structure of the domestic steel or television industries should be, but what the U.S. Congress sitting in 1897 intended in defining countervailable subsidies. Some foreign observers must find this behaviour hard to understand. To U.S. officials the practices and procedures seem logical and the idea of a more highly interventionist approach, such as assigning market shares, seems inappropriate.

It will be only partial comfort to any exporter to the U.S. market that the American approach is becoming at one and the same time more and less legalistic. The Trade Agreements Act of 1979 went far to establish more accessible and automatic remedies against foreign dumping and subsidization, as well as invocation of remedies provided by the new Geneva trade agreements. At the same time, however, the potential for disruption of trade through the use of those remedies has caused U.S. trade policy officials to examine more pragmatic alternatives such as seeking increased foreign investment in the United States, voluntary export restraint, and other less formal understandings about future conduct in the U.S. market.

The maintenance of the U.S. as an open market for other country's exports is highly dependent on the U.S. private sector's and the Congress' appreciation that there is equity provided in terms of market access abroad and that injurious and/or unfair foreign competition particularly in our own market will be remedied promptly. When the Executive Branch is seen as being unable to obtain these objectives, Congress' reaction has been to progressively narrow the scope for discretion allowed to the Executive in the administration of the trade laws. This lack of discretion can be viewed from abroad as resulting in increasingly arbitrary U.S. government interference in trade, because there is in fact less and less flexibility to work with other countries on mutually acceptable solutions.

These are currently numerous allegations of widespread dumping of steel into the United States market. Under U.S. law, if antidumping or countervailing duty cases are filed, a preliminary determination would have to be made by the administering authorities within less than a half year, at which point liability for additional duties would attach if the allegations are found to be supported by sufficient facts. If the margins are substantial, and material injury is found to exist or to be threatened, current patterns of trade would be very likely to be substantially altered immediately. The Trade Agreements Act of 1979 is designed to obtain this result if the facts warrant.

FORMATION OF THE INTERNATIONAL STEEL COMMITTEE

OECD member governments have, for better or worse, a high degree of involvement in the steel sector, ranging from ownership of productive facilities to an array of other activities that create many of the conditions under which steel is produced, sold and traded. Thus governments, to varying extents, regulate the conditions of employment in steel plants, the amount and kind of energy the plant uses, the location of new productive facilities, the limits of water and air pollution, the degree and type of permissible competition (and price policy), the ability to accumulate capital through tax and credit policies, and ultimately the amount of aggregate demand in the domestic economy (and to the extent that fiscal and monetary policies are successful this determines the domestic demand for steel).

These "internal" policies have traditionally not been considered part of trade policy, and were not matters for international discussion, although they profoundly affect the international competitiveness of national enterprises. With the beginning of the steel "crisis" of the mid-1970's, and the creation of the International Steel Committee in the fall of 1978, this perception began to change. Domestic policies as well as trade policies affecting the steel sector came to be regarded as legitimate causes for international concern. Restructuring plans, reductions in capacity, trade measures, government assistance to firms and workers, pricing policies, all were the subject of reporting, data-collection and international discussion, to an extent previously regarded as unacceptable. This was an important step forward in promoting international understanding, although no specific acts of cooperation were called for.

The International Steel Committee is now just over one year old. The impetus for its creation was the growing involvement of OECD member governments in steel sector issues, and the concern of each over the impact that others' measures might have on their own industry. The Committee provided more than a forum for data collection and exchange of information and views. Its charter listed a number of common objectives. It called upon members to work together to minimize trade restrictions in scope and duration, and to keep these measures within GATT rules.

But measures were considered inevitable. Governments would be enabled "to act promptly to cope with crisis situations in consultation with interested trading partners in conformity with agreed principles". Moreover, they should work together to "facilitate needed adaptations". Lastly, and perhaps most importantly, the Committees members recognized the need to *"facilitate multilateral cooperation consistent with the need to maintain, to anticipate and, to the extent possible, prevent problems"*.

Unfortunately, the crisis in the steel sector is not about to pass yet. Slow growth in the OECD members' economic appears to be all that can be anticipated for several more years, with continued relatively slack demand and excess capacity. The magnitude of the problem places Governments at something of a cross-roads for dealing with steel. For a reasonable period they could do without an international consensus on what amount of government intervention to bolster shares of world production and of trade patterns was acceptable, particularly if recovery, the end of a trough in the business cycle, was in sight. But the most serious problems confronting the steel sector are not just cyclical, major restructuring of a number of member countries' industries is required. Absent a substantial turnaround in the near term toward much higher demand and capacity utilization, the changes in comparative advantage, in terms of labor and raw material costs, and efficiency of plant and equipment, is increasingly going to have to show up in the market place.

NEXT STEPS IN INTERNATIONAL COOPERATION

If the factual assumptions listed above are accurate, a central question is how best can the transition be accomplished towards a world market shaped by international competitive factors freed to a maximum extent from distorsions introduced by governments. One approach, which now seems more likely to be used than any other, is the widespread resort to antidumping actions to test whether current trade patterns are appropriate.

The use of antidumping actions have full international sanction under the General Agreement on Tariffs and Trade, the new GATT Antidumping Code, and the International Steel Committee's charter.

There are numerous drawbacks to the use of Antidumping actions, however, from both a domestic producer's and a foreign exporter's point of view. The investigation is time-consuming (although less so under the 1979 Act), during which uncertainty continues. The measurements of prices and costs are difficult and often are necessarily imprecise and subject to dispute. The accumulation of the facts is burdensome for all concerned. The degree of injury is a relatively subjective and hard to predict judgment. In the judgment of any participant, the end result may be unsatisfactory, especially as it is arrived at without fully taking into account the interests of concerned parties. The impact of the result, after substantial delay, is likely to have a disruptive impact on trade.

Nevertheless, unilateral measures are generally preferred by trade-policy-makers. This seems to be due to a number of reasons: Unilateral measures do not have the appearance of a combination in restraint of trade. They can be condemned in the strongest terms by those adversely affected. This is felt not only to have a cathartic effect, but also to be likely to limit the duration of the measures. Failure to give consent to restrictions on one's exports also preserves all rights of response, as well as having the psychological benefit of appearing to retain sovereignty over one's trade unimpaired.

For a number of reasons, closer international cooperation in the steel sector would appear to be worth exploring. First, more is needed with respect to steel than just trade measures. What is needed primarily is structural adjustment. A trade measure which is not accompanied by an adjustment plan would of necessity have to be of indefinite duration. Therefore, what is needed is a comprehensive approach bringing in all relevant aspects of the problem. The trade measures that formed a part of this plan would, since they would have been part of an international agreement, be more acceptable to those affected, taking into account their legitimate interests and tailored to facilitate adjustment.

The results of international negotiation, where basic objectives are not sacrificed, are often superior to the use of unilateral measures. Negotiation is a moderating process, requiring a balancing of the interests affected and bringing about the desired results without confrontation. It is unclear why trade measures imposed pursuant to an international understanding should be considered to be more likely to be of excessive duration or restrictiveness as opposed to measures imposed unilaterally. There is no compelling logic to such a position.

The general principles to construct an appropriate international solution already exist as part of the OECD Council decision establishing the International Steel Committee such as: minimizing trade actions, conforming such actions to GATT, facilitating needed structural adaptations with the aim of achieving fully competitive industries, avoiding encouraging economically unjustified investments. The stated limitation: cooperation should be consistent with the need to maintain competition.

The adoption by the OECD Council in June 1978 of the "general orientations for adjustment policies" is a more elegant and detailed statement setting high standards for member governments. It states that intervention should be temporary and reduced progressively according to a prearranged time-table. The most important point perhaps is that:

"Such action should be integrally linked to the implementation of plans to phase out obsolete capacity and reestablish financially viable entities, without however, seeking to raise prices above levels providing an adequate return to efficient producers".

Absent an international understanding on steel, trade measures would be deprived of their most important function, to be part of a common scheme in which they are ancillary to a series of national adjustment programmes.

The suggestion then is to make the international steel Committee into something more than a shell in which national policies are discussed, and hopefully influenced, but to use it to foster the construction of a common international approach for steel, an international understanding which would avoid the potentially sharp and divisive confrontation that is likely to burden economic relations in the near future absent such an approach.

For the United States to participate in an international steel approach would require a level of conscious industrial sector analysis to which we are unaccustomed as a Government, but it could be done. It would also require that a positive adjustment package of domestic measures be fashioned for the U.S. steel industry (including tax, environmental policies, etc.). Historical trade patterns would be altered, but this could be accomplished over a reasonable period, as national adjustment programs took effect.

CHAIRMAN:

I would now like to call on Mr. Gadonneix. We have finished the more general presentations and we come to comments.

Mr. GADONNEIX:

Thank you Mr. Chairman.

We can see from what was said both yesterday and today how difficult it is to adapt the steel industry in depth to new world market conditions.

I would like to contribute to the present discussions by quoting a concrete example of adaptation, that of the French steel industry, and also to voice concern about various comments suggesting that there is a temptation to revert to protectionism.

The French Government and public are committed towards restructuring the industry. I would like to quote a few figures. In 1977, the French steel industry employed 145,000 persons and produced 22 million tonnes. In 1981, production will reach at least the same level, I hope, but with approximately 40,000 fewer persons employed.

Restructuring has, of course, required an exceptional financial effort, first, to help regions previously dominated by the steel industry to convert and, above all, to finance social measures. Agreements have been signed with the trade unions. They have helped to guarantee an income level for those made redundant and they have also enabled any steel worker out of work to be offered a job either within the same company or in another firm.

Finally, productivity investments of over Fr. 5,000 million are planned for the next few years without leading to any capacity increase.

This is an outline of the policy we have followed. But, I have heard a few suggestions that competitiveness could only be improved if protectionist barriers were strengthened. In the OECD Council Decision of 26th October, 1978 establishing a Steel Committee, paragraph A of the initial commitments indeed allows governments to lay down certain price restrictions in order to control unfair trading practices. As a result, two mechanisms have been adopted, one by the Community and the other by the United States.

In early 1978 the European Community set up an anticrisis plan with provisions concerning external policy aimed at stopping dumping practices while maintaining traditional trade flows.

Likewise, the United States has set up similar mechanism of trigger prices with the same purpose in mind. In practice, however, because European Community producers were not allowed to align their prices with those of American steel manufacturers, they have been considerably penalised by the system.

Furthermore, because there is no gap between trigger prices and American listed prices, there has been a lack of flexibility which does not arise in Europe where penetration margins are officially recognised in the agreements with third countries.

The Community is now relaxing its rules towards third countries. This is in line with the guidelines issued by the OECD in 1978, which stated that trade restriction measures should be expeditiously removed or liberalised.

And now, I hear requests for tighter protection and news that dumping complaints are going to be lodged against European producers. Such threats of this sort are groundless on three scores: first, it seems that as a whole American producers have had satisfactory operating results in 1979, even though the results were not as good as in 1978.

In the United States capacity utilisation rates were well above those still encountered in Europe. And I am not even considering the situation of special steel producers, whose profits tripled under the protection of import quotas.

Secondly, I would like to point out that the volume of European imports into the United States dropped by about 20 to 30 per cent in 1979 compared to the previous year. This is indeed largely due to less lively market activity, especially in the motor industry, but also and above all to sales below trigger prices by certain American steel manufacturers.

Thirdly, if I have understood it correctly, the complaints are to be primarily lodged against the Community alone, although the latter's share in American imports has dropped. Anyhow — this probably touches on issues that have been raised today — I wonder whether the trigger price mechanism can remain at all credible if anti-dumping complaints are actually to be lodged when producers sell below trigger prices.

It seems to me that these new obstacles to freedom of trade might once more seriously disrupt the steel market and this would be bound to lead to retaliatory measures.

Furthermore, I feel it is essential that the same conditions of access to the American market apply to all countries. I saw a reference today to differentiated trigger prices, a measure that would obviously run in a counter direction.

Finally, we are convinced that future will belong to the steel industries that are able to modernise and restructure their activities. After a year of painful rationalisation, the French steel industry is proof of this. I hope that, in spite of the various problems to which I have just referred commonsense will prevail and the liberalism to which we are all attached will continue to inspire our individual policies.

I am also convinced that the dialogue started within the OECD Steel Committee, especially with the American authorities, will continue with the same concern for understanding and cooperation as ever before.

Mr. ORBAN:

Everyone here agrees that the United States needs a strong and efficient steel industry. This does not mean necessarily that it supplies 10 per cent of the market.

Imports averaged from a little bit over ten per cent to under 20 per cent over the past 12 years, and actually dropped in 1979 both in terms of tonnes and percentage of the market. I don't see how an import share of ten to 20 per cent would hurt the US defence effort in any way at all. The drop in import tonnage was caused largely by the trigger price system. The current import sales for new orders, in keeping with the general slow down in steel business in the United States, is also quite slow. One problem we have in the statistics is that there is always a six month lead between the placing of import orders and actual arrivals, so that the statistics tend to overlap.

I would like to emphasise that the trigger price system has worked even though it has its faults, both from the viewpoint of the domestic industry and from the import trade. The advantage of the trigger price system is that it makes for predictability. It is preferable, in my opinion to dumping actions, the outcome of which is unpredictable for both sides.

I was very pleased to see the consumers represented at this meeting, which has not been the case in the steel argument in the United States before. I would like to stress that American metalworking industries employ more people and have a much greater output in terms of dollars than the steel industry itself. I think it is a good thing that General Motors is represented here. I think it is unfortunate that the smaller steel consumers are not represented. I have been asked by the Independent Wire Producers Association to point out that they desperately need free access to the world market in order to stay competitive and to stay in business.

I would like to stress that in the past imports have always grown when there was a threat of a steel strike. At the present time, whenever we get a good rate of steel activity in the United States with the mills operating at 85 per cent or better, steel salesmen for the integrated steelmakers threaten customers with allocations, so people will rush out and will place import orders which then come in six months later when often the activity has slowed down. This then gives rise to complaints by the domestic industry that too many imports are being sold in a slow market.

We quite agree that the depreciation rules should be changed for much shorter depreciation to bring them in line with modern technological facts. This in our opinion should be for all metalworking, not just for steel. The 7 billion dollar investment figure would require an increase of 15 per cent in the steel price in 1979 dollars in order to come out of the income of metalworking and ultimately from the consumer in order to permit the programme about which the steel industry is talking. Whether this is done in the form of subsidies or tax allowances or higher prices really makes no difference, because the consumer and the taxpayer will have to pay in the long run anyway.

We have calculated that the present differential between steel wages in the United States and average industrial wages comes to just about 70 dollars per ton, which is just about what it would take to make the U.S. steel industry competitive internationally. This is in line with Dr. Marshall's comments that Mr. McBride has perhaps been a little too successful in looking out for the interest of his members. If steel prices in the U.S. were to be raised drastically, metalworking industries would become uncompetitive both at home and abroad. There would be more jobs lost in metalworking than would be gained in the steel sector.

We also believe that a certain percentage of imports are definitely needed in the United States in order to keep the industry (which is basically an oligopoly) honest. Simply insulating the market and raising prices and wages indefinitely is not the answer, because of the situation of the metalworking industries.

As a solution to the current problem we can see that it would make sense for the short term to have:

— a combination of the trigger price system with the trigger prices not set so high as to result in a virtual embargo for certain products and for certain areas, as seems to be developing at the present time; and

— a flexible system of voluntary restraints such as we had in 1969 through 1973, and as is actually practiced on an informal basis by Japan.

Of course it would be necessary that this would include suppliers other than the Community and Japan, because those suppliers today account for more than one third of total imports. However, we should revert to a free and open market in two, or a maximum of three years. Also, any such system should be tailored in such a way as not to hurt the US consumer.

Mr. VONDRAN:

After two days of discussions in this room, I am forced to note that neither with regard to the diagnosis nor to the cure are we in agreement. But on one subject there can be no doubt, that freedom for our markets for steel is in real danger. This is something I can deduce from the various statements made here.

One other point is obvious to me: whatever form protectionist intentions may take, these intentions will not solve the problem, they will merely shift it; the international solidarity which is necessary in a critical period will thereby be destroyed. Countries which are protectionist do their own industries no good service.

If I can speak for the German steel industry, if I can preach against protectionist measures, it is in the following context: my country supplies 35 to 40 per cent of its domestic market from imports. I compare the American figure in 1979 — the figure was 15 per cent; this figure was 3 per cent below the level for 1978. I am forced to recognise with some bitterness that the level of tolerance varies considerably from one country to another.

The Americans as the most important trading partner in this world establish certain criteria. They also have a political role. We assume that these circumstances should take due account of her international responsibilities. If at present in the United States security reasons are mentioned in an attempt to justify self-sufficiency, I should say that the freedom and health of the Western world cannot be separated; for more than a quarter of a century no one disputed this. This idea appears in several international treaties, and it is part of the philosophy of OECD. If self-sufficiency in steel were to be adopted as an idea, this would be deviation from a successful concept. I think this is an important matter. I think we must sound a word of warning since the effects of policies of this sort cannot be assessed in advance.

One further point. The German steel industry is against unilateral measures of course, but it realises that the American steel industry requires time to adjust and adapt. If a certain period is necessary for a consolidation, my industry is willing to understand the initiatives taken by governments to accompany this process. But this assumes that traditional trade flows should be respected. Government initiatives should not create a situation where worldwide regulations are established. A minimum of intervention should be practiced and considerable freedom should be left to steel enterprises. OECD could perhaps further improve its instruments for analysis and should perhaps be available as a neutral observer for governments. It should not be an organisation which dictates policies.

Mr. ATKINSON:

I would like to speak from the standpoint of the automotive industry. I do not wish to speak specifically about trigger price mechanism, but rather of this as an example of an industry specific measure. It was brought up this morning initially by Mr. Florkoski. Measures of this type have been referred to frequently.

As a result of the U.S. trigger price mechanism, it has been calculated that imported automobiles in the U.S. have a $50 cost advantage over U.S. produced vehicles contained in the price of the steel within the car. This is despite the intention of either the steel industry people who urged a solution to their problems or the government officials. It has come at a time of crisis in the U.S. automotive industry, a crisis produced also by needs to adapt to a new fuel picture and to government mandated design changes. Together with these other factors, the higher price for materials is helping to build pressure for barriers against automotive imports. There is a lesson in this. First, an industry-specific measure against market disruption will be self-defeating. The steel will come into the country in finished form, if it does not come in semi-finished. Second, the protection will spread downstream to user industries. What is true in domestic measures on specific industry remedies, also must be true in the international arena. This, in spite of the fact that I am certain the industry does not wish to pass on its distress to others.

A second point: a procedural one. For the actions of the Steel Committee to be generally accepted, it cannot be seen as an arrangement limited to steel producers broadly defined, that is, including steel labour and the government officials concerned with the industry itself. The Committee must have wider participation.

Mr. SOHLMAN:

I would like to make a very brief comment. I would rather make the case for free trade here in the spirit of the symposium rather than as a statement in an international trade negotiation.

All here seem to agree that we are facing a prolonged period of over-supply. Several have stated that this calls for concertation, co-ordination between our nations in order to reduce capacity or at least avoid the build-up of further new capacity. At the aggregate level this seems to be a good idea. But looking at each country and each company, you find a more diversified and frequently more hopeful picture. Investments are being made in more productive capacity. Technological development is taking place. New products are being developed. These, I think, should be taken into account in our discussions.

On this score I think I am following the reasoning of Dr. Peco yesterday. The argument was made yesterday that if we all try to become more competitive, then none of us will be relatively more competitive. Our relative positions will not change. This is of course correct in a shorter time perspective. But in a somewhat longer perspective, this is perhaps valid too, but not very relevant. The more competitive we become, the better off we will all be. After all, this is the history of the development of our nations, that we in free competition — or more or less free competition — have been able to become more and more productive.

Viewed in this perspective, the statement has been made that market shares should not be increased in other countries. But I would suggest that this is an alarming statement. If any of our countries or companies finds a new a more efficient process or a new product, then this should logically lead to increased exports and inversely increased imports into other countries in this group. It is of course quite possible to shield one's own market from such influences, but this I suggest would not be to the benefit of either the potential importing country or the world economy as a whole. The net effect of such actions would in fact be to decrease the potential GNP for the whole of our group of countries and for the world as a whole.

Earlier in our meeting several speakers have emphasised the necessity for politicians to create the basis for long-term economic growth. I think it is unfortunately appropriate also to point out that future economic growth can be restricted by putting a straightjacket on foreign trade, including foreign trade in steel.

The portent of my intervention, is a call for great caution in the face of the different protectionistic tendencies evident in some papers and in some statements at this symposium. A retreat into protectionism in the face of the present problems would not only result in disastrous effects on steel trade, but also carry great risks of disrupting the whole trading system.

Following an earlier statement by my Canadian colleague, I would urge all governments to stick to the rules elaborated with great pain in the GATT. These rules should also be applied with great caution. At this point of time, with great uncertainty as regards the future economic development, it is imperative that no government take actions which would affect the atmosphere in our trading relations, thus precipitating a development that none of us desires.

Mr. SIMMONS:

I appreciate the opportunity to offer some comments. The perspective from which I speak is as a United States speciality steel producer who, with other speciality steel producers in the United States, has faced the problems caused by unrestrained and disruptive imports since 1962. These imports continued to increase without respite, including through the voluntary restraint agreement period of 1968 until the shortage period of 1973 and 1974, and were followed immediately by further sharp increases in 1975 and 1976 and then limited only by the speciality steel quotas imposed in June 1976 which, for the benefit of those who are not familiar, permitted record imports into the United States in 1979. Thus, we believe that there is no segment of the steel industry which has a longer familiarity with this problem.

If the thesis presented yesterday — that comparative advantage will favour developing nations for carbon flat rolled products — is true, the same thesis is not true for speciality steels such as stainless steel sheet, plate and rod. Yet in the past four years, major capacity has been installed for such stainless steel products, capacity which far exceeds the developing nations' own requirements. Thus it is clear to us that significant quantities of this capacity are intended for export. At the least it will divert other exports to the United States market.

With the end of speciality steel quotas last week, it is of vital importance to the United States speciality steel industry that market disruption from this new capacity, as well as new capacity from developed nations, does not occur.

In listening to the speakers both yesterday and today, an unknowledgable observer might conclude that all the recent and projected capacity additions were made within the disciplines of private sector economics. This, of course, is not the case. Specifically, the opposite is true. Most, if not all of the capacity recently installed or projected by developing nations is state-owned, thus permitting such companies to violate the disciplines of the private sector for extended periods of time, if such state-owned companies wish to export and have other higher priorities than profit. At any point in time, at the least, such state-owned companies create great investment uncertainty for private companies with whom they compete.

The point I would make is that I would hope that this forum and the international Steel Committee would address the issues of how private companies must co-exist with state-owned companies over the long term. Failure to resolve this issue will lead, in my opinion, to even greater protectionist pressures in those nations where private companies do flourish.

Finally, with the increasing formal and informal agreements which limit carbon steel trade flows, there are none which limit in any way speciality steel imports to the United States following the expiration of the speciality steel quotas last week, with the possible minor exception of stainless steel wire. I respectfully submit that the OECD should give close attention to this matter and to any resulting market disruption which may occur in the immediate future. The speciality steel industry, as has already been reported, is profitable, is efficient, and is technologically advanced. Yet even under these conditions, competition with state-owned companies, who do not have to meet our disciplines of capital formation and profit, make investment most uncertain in the private sector.

Mr. CASPARI:

Mr. Chairman, the contributions up till now and the discussion have shown how multiple are the problems and how multiple are the tasks that have to be accomplished by the steel industry. We have also seen that there is a plurality of problems and tasks for international trade, for the whole economy, for society and for politicians. To this extent, this symposium has been of great use to us within the Commission and within the EEC.

There are also a great many proposals concerning the further work of the Steel Committee and the working group of that Committee. I shall be glad if some governments, some states which are represented here but are not members of the Steel Committee would become members of the Committee. This is not just a formula of courtesy; this is really a word of thanks for all the stimulating proposals that have been made. However, I would like to make a few further remarks on the basis of previous contributions and on the basis of the discussion up till now. I will be able to keep it short because what I was going to say has, by and large, been said by my French colleague.

I would like to say that this whole debate, or the basic lines of this debate, corresponds to what was said in November 1977 in the Steel Committee and in its working party. That was the first time, I believe that the major countries of OECD sat down to work out a joint concept to counter the crisis. At that time the prime importance of restructuring was pointed out. It was said that this would be a very painful and long process and that it should be solved through international co-operation and that no one should try to pass on the burden of this to anyone else. The reply of the Commission to this was that we had embarked more deeply than anyone else on a plan for restructuring. I think Viscount Davignon pointed this out very clearly.

Today's discussion has also shown that we realise the social responsibilities that are linked up with this whole set of problems. Above all, the discussion and the figures that were put forward have shown that the EEC has not tried to push the burden onto other states. This is shown in figures on the use of capacity; it is shown in figures on unemployment; figures on imports and exports and on production. Import penetration, which up to 1975 was traditionally about 6 to 7 per cent within the Community, has moved up to 10 or 11 or 12 per cent. This means that import penetration into the market of the EEC has risen and has not gone down, as is the case in other parts of the world.

The share of exports to production has remained at 20 to 22 per cent for a few years and last year we exported less than in the past. All of these, I think, are signs which prove that we have not tried to pass the burden onto others, the burden imposed by restructuring measures.

The second point that was mentioned is that the problems of the steel industry cannot be solved by quantitative restrictions and protectionism. I think that everyone was unanimous on this point today. This also is the firm conviction of the Commission and of the EEC. Viscount Davignon stated all this much more eloquently than I ever could.

Now I come to my most important point. This is in connection with the question of prices. It has been stated clearly that we must pay close attention to the problem of prices, because as always in times of a weak economic situation, we find in the steel industry that there is a tendency towards making sales at a loss. All measures to counter this must, however, take into account traditional trade flows. This has always been our conviction because we must not try to pass on the burden to others. The Commission and the Community have borne this in mind. They have acted in harmony with the rules of the GATT and brought in a system of basic prices. I would like to add that these basic prices are lower than the American trigger prices.

We also believe that in this phase of restructuring and in this attempt to stop prices from sinking further, we ought to do everything in order to satisfy our traditional suppliers and protect our traditional trade flows. These measures must therefore be of a temporary nature, and these temporary measures should not mean that we violate the rules in the matter of traditional trade flows. We have been able to raise some prices but nonetheless, the Community prices are at the lower limit. They are lower than the U.S. prices. This does not mean that we have artificially fixed prices, but these prices remain quite clearly below U.S. prices. We have at the same time let quantities of imports go up, or at least remain stable, and these quantities have recently been going up.

To conclude, I would like to say that as against this position, our exports to the U.S. dropped by 2 million tons last year. Our share of the market in the U.S. has also gone down. I believe that this situation and the attitude of the Community should act as a stimulus to all of those who feel it is now necessary to make decisions. They too, I think, should bear in mind their responsibility for international co-operation. I feel that measures should not be taken now that would lead to a beggar-my-neighbour policy; measures should not be taken that can lead to distortion of international trade. I think that in this process the Community has made its contribution, and I hope that others will make a similar contribution.

Mr. ACLE:

It has been recognised that the steel industry is strategic for economic and political reasons. This is so because it is the base of any industrial process. This applies to all countries despite their degree of development. However, since the demand for steel is a derived demand, it tends to reflect the general economic situation of each country. Steel is demanded in the form of durable goods and capital goods. To have this point clear is very important in understanding why the situation of the steel industry in some developing countries differs radically from that prevailing in the developed countries. The former group is engaged in an active process of industrialisation with a deliberate policy of import substitution. For obvious and strategic reasons, this means the necessity to provide internally steel for the emerging capital and consumer goods industry. Therefore, the steel production in developing countries is domestic market-oriented; it doesn't have as its main purpose to export. In the case of Mexico, there is still exportation to the U.S.

market, but if you consider the American steel market as a whole or the trade deficit existing against Mexico with respect to the U.S., exports can be qualified as negligible.

In consequence, we believe that the fears of developed countries with respect to the actions of some developing countries are exaggerated. However, such fears are contagious. Therefore, developing countries are also afraid about the attitude of the developed countries which, I believe, puts too much emphasis on the emerging steel industry of LDCs as an explanation of their internal situation when this can be mainly explained either by domestic factors or by the interaction among developed economies. In fact, this fear was one of the reasons why Mexico has refrained from accepting the invitation to participate in the OECD Steel Committee. In other words, and to put it with a very Mexican example: we saw joining the Committee as entering into a bullring with the awkward task of being the bull.

In addition, we appreciate some problems with respect to our formal participation in an OECD Committee without being an active member of the OECD. That means, among other things, to be absent from the discussions on general economic policy, which in the end determine in great measures the situation of the steel industry.

Nevertheless, Mexico is well aware of its responsibility as an emerging industrial country. Therefore, we are willing to discuss at international levels and to co-operate at the international level in the solution of the problems of the steel industry.

Mr. HORMATS:

First, as Chairman of the OECD Steel Committee, I thank the OECD Secretariat for its hard work in preparing this excellent symposium. I know the many hours which went into this, and I think we all appreciate the superb job that has been done. Many constructive suggestions have been made here, including a number of proposals specifically directed to the OECD Steel Committee. I can assure you that at our next meeting in April we will begin serious and systematic consideration of these ideas. I am also gratified that many non-OECD countries are represented here and hope that this represents the beginning of a period of even closer co-operation with those countries.

Second, as an official of the U.S. Government, I should like to express my appreciation to Congressman Vanik, and to the business, labour leaders and private citizens from my country who have attended this symposium. As noted by the Congressman, Mr. Foy, Mr. McBride and others, the U.S. steel industry, like those of a number of other countries represented here, has undergone and is today undergoing important and painful adjustments to improve its productivity and its profitability. In the last several months, senior government, business and labour leaders have engaged in a especially intense discussions designed to address the problems of this industry. In recent years the government has responded positively to many of the industry's concerns, including institution of the Trigger Price Mechanism and improvements in the depreciation allowance. We are now carefully analysing proposals made in the "Orange Book", which was recently noted by Mr. Foy. Because we share the desire for a healthy U.S. steel industry, we are working to respond positively to these proposals within the context of our overall effort to increase the productivity of and curb inflation in the U.S. economy, and in keeping with our recognition of and our obligations to our trading partners.

Some have raised concerns that dumping cases might be filed by U.S. firms. I cannot of course predict with certainty what actions individual private parties in my country will take. I would, however, take this opportunity to stress that such filings, should they occur, would only be the beginning of a process. That process would be carried out with scrupulous fairness in a manner fully consistent with U.S. international obligations as negotiated recently in Geneva. We hope that others, sharing the desire to avoid actions which distort trade flows, would recognize this. I can also assure you that in the period ahead the U.S. Administration intends to continue working closely with the other parties concerned to find responsible solutions.

Mr. ALTMANN:

I am speaking on my own behalf but also on behalf of my colleagues here, who are steel merchants in the main OECD countries. Overall, we represent approximately 5,000 firms which, as you know, act as chief partners of the producing industry in the distribution of steel products on the various national and international markets.

On the occasion of this symposium, I should first like to convey my thanks for being invited to this two-day meeting — we were highly flattered. But at the same time, we are slightly disappointed because the problems of marketing of steel products have hardly been mentioned. If permitted to express our wish, we hope that things will be different in future and that not only products and production, but also distribution and marketing will be discussed.

Of the various papers, that of our friend Dr. Peco came closest to our area of interest. Dr. Peco referred to pathological disorders in the distribution trade. I should like to point out that if you decide to honour us with an invitation to another meeting, we will be only too glad to supply medical certificates to the effect that our profession is healthy in both mind and body.

Mr. CASTEGNARO:

I would like to thank you as a trade union representative of a small country, Luxemburg. We are sometimes suspected of having a decisive influence on worldwide policy. I would first of all like to thank the organisers of this Conference for having taken the initiative of calling the Conference. I would like to congratulate them too.

As trade unionists, we did not have any illusions when we came along here that within the framework of a single conference it would be possible to restore harmony to the steel industry at the world level. We feel that we should not leave things at this one single meeting. And we feel that in the course of future conferences of this type, we should clearly define what the prospects actually are, otherwise we shall not succeed in coping with the situation.

We agree with those who have said that forecasting instruments will have to be further developed and improved. This is a necessary condition before political discussions can take place, and political discussions will have to be continued in such fora as this one. For us, these data have to be established and presented in such a way that they can be translated into a plan of action that puts forward clear objectives. This plan of action must, of course, go beyond the limits of the steel sector, otherwise how can we deal with the matter of the developing countries whose situation was described yesterday by Herr Wienert, and how could we compensate the developments in these countries. There has to be compensation for having them give up the idea of building up their own steel industry.

This has to be seen in contrast to the economic and social preoccupations of the various countries. We know that the process of restructuring of the steel sector throughout the world is by no means completed; quite the contrary. We are aware that in other areas, in other sectors of the economy, developments are not very favourable either. And for this reason there is some danger of a further dramatic development in unemployment throughout the OECD. We can be sure too that social tensions will be aggravated in the future, and we could even arrive at some form of social explosion.

Basically, there are not many ways of getting out of this kind of vicious circle that we have got ourselves into. As trade unionists, we have to hope that it will be possible to solve these problems using other means and to, in some ways, step up economic growth. In the face of foreseeable developments in the field of employment, we hope that there will be an appropriate increase in growth rates, and we consider that this ought to be a priority goal of OECD countries. Other important priorities will have to be subordinated to this first goal. The measures taken within the framework of achieving this first priority will no doubt have favourable effects on the steel sector.

We are making this sort of demand in the full knowledge that adaptation is necessary within the steel sector. As trade unionists, we are ready to undertake our part of the responsibility in this whole process, but not at any price. It has been pointed out that the conditions in the steel sector have to take into account certain elements — elements connected with trade, financing and also social policy. Within the framework of these objectives, we have to have a plan of action that will lead back to full employment in the regions concerned. The trade unions are ready to play a positive role in the matter of restructuring but only if the other bodies involved in this process are prepared to try to exert their influence upon this goal of restoring full employment. Governments and representatives of industry will have to acknowledge that these goals are their goals too.

The fact that such a model can function has already been proved, although in a very small framework. For example, in Luxemburg, in 1975 and 1976, on the trade union side we agreed to accept restructuring of the steel industry but, beforehand, we took the opportunity of making the steel industry aware of its past errors. This positive reaction on the part of the trade unions was only possible because we had a tripartite conference on steel, and this put us in a position to exert an influence in our turn on developments in the steel industry. It was possible to have some influence and to get guarantees for income of steel workers and others. We have agreed to a programme to create replacement in the regions affected by restructuring. These agreements have been inserted into an overall agreement.

All of this was negotiated and examined in the tripartite set-up. We are convinced that this model could also be applied at the international level. And for this reason we support the idea and the claims of the International Federation of Metalworkers — demands that have been put forward in concrete form to the conference.

The trade unions can only accept their responsibilities in this matter of restructuring. They can only accept the idea of positive adjustments so long as these developments do not take place over their heads and without their participation. We have to see what the effects are going to be on us within the traditional steel regions. For this reason, the Steel Committee should play a more active role in the future and should take into account what is being asked for by the IMF and make sure that the programme asked for is applied within the short term.

CHAIRMAN'S, STATEMENT ON CONCLUSION
OF THE SECOND DAY SESSION

First, I should like to apologise for the ruthless way I handled those who wished to speak. In spite of this, we have gone beyond the time limit, which shows how useful the discussions were.

I think that it would be quite pointless to try to draw conclusions after this day of discussions. In a way, this would also distort the purpose of the Symposium, which was to bring together all those who in some capacity or another are concerned with steel trends within their own country or at international level. And we have tried to take into account viewpoints of governments, enterprises, trade unions and consumers. I think it was important to try to do so.

I feel it would be wrong to draw conclusions, but it would be equally wrong not to mention some of the salient points on which the discussions centred. The first point, which was a theme running throughout our discussions, is that there will be no healthy, competitive steel industry offering job security to its workers under the right sort of conditions unless the process of industrial adaptation is continued. This has been acknowledged by all speakers from all circles and, in my opinion, this is a vital point.

The responsibility for adjustment and adaptation is shared by states with a traditional steel industry and, above all, the industrialised countries. But, this responsibility is also shared by those countries setting up new steel industries. This is normal, necessary and healthy in terms of development of international trade, provided the general context of this development is clearly understood. If I may, I should like to tell our Mexican colleague that the OECD countries were delighted to see non-Member states participating in our work. On a lighter note, I would like to say that if he were the bull, I would be reluctant to be the matador.

The second important point is that the problem of restructuring cannot be considered within a purely macro-economic or purely trade context. Trade union representatives rightly emphasized that one of the consequences of restructuring was the effect of changes on the steelworkers. They stated in no uncertain terms that they felt that it was wrong for workers only to be involved when the time came to discuss the consequences of restructuring. The trade union representatives felt it would be essential to involve the workers in the adjustment process itself; this adjustment process requires consultation, concertation and enough time for the process to be applied under conditions which can be borne by those concerned.

In this context, the value of drawing up social policies favouring adaptation has also been recognised. This concerns vocational training, the provision of new jobs and certain other direct measures; governments and enterprises will have to do more in this particular area. I think it was important that we heard an industry spokesman who acknowledged that industry itself is responsible for creating alternative jobs, and he gave us certain examples in this connection.

Many spokesmen stressed the strategic nature of maintaining a steel industry in the context of the economic policy of the various countries concerned. They added, however, that this requirement also depended on the existence of a healthy industry, i.e. an industry competitive in relation to world trade standards. This is yet another instance of the link between the need for restructuring and the harmonious development of industry.

As I see it, these comments were made with varying emphasis but each reflected the same general recognition.

We then heard a very important and interesting paper on the impact of consumer attitudes on the development of industrial policy. I feel this is a useful and necessary dimension. It is also a new dimension and must become part of our reflection on economic problems. I am convinced that we shall long remember Ms. O'Reilly's paper, which met with spontaneous applause both from those who agreed and from those who did not agree. Clearly, in our future discussions, the role of consumers will be more, not less, important; I believe this is one of the considerations we ought to bear in mind for the future.

Finally, we examined the subject of protectionism from many different angles. One of the participants said: "We are all against protectionism, against the principle of protectionism." But the problem really begins from the time these statements of principle are made. Once you start making these statements, you can be sure the problem really exists.

I think that this is a very fundamental issue, an issue requiring powers of imagination, goodwill, courage and resolution on the part of all participants, because we have recognized two facts:

— First, no market, however vast it might be — and there are many large markets represented around this table — can risk dealing with its problems on a unilateral basis. Otherwise, there will be contagion throughout the whole economic system.

— Around this table, we also acknowledged the fact that to the extent that genuine adjustment policies are being implemented, this imagination and resolution which I just mentioned must also be used not to create new international trade rules but to interpret those rules which already exist with commonsense, with a sense of fairness and justice, so that whenever it is necessary to allow further time for the restructuring, we will not avoid keeping the rules but will try to promote the deeper forms of co-operation that should exist among us to solve human problems.

Various appeals were made and these appeals were of importance. I very much hope that each person will leave this symposium asking himself in what way it is possible to take account of all these legitimate forms of interest without calling into question the legitimate interests of anyone.

I am like Mr. Wolff; I am an optimist. I think that in spite of the difficulties, in spite of low growth, in spite of the extraordinarily complicated nature of the adjustment process and all its implications — if each of us accepts the idea of not making use of his position of force, I believe that it is possible to find solutions to the problems we face in such a way that the legitimate interests of some are not out of harmony with the legitimate interests of others.

Finally, I would like to make two comments of a procedural nature.

First, I should like to thank Mr. Hormats, Chairman of the Steel Committee, for his statement that at his group's next meeting in April he would reflect on the priorities for the OECD Committee in the light of what has been said today, so that the work we unanimously agreed is not only useful but necessary can be developed for the benefit of all of us here. It must be useful from the standpoint of statistics. Mr. Foy and I discussed this problem this morning, and others mentioned it later. It is very important for us to know exactly what we are talking about. Otherwise, we might simply seem to be pleading on every occasion on behalf of our own interests. Then there is the social dimension. We discussed that. There is the matter of reflecting on the experience gathered in various quarters. I think the Steel Committee has an important task ahead of it.

The second comment is that we would not still be here to nearly 6.00 p.m. on the second day of this meeting if we did not feel that it would be useful to hold this discussion. I would like to thank the Secretary-General of the OECD, Mr. Van Lennep, and Mr. Wootton and his assistants for their untiring efforts. Although many of us were in favour of the meeting, without their work this meeting could not have been of the quality that it has been. I feel certain that I am speaking on behalf of everyone when I thank the Secretary-General for the work that has been done and I ask the participants at the meeting to begin immediately preparing for the next Steel Symposium.

I consider that the first Symposium meeting is now over. I would like to thank everyone for having taken part, and wish you a pleasant journey home.

Mr. Van LENNEP:

Just one second to say one or two words at the end of this Symposium. I think indeed we at OECD will look at this Symposium as a first step and as an important step. We are very grateful to all who have participated in the Symposium. We will carefully reflect on the many suggestions that have been made; they will be a real impulse — a positive impulse in our work.

But I would hope that we will not separate before having said a word of sincere thanks to the Chairmen of this Symposium, Mr. Hodges yesterday and Viscount Davignon today. We certainly are extremely grateful to them for having conducted a rather complicated Symposium in such an effective, efficient and agreeable manner. You, Mr. Davignon, have picked up today what we concluded yesterday and you have with firmness, with your sense of humour and the dynamism we know, led this Conference to a very successful end. We are very grateful to you indeed. Thank you very much.

LISTE DES PARTICIPANTS AU SYMPOSIUM SUR
«L'INDUSTRIE DE L'ACIER DANS LES ANNÉES 1980»

les 27 et 28 février 1980
au Château de la Muette, Paris

LIST OF PARTICIPANTS IN THE SYMPOSIUM ON
"THE STEEL INDUSTRY IN THE 1980s"

27th and 28th February, 1980
Château de la Muette, Paris

PRÉSIDENTS/MODÉRATEURS

CHAIRMEN/MODERATORS

Mr. Luther Hodges, Jr.
Deputy Secretary of Commerce, U.S. Department of Commerce, Washington,
D.C.
Vicomte Etienne Davignon
Membre de la Commission des Communautés Européennes, Bruxelles

PARTICIPANTS INDIQUÉS SELON LEUR NATIONALITÉ

PARTICIPANTS LISTED BY NATIONALITY

ORATEURS ET COMMENTATEURS

SPEAKERS AND PANELLISTS

Ils figurent parmi les participants
et leurs noms sont indiqués du signe*

These are included among participants;
their names are indicated by an asterisk

ALLEMAGNE/GERMANY

H.E. Dr. Horst-Krafft Robert
Head of the German Delegation to the OECD, Paris
Monsieur Dieter Henning
Vorstand IG Metall, Düsseldorf
Dr. Samuel Karres
Friedrich Ebert Stiftung, Bonn
Dr. Werner Kaufmann-Buhler
Counsellor German Delegation to the OECD, Paris
Dr. Ing. E.h. Willy Korf
President, Korf-Stahl AG, Baden-Baden
Dr. Herbert Kriegbaum
Verein Deutscher Maschinenbau-Anstalten E.V. (VDMA), Frankfurt am Main
Dr. Bernhard Molitor
Ministerialdirektor Federal Ministry of Economic Affairs, Bonn
Dr. Gerhard Ollig
Ministerialrat, Federal Ministry of Economic Affairs, Bonn
Dr. Ruprecht Vondran
Hauptgeschäftsführer, Wirtschaftsvereinigung Eisen- und Stahlindustrie, Düsseldorf
Dr. H.G. Vorwerk
Director General, EUROFER, Brussels
Monsieur Heinz H. Wersig
Geschäftsführendes Mitglied des Vorstandsrates, Bundesverband Deutscher Stahlhandel, Düsseldorf
Monsieur Helmut Wienert*
Research Economist Rheinisch-Westfälisches Institut für Wirtschaftsforschung (Institute of Rhine-Westphalia for Economic Research), Essen, Germany

ARGENTINE/ARGENTINA

S. Exc. M. Tomas J. de Anchorena
Ambassadeur d'Argentine en France
Mme Ileana Di Giovan de Suarez
Secrétaire, Ambassade d'Argentine en France
M. Gustavo Menendez
Fabricaciones Militares

AUSTRALIE/AUSTRALIA

Mr. R. Crawford
Commercial Counsellor Australian Delegation to the OECD Paris
Mr. J. Brunner*
Chief Economist, Broken Hill Proprietary Co. Ltd., Melbourne, Australia

AUTRICHE/AUSTRIA

S. Exc. le Dr. Peter Jankowitsch
Chef de la Délégation de l'Autriche auprès de l'OCDE Paris
Dipl. Ing. V.D. Josef Fegerl
Direktor, Voest Alpine A.G., Linz

Ing. Dkfm. Dr. Gottfried Gröbl
Ministerialrat Ministère Fédéral du Commerce et de l'Industrie, Vienne
Mr. Michael Sagmeister
Zentralsekretär Österreichischer Gewerkschaftsbund, Vienne
Dr. Dkfm. Helmut Schmid
Direktor Vereinigte Edelstahlwerke Aktiengesellschaft (VEW), Vienne

BELGIQUE/BELGIUM

S. Exc. Monsieur Hervé Robinet
Chef de la Délégation de la Belgique auprès de l'OCDE, Paris
Monsieur Louis Coosemans
Directeur FABRIMETAL, Bruxelles
Monsieur J. Doyen
Secrétaire général, Centrale Chrétienne des Métallurgistes de Belgique, Bruxelles
Monsieur Germain Duhin
Secrétaire général, Fédération Générale du Travail de Belgique, Bruxelles
Monsieur Jean Ghislain
Directeur de la Division du Marché, Groupement des Hauts Fourneaux et Aciéries Belges, Bruxelles
Monsieur Christian Oury
Président Directeur général, Groupement des Hauts Fourneaux et Aciéries Belges, Bruxelles
Monsieur K. Peerenboom
Directeur général de l'Industrie, Ministère des Affaires Economiques, Bruxelles

BRÉSIL/BRAZIL

Mr. Henrique Brandao Cavalcanti
President, Siderbrás S.A., Brasilia
Mr. Jorge Gerdau Johannpeter
President, Instituto Brasileiro de Siderurgia (IBS), Porto Alegre
Mr. Fred Woods de Lacerda
Secretary-General, Instituto Brasileiro de Siderurgia, Rio de Janeiro
Mr. Aluisio Marins
Ministry for Industry and Trade, Brasilia

CANADA

Mr. Stephen H. Heeney
Acting Head of the Canadian Delegation to the OECD, Paris
Mr. W. Black
Department of Industry, Trade and Commerce, Resource Industries Branch, Iron and Steel Division, Ottawa
Mr. Stewart Cooke
Director, District 6, United Steelworkers of America, Toronto
Mr. E. Gerard Docquier
National Director, United Steelworkers of America, Toronto
Mr. G. Elliot
Office of General Relations, Department of Industry, Trade and Commerce, Ottawa

Mr. Robert C. Varah
Director, Commercial Development, Dominion Foundaries & Steel Ltd., Hamilton
Mr. G. L. Waters
General Marketing Manager, Commercial Planning, Steel Company of Canada Ltd., West Hamilton, Ontario

CORÉE/KOREA

Mr. Byung Wha Ahn
Executive Vice President Pohang Iron and Steel Co., Ltd., Seoul
Mr. Sang Soo Lee
Managing Director Korea Iron Industry Association, Seoul
Dr. Chong Hyun Nam
Senior Fellow Korea Development Institute, Seoul

DANEMARK/DENMARK

H. E. Mr. Hans Tabor
Head of the Danish Delegation to the OECD, Paris
Mr. Arne Hansen
Chairman, Lemvigh-Müller & Munck A/S, Copenhagen
Mr. N. B. Hansen
Director, Ministry of Industry, Copenhagen
Mr. J. Harne
Deputy Director, Federation of Danish Mechanical Engineering Industries, Copenhagen
Mr. Steffen Möller
Chief Economist, Danish Metalworkers Union, Copenhagen
Mr. B. Dan Nielsen
Director, Ministry of Foreign Affairs, Copenhagen

ESPAGNE/SPAIN

S. Exc. Monsieur Tomas Chavarri y del Rivero
Chef de la Délégation de l'Espagne auprès de l'OCDE, Paris
Monsieur Guillermo d'Aubarede Navia-Ossorio
Conseiller, Délégation de l'Espagne auprès de l'OCDE, Paris
Monsieur Jose Luis Corcuera
Secrétaire, Fédération métallurgique, Union Générale des Travailleurs, Madrid
Monsieur Ricardo Diez Serrano
Président du Conseil d'Administration, Altos Hornos del Mediterraneo S.A., Madrid
Monsieur Julian Garcia Valverde
Sous-directeur général des Etudes, Ministère de l'Industrie et de l'Energie, Madrid
Monsieur Luis Guereca Tosantos
Directeur général, UNESID, Madrid
Monsieur Emilio Lopez Torres
Sous-directeur général des Industries de Base, Madrid
Monsieur Miguel Salis Balzola
Président, UNESID, Madrid

ÉTATS-UNIS D'AMÉRIQUE/UNITED STATES OF AMERICA

H. E. Mr. Herbert Salzman
Head of the United States Delegation to the OECD, Paris
Mr. Thomas Atkinson
Director, International Economics Group General Motors Corporation, New York
Mr. Paul Bernish
Special Assistant to the Deputy Secretary of Commerce, Department of Commerce, Washington, D.C.
Mr. David Biltchik
Acting Deputy Assistant Secretary for International Economic Policy, Department of Commerce, Washington, D.C.
Mr. Lawrence S. Campbell
International Economist Department of Commerce, Washington, D.C.
Mr. Dean K. Clowes
Deputy Under Secretary for International Affairs U.S. Department of Labor, Washington, D.C.
Mr. James F. Collins
Executive Vice President American Iron and Steel Institute, Washington, D.C.
Dr. R. W. Crandall*
Senior Fellow The Brookings Institution, Washington, D.C.
Mr. Philmore H. Dennen
President, Association of Steel Distributors and Chairman of the Board, Dennen Steel Corporation, Grand Rapids, Michigan
Mr. George A. Ferris
Vice-President, Basic Products Operations, Ford Motor Company, Dearborn, Michigan
Mr. Lewis W. Foy*
Chairman, Bethlehem Steel Corporation, Bethlehem
Mr. Richard W. Heimlich
Assistant U.S. Trade Representative, Office of the U.S. Trade Representative, Washington, D.C.
H. E. Mr. Robert D. Hormats
Deputy U.S. Trade Representative, Office of the U.S. Trade Representative, Washington, D.C.
Mr. Lloyd McBride*
President, United Steelworkers of America, Pittsburgh
Dr. Paul W. Marshall*
Director, Putnam Hayes and Bartlett Inc., Newton, Massachusetts
Mr. Albert A. Monnett, Jr.
Vice-President and Assistant to the Chairman, United States Steel Corporation, Pittsburgh
Mr. Kurt Orban
President, American Institute for Imported Steel, New Jersey
Ms. Kathleen O'Reilly*
Executive Director, Consumer Federation of America, Washington, D.C.
Mr. R. Peabody
President, American Iron and Steel Institute, Washington, D.C.
Mr. Roger R. Regelbrugge
President, Korf Industries, Inc., Charlotte, North Carolina
Mr. David M. Roderick (for 27th only)
Chairman and Chief Executive Officer, U.S. Steel Corporation, Pittsburgh

Mr. Richard F. Schubert
President, Bethlehem Steel Corporation, Bethlehem
Mr. John J. Sheehan
Director of Legislative Affairs, United Steelworkers of America, Pittsburgh
Mr. Richard P. Simmons
President, Allegheny Ludlum Steel Corporation, Div. Allegheny Ludlum Industries, Inc., Pittsburgh
Commissioner Paula Stern
United States International Trade Commission, Washington, D.C.
Representative Charles Vanik*
House of Representatives, Washington, D.C.
Mr. Robert G. Welch
President, Steel Service Center Institute, Cleveland, Ohio
Mr. Alan Wolff*
Verner, Liipfert, Bernhard and McPherson, Washington, D.C.

FINLANDE/FINLAND

H. E. Mr. Pekka Malinen
Head of the Finnish Delegation to the OECD, Paris
Mr. Christian Andersson
Ministry of Trade and Industry, Helsinki
Mr. Helge Haavisto
Chairman of the Board, Rautaruukki Oy, Helsinki
Mr. Juhani Linna
Executive Director, Association of Finnish Steel and Metal Producers, Helsinki
Mr. Eino Yrjönen
Finnish Metalworkers' Union, Helsinki

FRANCE

Monsieur Robert Altmann
Directeur général, Société Lorraine des Produits Métallurgiques, La Plaine-Saint-Denis
Monsieur Michel Collas
Président, Chambre syndicale de la Sidérurgie française, Paris
Monsieur Claude Etchegaray
Président-directeur général, USINOR, Paris
Monsieur Pierre Gadonneix
Directeur des industries métallurgiques, mécaniques et électriques, Ministère de l'Industrie, Paris
Monsieur René Lapierre
Administrateur civil Direction des Relations économiques extérieures, Ministère de l'Economie, Paris
Monsieur Claude Leroy
Directeur, SACILOR
Monsieur J. Mayoux
Président du Directoire de SACILOR, Paris
Monsieur Bernard Mourgues
Secrétaire général, Fédération des Industries métallurgiques, Force Ouvrière, Paris
Monsieur Jean Reinhold
Directeur général, Ugine Aciers, Paris

Monsieur Pierre Robert
Secrétaire national à la Fédération générale de la Métallurgie, Paris
Monsieur Yves Pierre Soulé
Délégué général, Chambre syndicale de la Sidérurgie française, Paris
Monsieur Gérard Valluet
Conseiller commercial, Direction des Relations économiques extérieures, Ministère de l'Economie, Paris
Monsieur André Vernier
Président, Fédération Sidérurgie pour la C.F.T.C. (Confédération Française des Travailleurs Chrétiens), Paris
Monsieur André Wille
Premier vice-président, Métallurgie, C.G.C. (Confédération Générale des Cadres), Paris

GRÈCE/GREECE

S. Exc. Monsieur Dimitri Athanassopoulos
Chef de la Délégation de la Grèce auprès de l'OCDE, Paris
Monsieur Dimitris Th. Angelopoulos
Président, directeur général, Halyvourgiki S.A., Athènes
Monsieur Théodore Angelopoulos
Halyvourgiki S.A., Athènes
Monsieur Georges Loucopoulos
Conseiller, Délégation de la Grèce auprès de l'OCDE, Paris

IRLANDE/IRELAND

Mr. Donal Hurley
Deputy Head of the Irish Delegation to the OECD, Paris
Mr. Michael Mullan
General Secretary, Irish Transport and General Workers Union, Dublin
Mr. J. Mullen
Deputy Director for E.C.S.C. Affairs and Economic Planning, Irish Steel Holdings, Dublin

ISLANDE/ICELAND

Mr. Sveinn Björnsson
Counsellor, Icelandic Delegation to the OECD, Paris

ITALIE/ITALY

Son Exc. Monsieur Fausto Bacchetti
Chef de la Délégation de l'Italie auprès de l'OCDE, Paris
Monsieur Mario Casadio
Ministère de l'Industrie, Rome
Monsieur Mario Gerbino
Conseiller, Délégation de l'Italie auprès de l'OCDE, Paris
Dr. Gian Carlo Guiducci
Italsider S.P.A., Gênes

Dr. Gaetano Moro
Dirigente Settore Siderurgico, Ministère des Participations d'Etat, Rome
Monsieur Ferdinando Palazzo
Administratore Delegato de la Société TEKSID, Torino
Dr. F. Peco*
Delegato della Presidenza, Associazione Industrie Siderurgiche Italiano (ASSIDER), Milan

JAPON/JAPAN

Mr. Hiromu Fukada
Minister, Deputy Head of the Japanese Delegation to the OECD, Paris
Mr. Naohiro Amaya
Vice-Minister for International Affairs, Ministry of International Trade and Industry, Tokyo
Mr. Akira Chihaya
Deputy General Manager of the President's Office, Nippon Steel Corporation, Tokyo
Mr. Masahiro Fukukawa
Third Secretary, Japanese Delegation to the OECD, Paris
Mr. Mitsuhiko Hazumi
Deputy Director-General Economic Affairs Bureau, Ministry of Foreign Affairs, Tokyo
Mr. Tsutomu Kono*
General Manager, Corporate and Economic Research Department, Nippon Steel Corporation, Tokyo
Mr. Hiroshi Miwa
Executive Vice-President, Kawasaki Steel Corporation, Tokyo
Mr. Teruyoshi Murozumi
General Manager, Planning and Business Research Department, Kawasaki Steel Corporation, Tokyo
Mr. Takuhiko Nakamura
Chairman, Japanese Federation of Iron and Steel Workers' Unions, Tokyo
Mr. Masahisa Naitoh
Director, The Americas-Oceania Division International Trade Policy Bureau Ministry of International Trade and Industry, Tokyo
Mr. Kunio Okabe
Deputy General Manager, Corporate Planning – International Department, Nippon Steel Corporation, Tokyo
Mr. Eishiro Saito
President Nippon Steel Corporation, Tokyo
Mr. Masuo Shibata
Director General for International Economic Affairs Department, International Trade Policy Division, MITI, Tokyo
Mr. Hajime Tahara
Director, European Office, Japan Iron and Steel Federation, Brussels
Mr. Hiroshi Takano
Executive Vice-President, Nippon Kokan K.K., Tokyo
Mr. Chikao Tsukuda
Minister (Industry and Trade), Japanese Delegation to the OECD, Paris

LUXEMBOURG

S. Exc. Monsieur André Philippe
Ambassadeur du Luxembourg en France, Chef de la Délégation du Luxembourg
    auprès de l'OCDE, Paris
Monsieur Lucien Bisenius
Premier vice-président, Fédération des employés privés, Helmsange
Monsieur Mario Castegnaro
Membre du Comité exécutif, O.G.B.L., Eschlalzette
Monsieur Paul Metz
Directeur général, ARBED, Luxembourg
Monsieur André Robert
Directeur, Groupement des Industries Sidérurgiques Luxembourgeoises,
    Luxembourg
Monsieur Armand Simon
Secrétaire général, Ministère de l'Economie et des Classes moyennes,
    Luxembourg
Monsieur Emmanuel Tesch*
Président, EUROFER, Bruxelles, Président, ARBED, Luxembourg

MEXIQUE/MEXICO

Mr. Alfredo Acle
Director General, Comision Coordinadora de la Industria Siderurgica, Mexico
Mr. Daniel Dultzin Dubin
Third Secretary (Economic Affairs), Embassy of Mexico, Paris

NIGERIA

Mr. S. O. Ogundele
Minister, Embassy of Nigeria, Paris

NORVÈGE/NORWAY

Mr. Torolf Raa
Deputy Head of the Norwegian Delegation to the OECD, Paris
Mr. Per Blidensol
Managing Director, A/S Norsk Jernverk, Oslo
Mr. Hans Breder
Technical Director, A/S Sønnichsen Rørvalseverket, Oslo
Mr. Odd Göthe
Deputy Secretary-General Ministry of Industry, Oslo
Mr. A. Nordli
Secretary, Norsk Jern- og Metallarbeiderforbund, Oslo
Mr. K. A. Sanden*
Editor, Norsk Jern- og Metallarbeiderforbund, Oslo

PAYS-BAS/NETHERLANDS

Mr. R. B. P. de Brouwer
Ministry of Economic Affairs, The Hague
Mr. L. D. Donk
General Director, Algemene Vereniging voor de Ijzerhandel (Steel Traders Association), The Hague
Mr. J. Heusdens
Director of the Dutch Iron and Steel Association, Deputy Director, ESTEL NV, Nijmegen
Drs. J. D. Hooglandt
Chairman, Board of Management, ESTEL NV Hoesch-Hoogovens, Nijmegen
Mr. H. Krul
Federation of Netherlands Trade Unions (FNV), Velsen Noord
Drs. S. J. G. Wijnands
Head of the Economic Department, Vereniging voor de metaal – en de elektro-technische Industrie (FME), Zoetermeer

PORTUGAL

S. Exc. Monsieur Henrique Granadeiro
Chef de la Délégation du Portugal auprès de l'OCDE, Paris
Monsieur Joâo Antunes Bártolo
Directeur général des Industries chimiques et métallurgiques, Ministère de l'Industrie et de l'Energie, Lisbonne
Monsieur José Diogo Costa
Département de l'Industrie métallurgique, Ministère de l'Industrie et de l'Energie, Lisbonne
Monsieur Fernando A. Dos Santos Martins
Président du Conseil de gérance, EQUIMETAL, Lisbonne
Monsieur Victor Rodrigues Pessoa
Directeur, Département central de Planification, Lisbonne
Monsieur Fernando Teixeira Gomes
Conseiller, Délégation du Portugal auprès de l'OCDE, Paris

ROYAUME-UNI/UNITED KINGDOM

H. E. Mr. A. F. Maddocks, C.M.G.
Head of the United Kingdom Delegation to the OECD, Paris
Mrs. Janet Cohen
Assistant Secretary, Iron and Steel Division, Department of Industry, London
Mr. S. J. Gross, C.M.G. (for 27th only)
Head of the Iron and Steel Division, Department of Industry, London
Mr. G. H. Laird
Executive Councilman, Amalgamated Union of Engineering Workers (Engineering Section), London
Sir Richard Marsh, P.C.
Chairman, British Iron and Steel Consumers' Council, Richmond
Mr. Michael Marshall, M.P.
Parliamentary Under-Secretary of State for Industry, Department of Industry, London

Mr. P. G. Naylor
Chief Executive, British Steel Corporation (Industry) Ltd., London
Mr. G. H. Sambrook
Managing Director Commercial, British Steel Corporation, London
Mr. W. Sirs*
General Secretary of the Iron and Steel Trades Confederation, President of the
    International Metalworkers' Federation's Iron, Steel and Non-Ferrous Metal
    Department
Mr. H. C. Smith
General Secretary, National Union of Blast Furnacemen, Middlesbrough, Cleve-
    land

SUÈDE/SWEDEN

H. E. Mr. Hans Colliander
Head of the Swedish Delegation to the OECD, Paris
Dr. John Ekström
Ministry of Industry, Stockholm
Professor Erik Höök
Managing Director, Jernkontoret, Stockholm
Professor Erik Ruist*
Handelshögskolan (Stockholm School of Economics), Stockholm
Mr. Olof Rydh
Secretary, Sv. Metallindustriarb. förbundet, Stockholm
Mr. B. Särhagen
Swedish Union of Clerical and Technical Employees in Industry, Stockholm
Mr. Staffan Sohlman
Deputy Director General, Board of Commerce, Stockholm
Mr. Björn Wahlström
Managing Director, SSAB Svenskt Stål AB, Stockholm

SUISSE/SWITZERLAND

Monsieur Paul Änishänslin
Office fédéral des Affaires économiques extérieures, Berne
Monsieur Heinz W. Frech
Directeur général, von Roll Ltd., Gerlafingen
Monsieur Alexander Lamparter
Président, Association suisse d'Industries de transformation d'acier laminé à
    chaud, Evilard
Dr. Ferdinand Oehen
Von Moos Stahl AG., Lucerne
Monsieur Agostino Tarabusi
Secrétaire central, Fédération des travailleurs de la métallurgie et de l'horlogerie,
    Berne

TURQUIE/TURKEY

Monsieur Isin Çelebi
Directeur du groupe de la Métallurgie et des Biens d'équipement, Organisation
    d'Etat de Planification, Ankara

Monsieur Ayhan Karlidag
Adjoint au sous-secrétaire d'Etat, Ministère de l'Industrie et de la Technologie, Ankara
Monsieur Ferruh Katrancigil
Directeur général adjoint S.A. Complexe Sidérurgique d'Ereğli, Ankara
Monsieur Mete Sayici
Conseiller, Organisation d'Etat de Planification, Ankara
Monsieur Memis Ali Usta
Directeur général adjoint, Entreprise d'Etat du Fer et de l'Acier, Ankara

VENEZUELA

Monsieur Alvaro Atencio
Conseiller économique, Ambassade du Venezuela, Paris
Monsieur E. Efrain Pico-Ponte
Attaché scientifique, Ambassade du Venezuela, Paris

YOUGOSLAVIE/YUGOSLAVIA

H. E. Mr. Gavro Cerović
Head of the Permanent Delegation to the OECD, Paris
Mr. Aleksandar Čavić
Technical and RD Manager, Yugoslav Iron and Steel Federation, Belgrade
Mr. Božidar Martinović
Vice-President, Executive Board, RMK Zenica, Zenica
Mrs. Mileva Stefanović
Counsellor of the Federal Institute for Social Planning, Belgrade

PARTICIPANTS MEMBRES D'ORGANISMES INTERNATIONAUX

PARTICIPANTS MEMBERS OF INTERNATIONAL BODIES

LES COMMUNAUTÉS EUROPÉENNES

THE EUROPEAN COMMUNITIES

La Commission/The Commission

Monsieur Jean-Pierre Leng
Délégué permanent de la Commission auprès de l'OCDE, Paris
Dr. Manfred Caspari
Directeur général adjoint des Relations extérieures, Bruxelles
Monsieur Pierre Defraigne
Cabinet du vicomte Davignon
Monsieur György von O'Svath
Administrateur principal, Direction des Relations extérieures, Bruxelles
Monsieur Hugo Paemen
Chef de Cabinet du vicomte Davignon

Monsieur Maurice Schaeffer
Directeur, Direction Acier, Direction générale du Marché intérieur et des Affaires
    industrielles, Bruxelles
Mr. Reginald Spence
Administrateur principal, Direction Acier, Direction générale du Marché intérieur
    et des Affaires industrielles, Bruxelles

Comité consultatif de la Communauté Européenne du charbon et de l'acier

Monsieur Rudolf Judith
Président du Comité exécutif, Bruxelles et Vorstand IG Metal, Düsseldorf
Monsieur Jean Danis
Secrétaire du Comité consultatif, Bruxelles

COMMISSION ÉCONOMIQUE POUR L'EUROPE
DES NATIONS UNIES – Genève
UNITED NATIONS ECONOMIC COMMISSION
FOR EUROPE – Geneva

Monsieur Jean-Roger Messy
Chef de la Division de l'Industrie

COMITÉ CONSULTATIF ÉCONOMIQUE ET INDUSTRIEL
AUPRÈS DE L'OCDE – Paris
BUSINESS AND INDUSTRY ADVISORY COMMITTEE
TO THE OECD – Paris

Mr. K. H. J. Clarke
Vice-President of B.I.A.C., Inco Limited, Toronto
Mademoiselle Yolande Michaud
Secrétaire générale du B.I.A.C., Paris
Monsieur Pol Provost
Vice-président du B.I.A.C., Bruxelles

COMMISSION SYNDICALE CONSULTATIVE
AUPRÈS DE L'OCDE – Paris
TRADE UNION ADVISORY COMMITTEE
TO THE OECD – Paris

Mr. Karl Casserini
Assistant General Secretary, International Metalworkers' Federation, Geneva
Mr. Herman Rebhan
General Secretary, International Metalworkers' Federation, Geneva
Mr. F. W. M. Spit
General Secretary, World Federation for the Metallurgic Industry (W.C.L.),
    Brussels
Mr. Kari Tapiola
Seneral Secretary of the T.U.A.C., Paris

INTERNATIONAL IRON AND STEEL INSTITUTE – Brussels

Mr. D. F. Anderson*
Director, Department of Economic Affairs, International Iron and Steel Institute, Brussels,

ORGANISATION DES NATIONS UNIES POUR LE DÉVELOPPEMENT INDUSTRIEL (ONUDI) – Vienne UNITED NATIONS INDUSTRIAL DEVELOPMENT ORGANISATION (UNIDO) – Vienna

Dr. B. R. Nijhawan*
Senior Adviser (Inter-Regional)

OECD SALES AGENTS
DÉPOSITAIRES DES PUBLICATIONS DE L'OCDE

ARGENTINA – ARGENTINE
Carlos Hirsch S.R.L., Florida 165, 4° Piso (Galería Guemes)
1333 BUENOS-AIRES, Tel. 33-1787-2391 Y 30-7122

AUSTRALIA – AUSTRALIE
Australia & New Zealand Book Company Pty Ltd.,
23 Cross Street, (P.O.B. 459)
BROOKVALE NSW 2100   Tel. 938-2244

AUSTRIA – AUTRICHE
OECD Publications and Information Center
4 Simrockstrasse 5300 BONN   Tel. (0228) 21 60 45
Local Agent:
Gerold and Co., Graben 31, WIEN 1. Tel. 52.22.35

BELGIUM – BELGIQUE
LCLS
44 rue Otlet, B 1070 BRUXELLES -Tel. 02-521 28 13

BRAZIL – BRÉSIL
Mestre Jou S.A., Rua Guaipà 518,
Caixa Postal 24090, 05089 SAO PAULO 10. Tel. 261-1920
Rua Senador Dantas 19 s/205-6, RIO DE JANEIRO GB.
Tel. 232-07. 32

CANADA
Renouf Publishing Company Limited,
2182 St. Catherine Street West,
MONTREAL, Quebec H3H 1M7 Tel. (514) 937-3519

DENMARK – DANEMARK
Munksgaards Boghandel,
Nørregade 6, 1165 KØBENHAVN K. Tel. (01) 12 85 70

FINLAND – FINLANDE
Akateeminen Kirjakauppa
Keskuskatu 1, 00100 HELSINKI 10. Tel. 65-11-22

FRANCE
Bureau des Publications de l'OCDE,
2 rue André-Pascal, 75775 PARIS CEDEX 16. Tel. (1) 524.81.67
Principal correspondant :
13602 AIX-EN-PROVENCE : Librairie de l'Université.
Tel. 26.18.08

GERMANY – ALLEMAGNE
OECD Publications and Information Center
4 Simrockstrasse 5300 BONN   Tel. (0228) 21 60 45

GREECE – GRÈCE
Librairie Kauffmann, 28 rue du Stade,
ATHÈNES 132. Tel. 322.21.60

HONG-KONG
Government Information Services,
Sales and Publications Office, Baskerville House, 2nd floor,
13 Duddell Street, Central.   Tel. 5-214375

ICELAND – ISLANDE
Snaebjörn Jönsson and Co., h.f.,
Hafnarstraeti 4 and 9, P.O.B. 1131, REYKJAVIK.
Tel. 13133/14281/11936

INDIA – INDE
Oxford Book and Stationery Co.:
NEW DELHI, Scindia House. Tel. 45896
CALCUTTA, 17 Park Street. Tel. 240832

INDONESIA – INDONÉSIE
PDIN LIPI, P.O. Box 3065/JKT., JAKARTA, Tel. 583467

ITALY – ITALIE
Libreria Commissionaria Sansoni:
Via Lamarmora 45, 50121 FIRENZE. Tel. 579751
Via Bartolini 29, 20155 MILANO. Tel. 365083
Sub-depositari:
Editrice e Libreria Herder.
Piazza Montecitorio 120, 00 186 ROMA. Tel. 6794628
Libreria Hoepli, Via Hoepli 5, 20121 MILANO. Tel. 865446
Libreria Lattes, Via Garibaldi 3, 10122 TORINO. Tel. 519274
La diffusione delle edizioni OCSE è inoltre assicurata dalle migliori
librerie nelle città più importanti.

JAPAN – JAPON
OECD Publications and Information Center,
Landic Akasaka Bldg., 2-3-4 Akasaka,
Minato-ku, TOKYO 107 Tel. 586-2016

KOREA – CORÉE
Pan Korea Book Corporation,
P.O.Box n° 101 Kwangwhamun, SÉOUL. Tel. 72-7369

LEBANON – LIBAN
Documenta Scientifica/Redico,
Edison Building, Bliss Street, P.O.Box 5641, BEIRUT.
Tel. 354429–344425

MALAYSIA – MALAISIE
and/et SINGAPORE-SINGAPOUR
University of Malaya Co-operative Bookshop Ltd.
P.O. Box 1127, Jalan Pantai Baru
KUALA LUMPUR Tel. 51425, 54058, 54361

THE NETHERLANDS – PAYS-BAS
Staatsuitgeverij
Verzendboekhandel Chr. Plantijnstraat
S-GRAVENHAGE Tel. nr. 070-789911
Voor bestellingen: Tel. 070-789208

NEW ZEALAND – NOUVELLE-ZELANDE
The Publications Manager,
Government Printing Office,
WELLINGTON: Mulgrave Street (Private Bag),
World Trade Centre, Cubacade, Cuba Street,
Rutherford House, Lambton Quay, Tel. 737-320
AUCKLAND: Rutland Street (P.O.Box 5344), Tel. 32.919
CHRISTCHURCH: 130 Oxford Tce (Private Bag), Tel. 50.331
HAMILTON: Barton Street (P.O.Box 857), Tel. 80.103
DUNEDIN: T & G Building, Princes Street (P.O.Box 1104),
Tel. 78.294

NORWAY – NORVÈGE
J.G. TANUM A/S Karl Johansgate 43
P.O. Box 1177 Sentrum OSLO 1   Tel (02) 80 12 60

PAKISTAN
Mirza Book Agency, 65 Shahrah Quaid-E-Azam, LAHORE 3.
Tel. 66839

PORTUGAL
Livraria Portugal, Rua do Carmo 70-74,
1117 LISBOA CODEX. Tel. 360582/3

SPAIN – ESPAGNE
Mundi-Prensa Libros, S.A.
Castello 37, Apartado 1223, MADRID-1. Tel. 275.46.55
Libreria Bastinos, Pelayo, 52, BARCELONA 1. Tel. 222.06.00

SWEDEN – SUÈDE
AB CE Fritzes Kungl Hovbokhandel,
Box 16 356, S 103 27 STH, Regeringsgatan 12,
DS STOCKHOLM. Tel. 08/23 89 00

SWITZERLAND – SUISSE
OECD Publications and Information Center
4 Simrockstrasse 5300 BONN   Tel. (0228) 21 60 45
Agent local :
Librairie Payot, 6 rue Grenus, 1211 GENÈVE 11. Tel. 022-31.89.50

TAIWAN – FORMOSE
National Book Company,
84-5 Sing Sung South Rd., Sec. 3, TAIPEI 107. Tel. 321.0698

THAILAND – THAILANDE
Suksit Siam Co., Ltd.,1715 Rama IV Rd.,
Samyan, BANGKOK 5   Tel. 2511630

UNITED KINGDOM – ROYAUME-UNI
H.M. Stationery Office, P.O.B. 569,
LONDON SE1 9 NH. Tel. 01-928-6977, Ext. 410 or
49 High Holborn, LONDON WC1V 6 HB (personal callers)
Branches at: EDINBURGH, BIRMINGHAM, BRISTOL,
MANCHESTER, CARDIFF, BELFAST.

UNITED STATES OF AMERICA – ÉTATS-UNIS
OECD Publications and Information Center, Suite 1207,
1750 Pennsylvania Ave., N.W. WASHINGTON, D.C. 20006.
Tel. (202)724 1857

VENEZUELA
Libreria del Este, Avda. F. Miranda 52, Edificio Galipàn,
CARACAS 106. Tel. 32 23 01/33 26 04/33 24 73

YUGOSLAVIA – YOUGOSLAVIE
Jugoslovenska Knjiga, Terazije 27, P.O.B. 36, BEOGRAD.
Tel. 621-992

Les commandes provenant de pays où l'OCDE n'a pas encore désigné de dépositaire peuvent être adressées a :
OCDE, Bureau des Publications, 2 rue André-Pascal, 75775 PARIS CEDEX 16.
Orders and inquiries from countries where sales agents have not yet been appointed may be sent to:
OECD, Publications Office, 2 rue André-Pascal, 75775 PARIS CEDEX 16.

OECD PUBLICATIONS, 2 rue André-Pascal, 75775 Paris Cedex 16 - No. 41 565 1980
PRINTED IN FRANCE
(2250 N- 58 80 02 1) ISBN 92-64-12081-5